simple methods for aquaculture

POND CONSTRUCTION
for freshwater fish culture

pond-farm structures and layouts

Text: A.G. Coche
J.F. Muir
Illustration designs and layout: T. Laughlin

Reprinted by arrangement with the
Food and Agriculture Organization of the United Nations

by

Daya Publishing House®
A Division of
Astral International Pvt. Ltd.
New Delhi - 110 002

"This book was originally published by the Food and Agriculture Organization of the United Nations (FAO) as Fish and fishery products, FAO Fisheries circular No. 821".

"The designations employed and the presentation of material in this publication do not imply the expression of any opinion, whatsoever on the part of the Food and Agriculture Organization of the United Nations concerning the legal status of any country, territory, city or area or of its authorities, or concerning the delimitations of its frontiers or boundaries."

'The designations 'developed' and 'developing' economies are intended for statistical convenience and do not necessarily express a judgment about the stage reached by a particular country, territory or area in the development process.'

First Indian Edition, 2001
Reprinted, 2016
ISBN: 978-81-7035-253-2

Published by : **Daya Publishing House®**
A Division of
Astral International Pvt. Ltd.
– ISO 9001:2008 Certified Company –
4760-61/23, Ansari Road, Darya Ganj
New Delhi-110 002
Ph. 011-43549197, 23278134
E-mail: info@astralint.com
Website: www.astralint.com

Printed at : **Chawla Offset Printers,**
Delhi-110 052

THE AQUACULTURE TRAINING MANUALS

ıe training manuals on simple methods for aquaculture published in .e FAO Training Series are prepared by the Inland Water Resources ıd Aquaculture Service of the Fishery Resources and Environment vision, Fisheries Department. They are written in simple language and esent methods and equipment useful not only for those responsible for ld projects and aquaculture extension in developing countries but also r use in aquaculture training centres.

ey concentrate on most aspects of semi-intensive fish culture in fresh ıters, from selection of the site and building of the fish farm to the sing and final harvesting of the fish.

O would like to have readers' reactions to these manuals. Comments, ticism and opinions, as well as contributions, will help to improve ure editions. Please send them to the Senior Fishery Resources ficer (Aquaculture/Publications), FAO/FIRI, Viale delle Terme di Carac- a, 00100 Rome, Italy.

The following manuals on simple methods of aquaculture have been published in the FAO Training Series:

Volume 4 — Water for freshwater fish culture
Volume 6 — Soil and freshwater fish culture
Volume 16/1 — Topography for freshwater fish culture: Topographical tools
Volume 16/2 — Topography for freshwater fish culture: Topographical surveys
Volume 20/2 — Pond construction for freshwater fish culture: Pond-farm structures and layouts

Three final volumes are being prepared:

Volume 20/1 — Pond construction for freshwater fish culture: Building earthen ponds
Volume 21/1 — Management for freshwater fish culture: Ponds and water practices
Volume 21/2 — Management for freshwater fish culture: Farms and fish stocks

HOW TO USE THIS MANUAL

The material in the two volumes of this manual is presented in sequence, beginning with basic definitions. The reader is then led step by step from the easiest instructions and most basic materials to the more difficult and finally the complex.

The most basic information is presented on white pages, while the more difficult material, which may not be of interest to all readers, is presented on pages with a grey or a light blue background.

Some of the more technical words are marked with an asterisk (*) and are defined in the Glossary on page 213.

For more advanced readers who wish to know more about fish-farm construction, a list of specialized books for further reading is suggested on page 214.

CONTENTS

CONTENTS, continued

CONTENTS, continued

CONTENTS, continued

CONTENTS, continued

TABLES

GRAPHS

7 MAIN WATER INTAKE STRUCTURES

70 Introduction

1. Water intake structures depend on the type of pond you have. You learned earlier that a fish pond can be supplied with water from different sources (see Chapter 1 of **Construction for freshwater fish culture**, *FAO Training Series*, **20/1**). **Several types of pond** were defined by their **intake structures:**

- **sunken pond:** no intake required;
- **barrage pond without diversion canal:** no intake required;
- **barrage pond with diversion canal:** main water intake combined with a diversion structure in the diversion canal;
- **diversion pond:** main water intake with or without a separate diversion structure downstream to raise the water level in the stream.

Types of ponds

Sunken pond
(no structure required)

Diversion pond

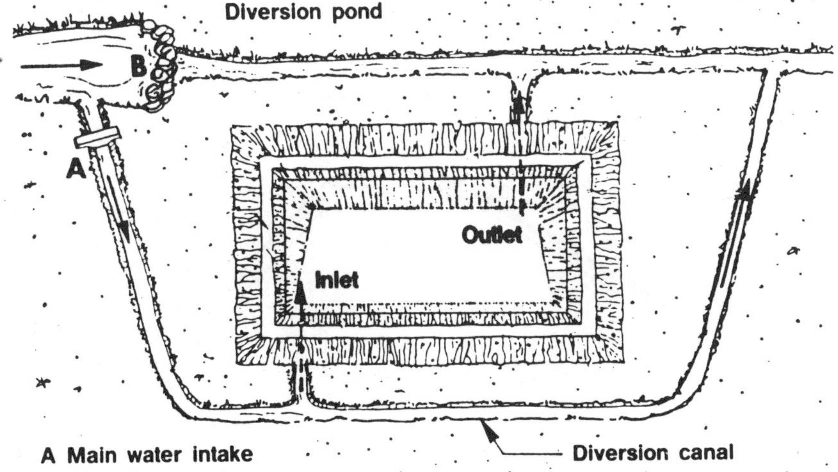

A Main water intake
B Diversion structure

Barrage pond
without diversion canal
(no structure required)

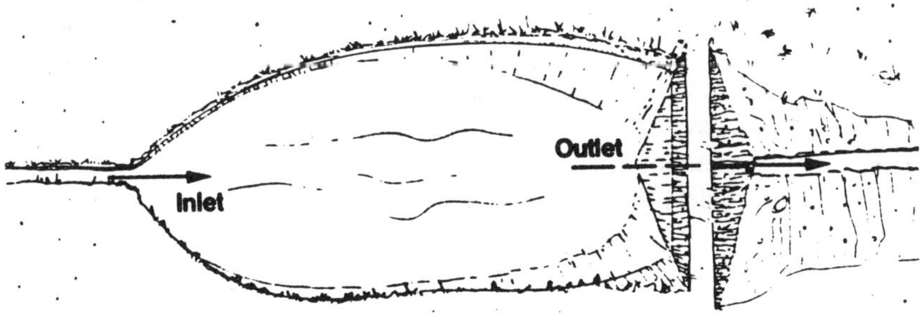

Barrage pond
with diversion canal

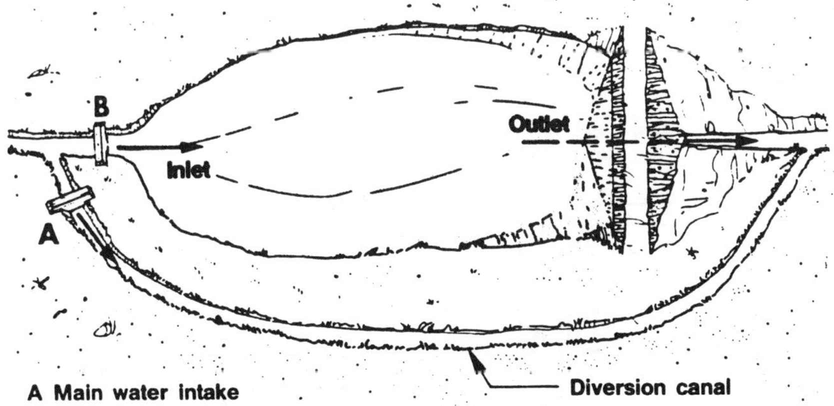

A Main water intake
B Diversion structure

Water supply provided from a reservoir

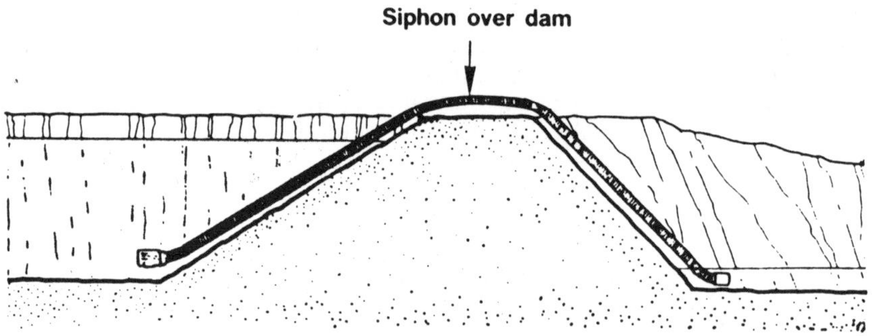

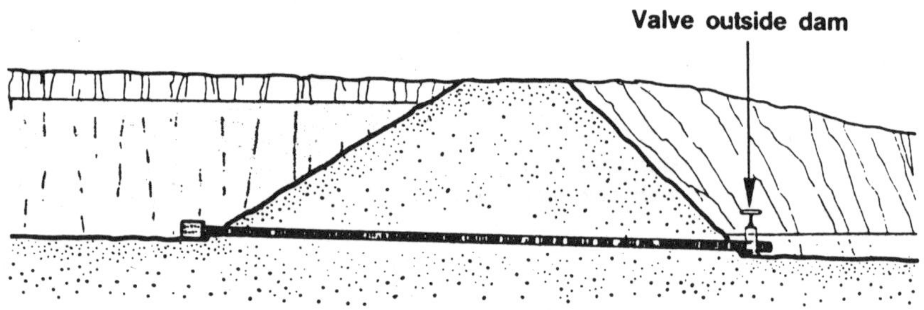

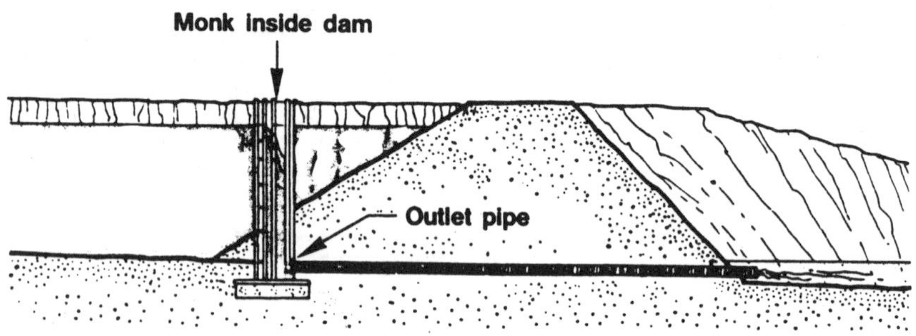

Note: if the water supply is provided from **a reservoir,** the intake structures are usually part of the system that releases the impounded water into the pond feeder canal. They may consist of:

- a **siphon** placed over the dam;
- the **reservoir bottom valve** at the downstream side of the dam;
- an **outlet sluice** at the upstream side of the dam such as a monk.

Selecting the water intake structures

2. The main elements of a water intake are:

- **a diversion structure**, to control the water level in the stream and to ensure it is sufficient to supply the intake but not to flood it (see Sections 73 to 75);
- **inlet level (and flow) control** in the intake structure itself, to control water supply to the ponds (see Section 76). It is usually connected to the water transport structure (see Chapter 8);
- **entrance protection**, such as coarse bars or piling or a range of screens to protect the intake from debris and scour damage (see Section 77).

3. There are many designs for water intake structures, some of which can be quite complex and require specialized design and construction. This manual concentrates on relatively simple designs (see **Table 31**) that you can build by yourself or with the assistance of a good mason.

TABLE 31.

Diversion structures to control stream water levels

Type of stream	*Structures required*		*Section*
SMALL ● water flow less than 10 l/s ● no significant flood conditions therefore: divert all available water by damming stream	Diversion structure (not to be submerged)	Earthen barrage Bamboo barrier Wooden barrier	73 73 73
	No need for water intake structure; excess water should be drained away from the pond		—
LARGE ● water flow at least twice the flow required ● significant flood conditions therefore: use only part of available water	Diversion structure required to raise water level (stream flow less than 100 l/s; structure can be submerged)	Pole barrier Stone barrier Gabion barrier Adjustable barrage (concrete pillars with boards)	74 74 74 75
	No need for diversion structure; water intake structure adjustable		76

Main water intakes

4. **Main water intakes** are used for the **overall regulation and diversion of water supplies** to a pond or group of ponds. In many cases, they are distinct from water transport structures, which are discussed in Chapter 8, and from smaller pond inlet structures, discussed in Chapter 9, which supply and control water flow into individual ponds.

5. The main purpose of an intake is to ensure a constant water supply that can be adjusted to suit local conditions.

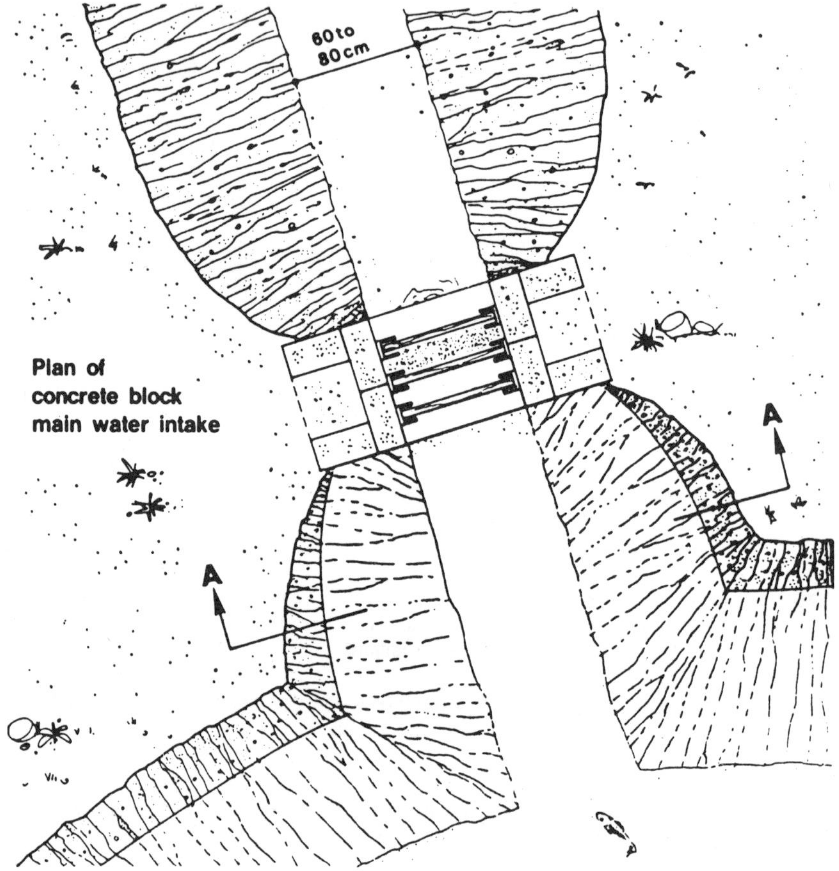

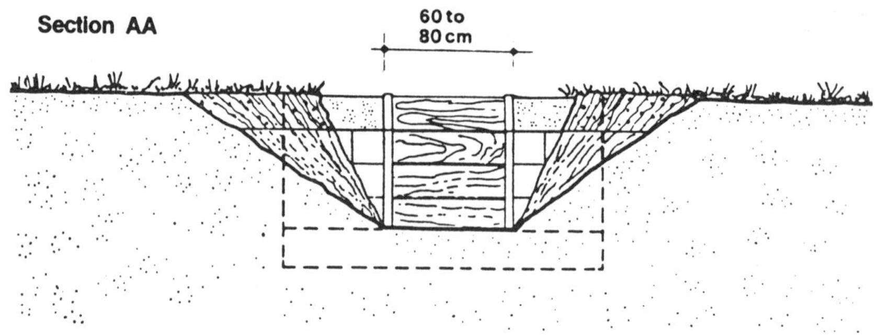

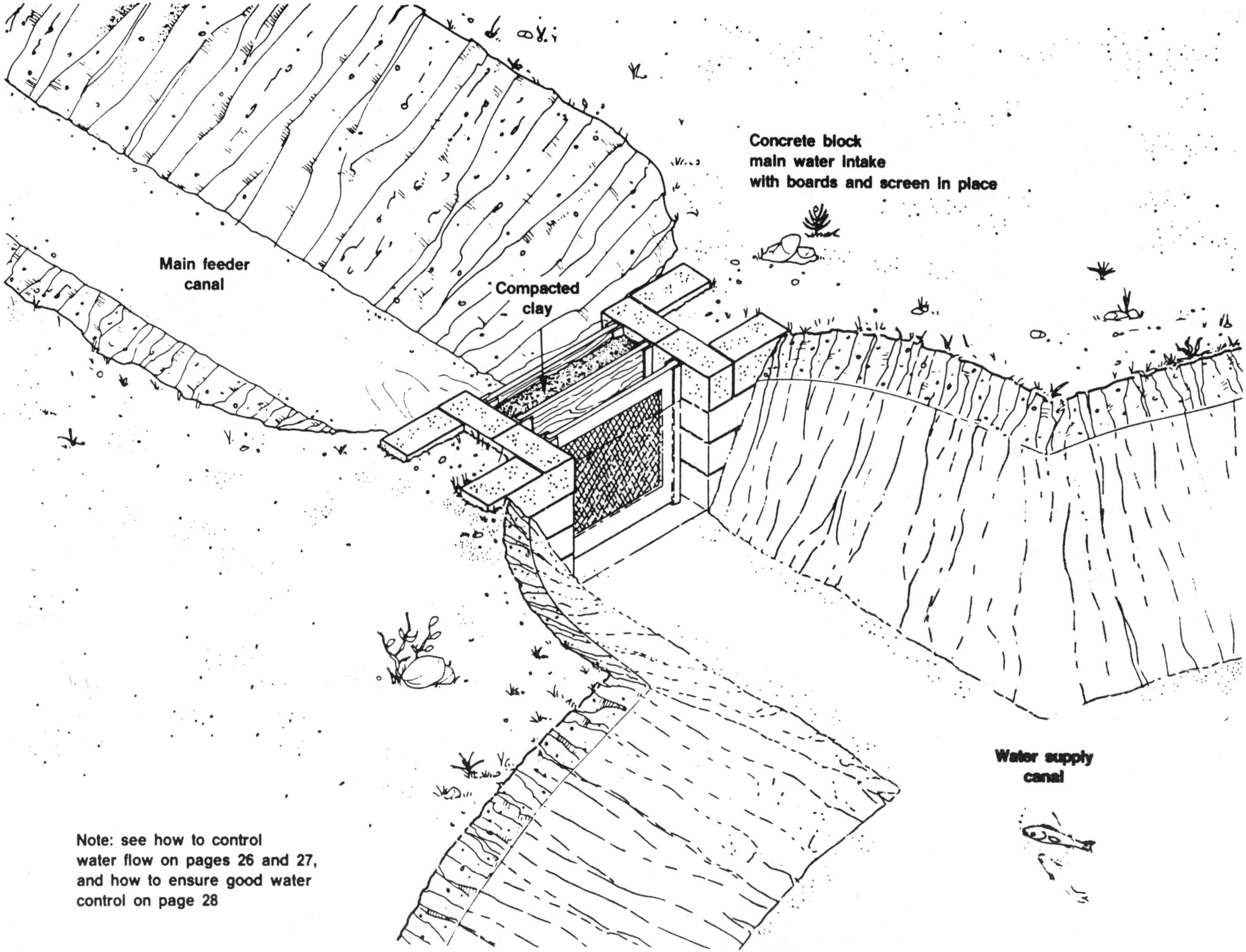

Note: see how to control
water flow on pages 26 and 27,
and how to ensure good water
control on page 28

Building a main water intake

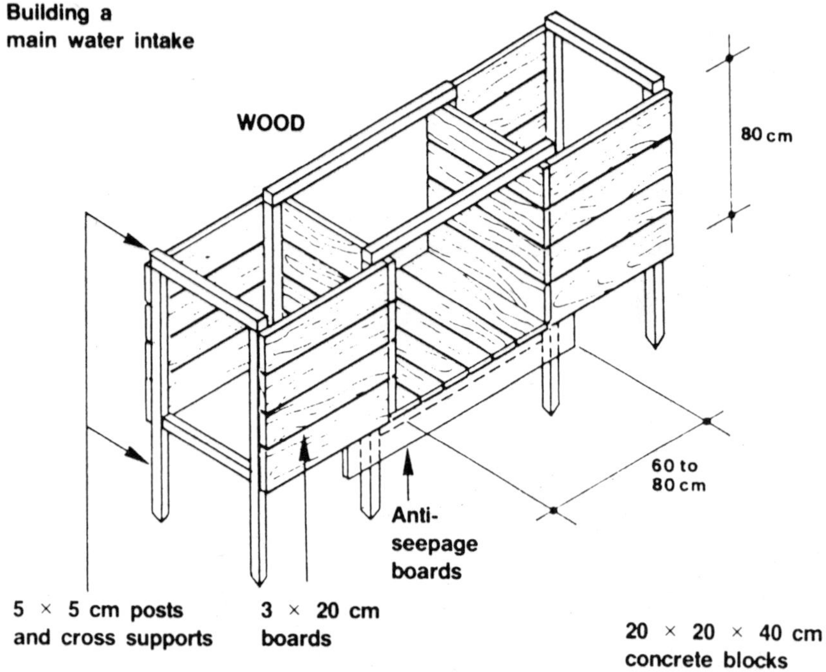

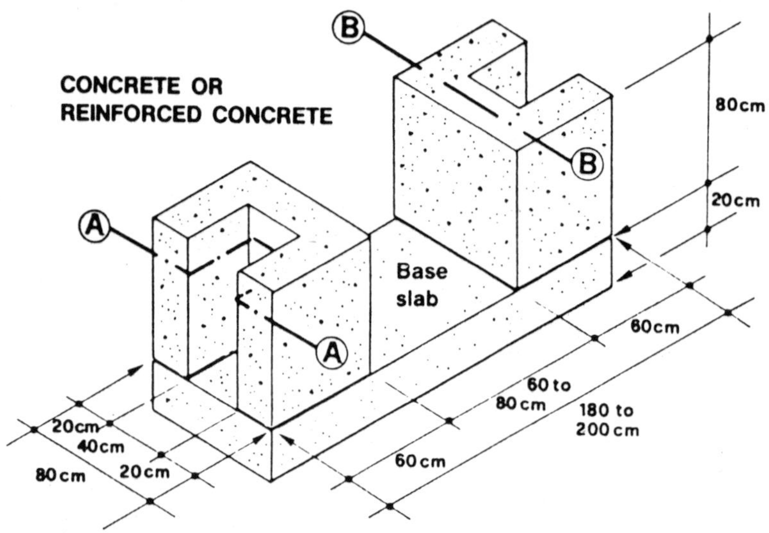

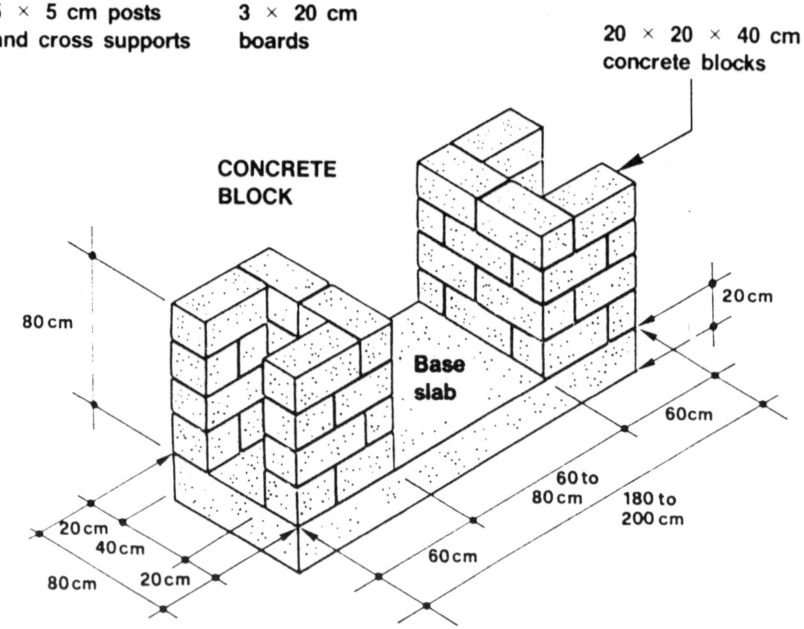

Placement of steel bars for reinforced concrete

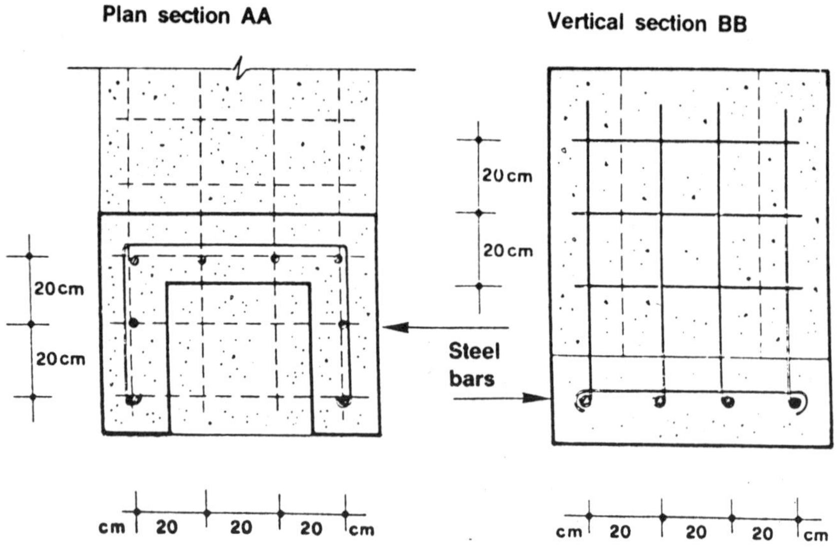

Note: the dimensions shown are suitable for structures for a medium-size pond system

Locating the main water intake along a stream

6. The pond site and its water feeder canal usually determine the location of the main water intake. If the pond is to be built along a stream, it is better to **select a site that has:**

- valley sides that are not too steep;
- a relatively level, stable and smoothly flowing section of the stream, reasonably free of debris and moving silt;
- no excessive forest over and around the feeder canal;
- a straight stretch of the stream.

Note: avoid large rivers with a fluctuating water level. Be very careful to make sure the intake is not set above the minimum water level of the river.

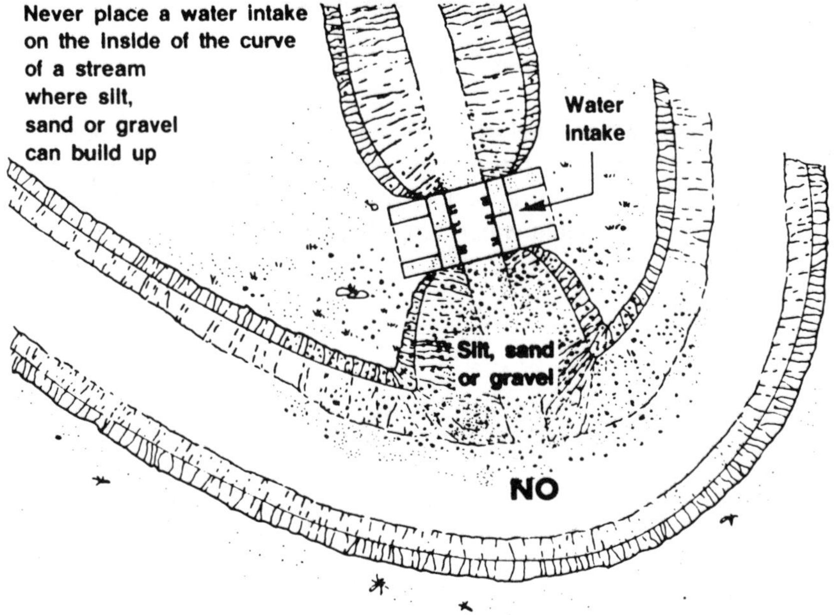

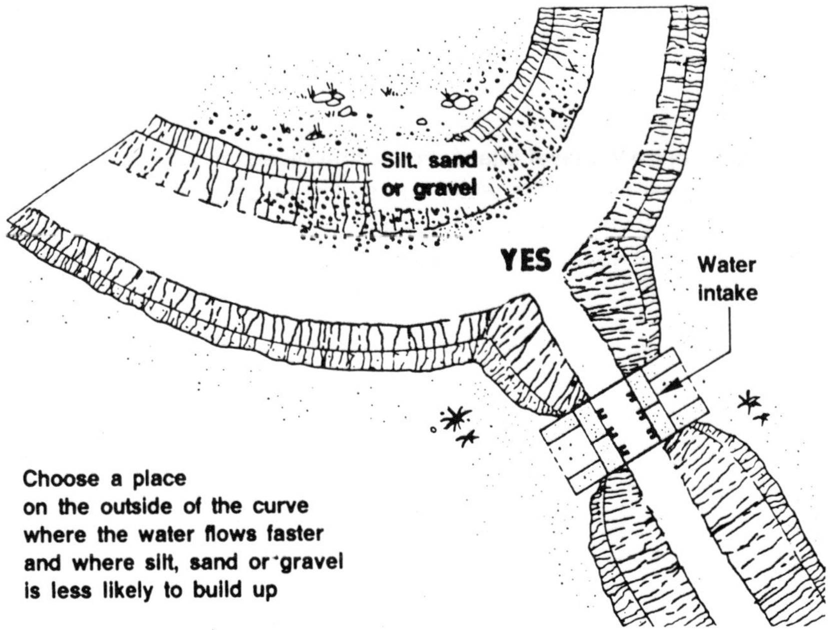

71 How to define the level of the water intake

1. There are two main types of intake:

- **an open or free-level intake,** in which the water supply levels are uncontrolled, and the intake operates in all water flow conditions. This system is simple and relatively cheap, but usually requires a reliable water supply that does not fluctuate excessively;

Free-level intake canal

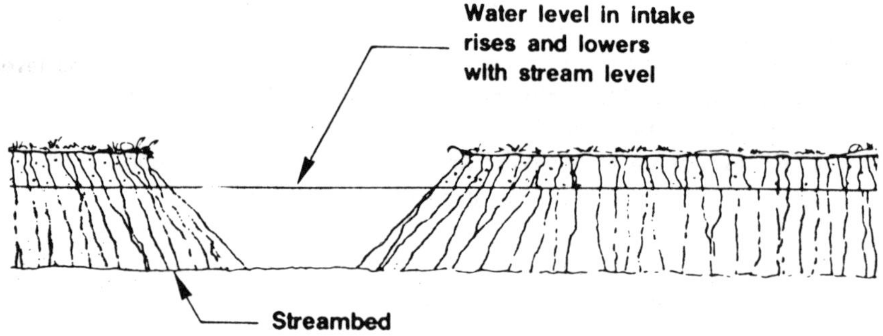

Controlled-level intake canal

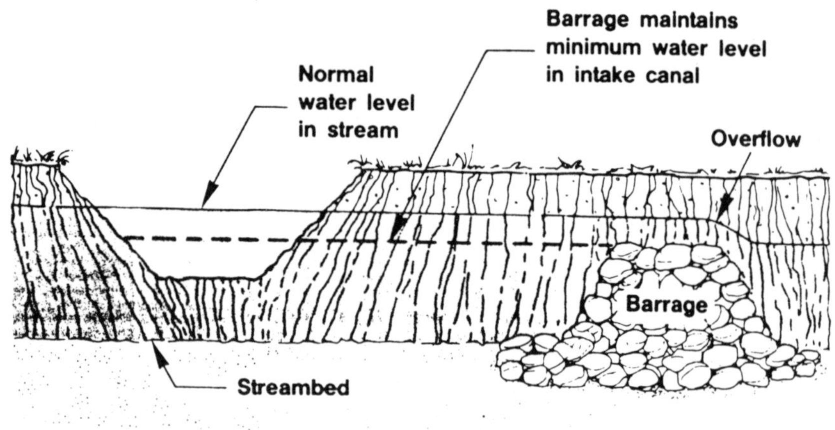

- **a controlled level intake,** in which a diversion structure is set up downstream in the water course for the purpose of maintaining water levels throughout a range of flow conditions. This system is more expensive but more reliable and provides a constant supply.

2. In both cases, **the important points** to consider are:

- the levels of the water source (river, stream, etc.) in relation to the water supply structure and the ponds themselves (see Section 19, **Construction, 20/1**);
- the depth from which you want to take the water (surface, lower levels or the complete depth of the water supply source).

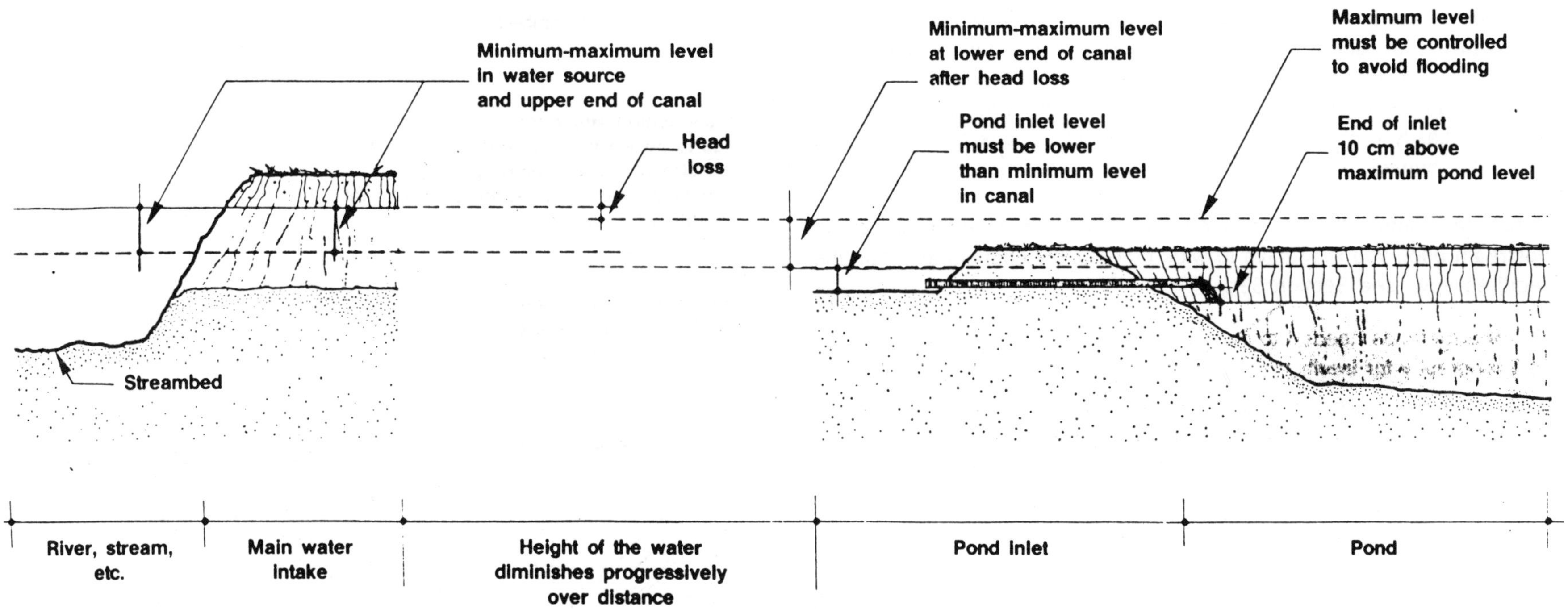

3. In the case of **an open intake system,** you must make sure that the level of the water in the supply source is sufficient at all times to allow you to take the water to the depth you need. You must also make sure that there is no risk of flooding the intake. As will be shown later, you can use an intake gate to control the incoming water supply.

4. In the case of **a controlled level system,** you can define the water level by setting the level of the diversion structure. The following points are very important.

(a) Check the longitudinal and cross-section profiles of the valley upstream of the structure to calculate the size of the flooded area that would be created behind the proposed structure (see Chapter 8, **Topography for freshwater fish culture,** *FAO Training Series,* **16/2**).
(b) Aim to set the diversion structure at the approximately minimum water level required for water flowing in the supply channel.
(c) Make sure that **flood water** can be removed, either over a weir or through a side channel (see Chapter 11). If the structure is made of soft, easily erodible materials (earth or clay), it is better to use a side channel.

Note: if the control structure has to be set lower to reduce **the size of the upstream pool,** you may have to widen the supply channel to obtain the required flow (see Section 82).

5. The methods needed to determine the relative levels are described in **Topography for freshwater fish culture,** *FAO Training Series,* **16/1** and **16/2**. Where possible, make use of local information. Ideally you should set up flow gauges and water-level stations. (See for example, Section 36 in **Water for freshwater fish culture,** *FAO Training Series,* **4**).

Make sure that flood levels do not overtop the control structure

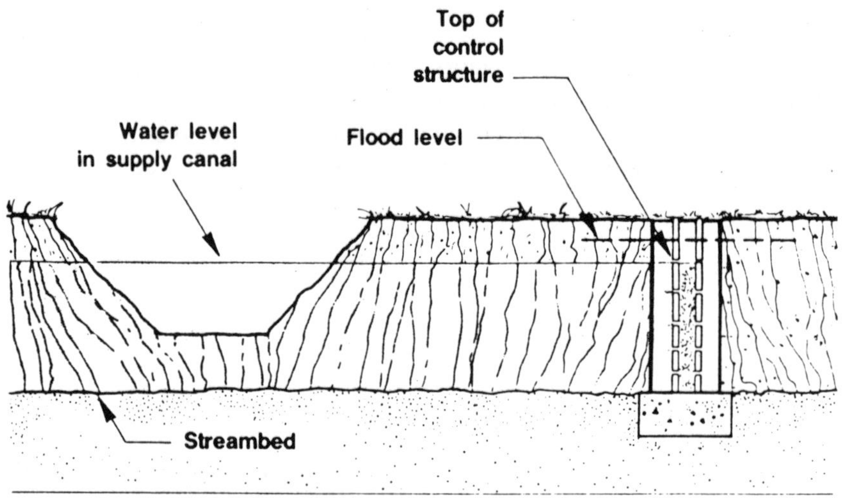

If the control structure has to be set lower to reduce the size of the upstream "pool", you may have to widen the supply canal to obtain the required flow

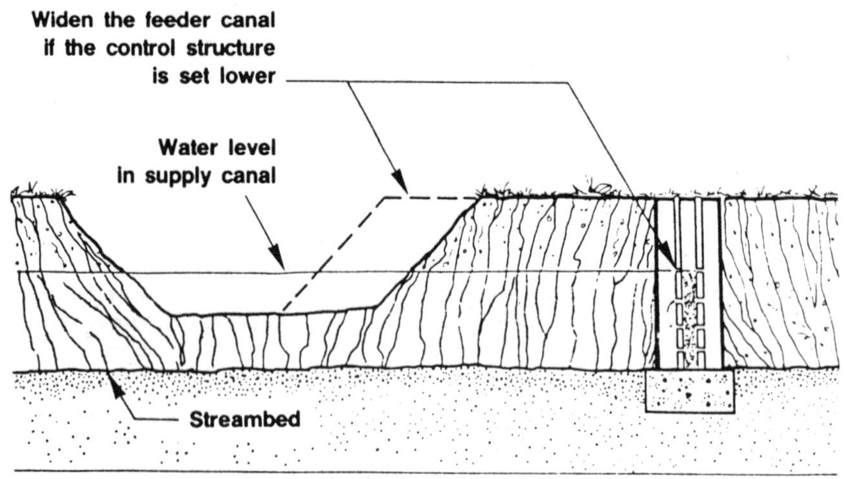

72 The size of the water intake

1. The wider the water intake area, the less will be the **head loss*** as the water flows to the ponds. This factor may become important when there is very little head available.

2. In most cases, however, **the water intake is about the same width as the supply canal** connected to it. The size of the supply canal is chosen according to the flow required (see Section 82). If the supply canal is particularly wide, or if you want to increase the head loss at the water intake (for example, when the external water level is much greater than the level required within the supply canal), the intake can be made narrower than the supply canal. Generally, **a narrow intake is easier to control,** as the sluice boards or gate controls are easier to move.

3. As an approximate guide, **Graph 6** gives typical flow rates through intake structures at different head loss. This head loss should be added to the supply canal head loss (Section 82) to define the relative levels of the intake and the ponds.

Example

If 0.20 m is available between the minimum intake water level and the pond supply, a flow of 0.25 m^3/s is required. It is calculated that head loss in the supply canal due to its bottom slope (see Section 82, paragraph 8) is 0.15 m. Possible head loss through intake is therefore limited to 0.20 − 0.15 m = 0.05 m or 5 cm. To ensure the required flow rate, the intake width would have to be at least 0.40 m or 40 cm (**Graph 6**).

4. The intake control structures are described later (see Sections 76 and 77). First, you will learn about the diversion structures that are used for intakes (**Table 31**).

GRAPH 6

Water flow through sluices

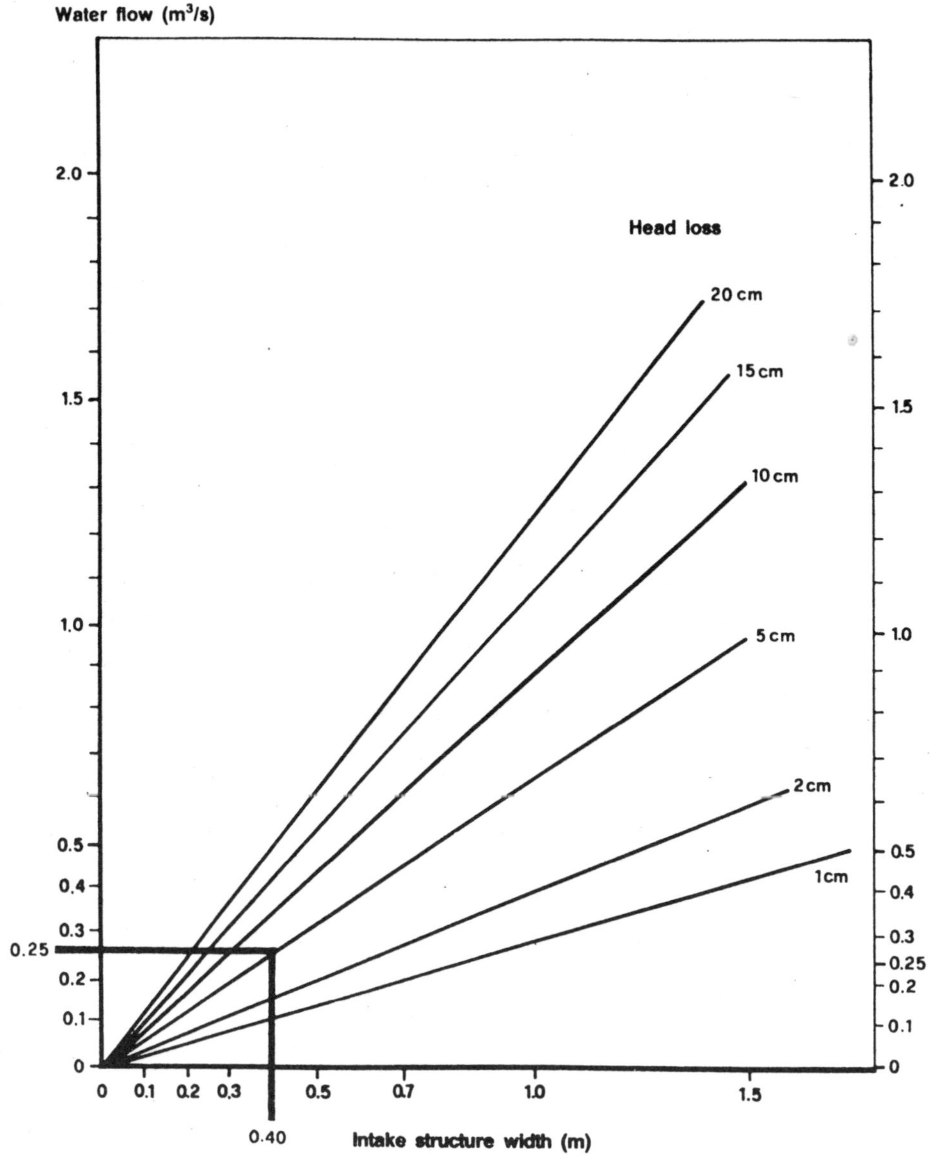

73 Simple diversion structures

1. Simple diversion structures can be constructed from **a range of materials**. These materials are suitable for holding back water, but should not be used where water regularly overflows.

Earthen barrage dam

2. You can **totally block the channel** of a small stream with an earthen dam (see Sections 61 and 66, **Construction, 20/1**). Proceed in the following way:

(a) **Design the dam** to be built as if it were for a barrage pond (see Sections 61 and 66, **Construction, 20/1**).

(b) **Divert the stream** around the construction site. It is easiest to do this when the stream flow is low, for example, toward the end of the dry season.

(c) Stake out the dam base, set out the earthwork and **build the dam across the stream channel** (see Section 66, **Construction, 20/1**).

(d) Construct **the intake structure, the water feeder canal and its overflow** away from the ponds.

(e) Gradually remove the temporary diversion, letting the stream establish itself in its original channel and fill the feeder canal with water.

Note: if necessary, protect the wet side of the new dam with rocks or stones.

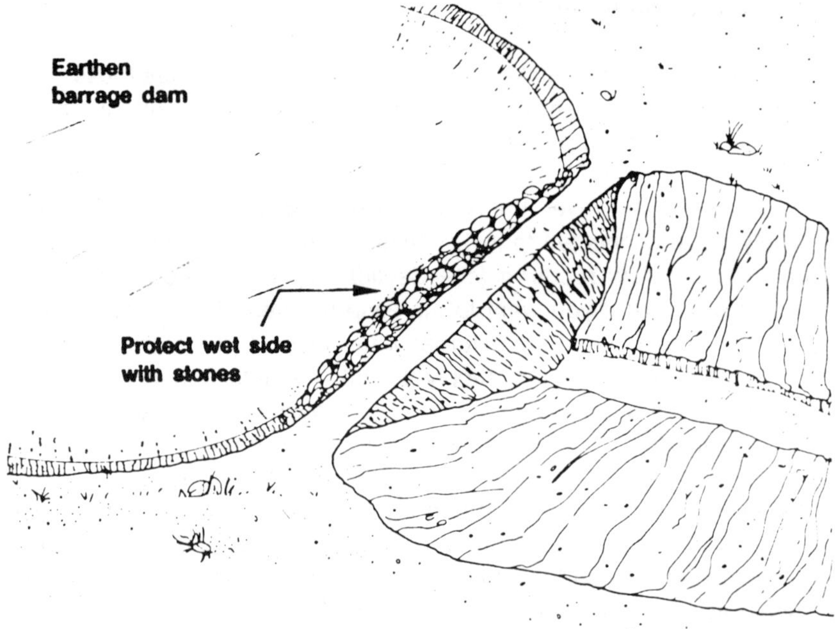

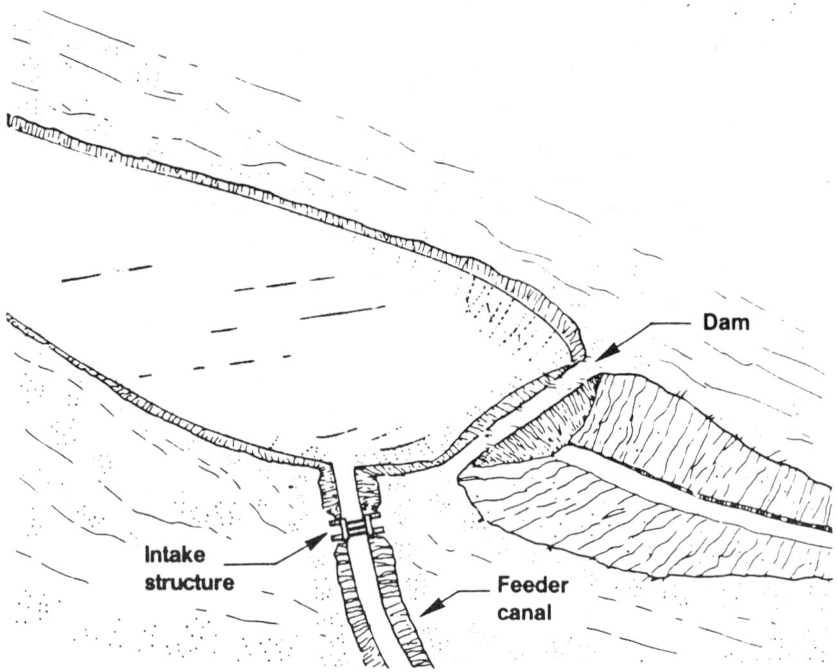

Bamboo or wooden pole barrier

3. You can also block the channel of a small stream using a double row of wooden or bamboo poles lashed together with flexible lianas or vines, and packed with clay soil between the poles to prevent water seepage.

4. Remember that:

- the double row of poles should be placed side by side and driven vertically into the ground;
- the barrier should extend well into each of the stream banks; and
- the barrier will be stronger if you build it curving against the flow of the stream.

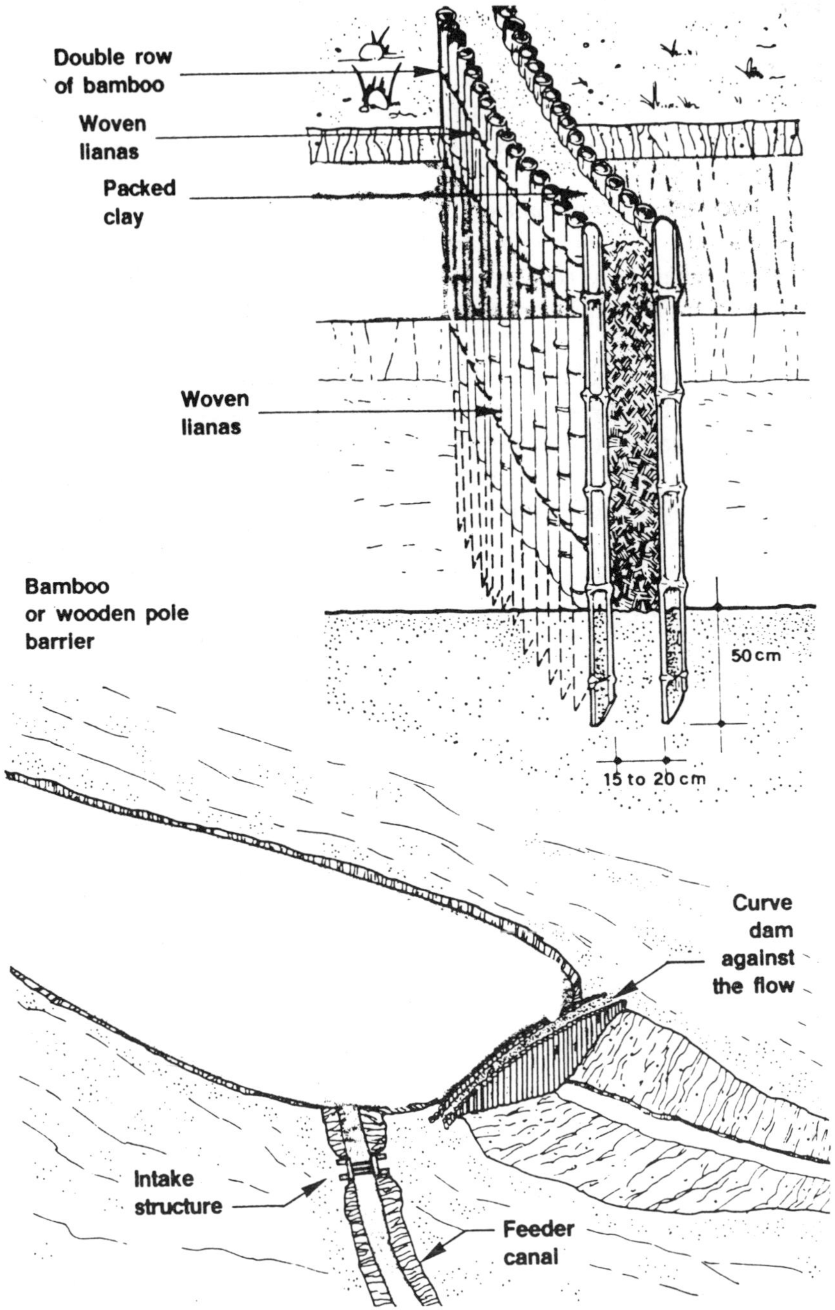

Wooden plank barrier

5. There are other ways you can build a barrier using planks and wooden poles. This kind of barrier **can easily be removed** in the rainy season when the water level begins to rise in the stream channel.

6. Two kinds of plank barriers are shown here. In the first, the planks are placed at a slight angle and braced by timbers. In the second, the planks are held in place between a light structure of logs and can be removed by lifting out one plank at a time.

(a) The planks should be well driven into the ground next to each other.
(b) The joints between the planks may, if necessary, be filled with heavy clay to make the barrier more impervious.
(c) You can also use medium- to heavy-weight polythene sheeting, overlapping bags, old inner-tubes or tarred felt or sacking to reduce seepage.
(d) The water level in the stream channel can be raised to reach a depth of 0.8 to 1 m.

Angled-plank barrier

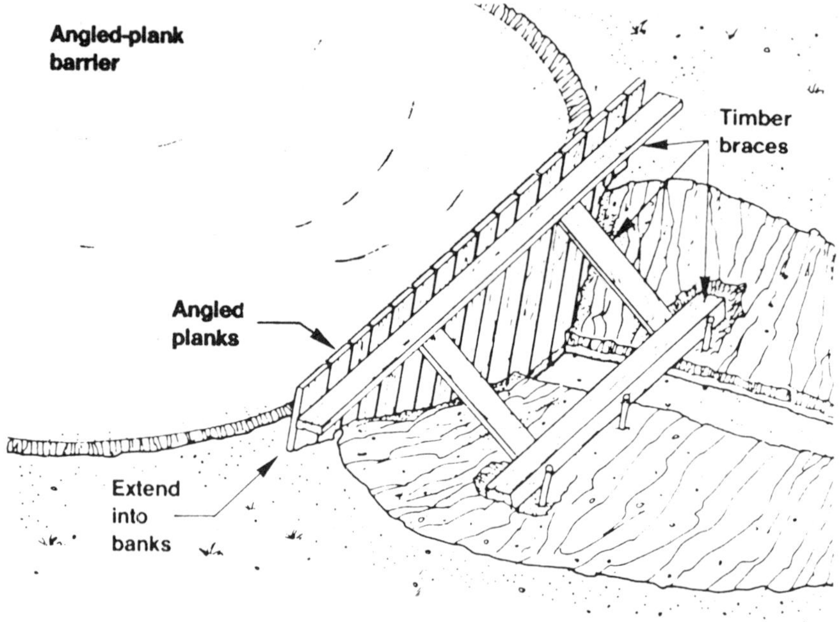

Vertical plank barrier

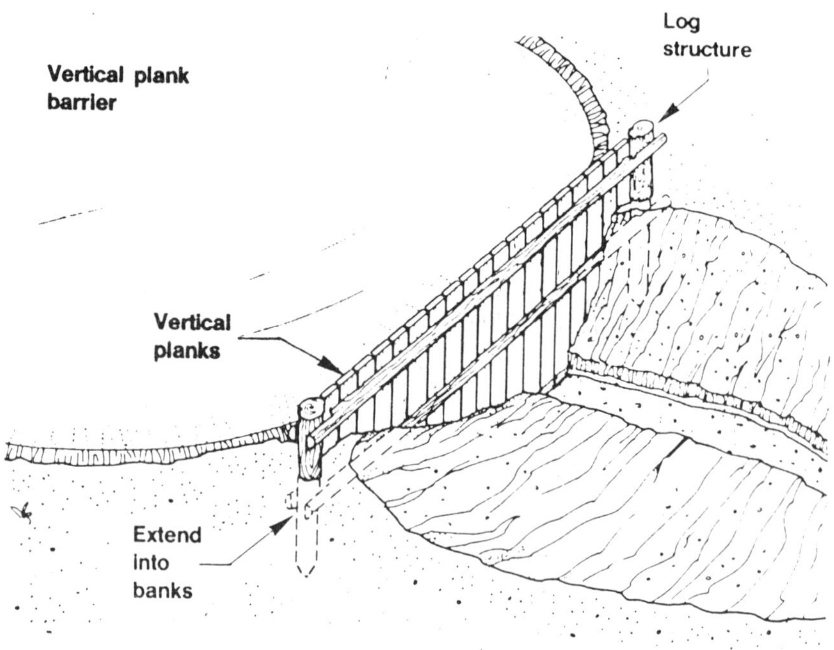

74 Submergible diversion structures

1. These structures can be used both for holding back water and for overflows.

Wooden pole barrier

2. The purpose of this type of barrier is only **to raise the water level** in the stream channel **without blocking the water flow completely**. Some water can escape through the pervious barrier, while the rest flows over the barrier.

3. The barrier is made of **two rows of wooden poles** driven vertically into the streambed and closely tied together with ropes or lianas. Gravel or rock can be placed downstream of the barrier base to reduce bottom erosion. The barrier should extend well into both stream banks.

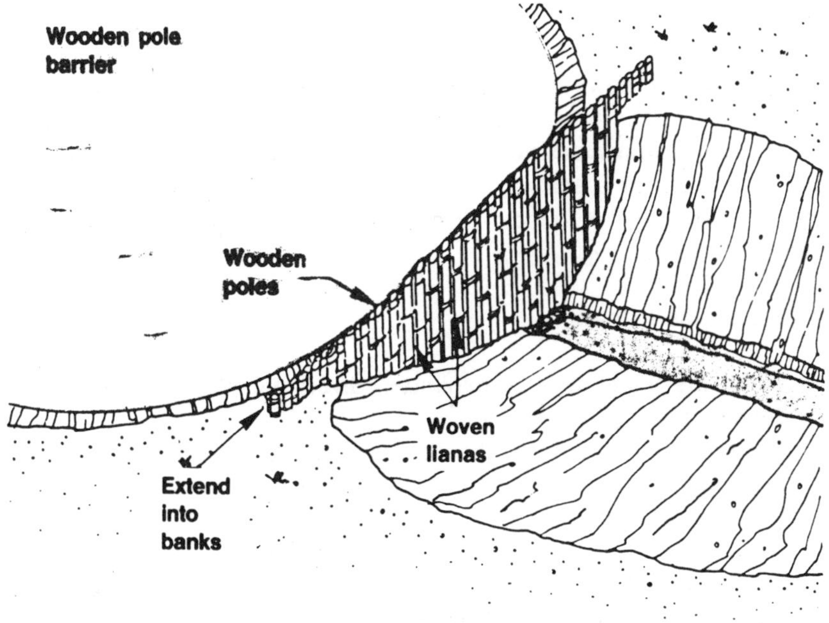

Rock barrier

4. This is a very simple submergible structure made **by piling rocks across the streambed** and forming a small porous barrier. You should build this barrier in layers. For each layer, use relatively large rocks first, and then fill the gaps with smaller rocks. The base width of the barrier depends on its final height, which **should not exceed one metre**. If you work carefully, you can build side slopes with a 1:1 ratio to save additional work. With this method, a one-metre-high barrier requires a base width of about 2.5 m to give a top width of 0.5 m.

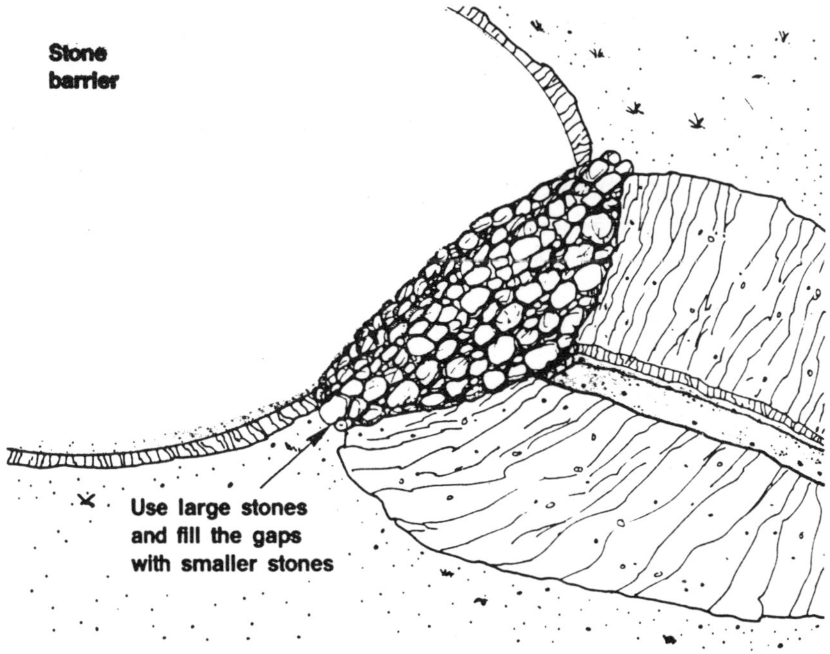

Gabion barrier

5. You learned how to construct gabions earlier (see Section 47, **Construction, 20/1**). These baskets can be used very effectively in small streams with **a maximum flow of less than 100 l/s** to divert part of the water and to act as a spillway when floods occur. They are particularly suitable when gravel is found on the streambed and when the stones can be found locally.

6. Proceed as follows:

(a) When the water flow is minimum, **divert the stream** around the construction site.

(b) Stake out the base of the barrier you wish to build, for example, a rectangular area 3 m wide across the streambed, at a **right angle to the flow direction**.

(c) Across this area, prepare **a horizontal platform** at a depth of about 0.5 m below the streambed level.

(d) Build **the foundation of the barrier** on the horizontal platform, using one layer of **thin gabions** (2 m × 1 m × 0.5 m), as shown in Section 37, **Construction, 20/1**.

(e) On top of this foundation build **the body of the weir** using two layers of **thin gabions** placed across and on the upstream part of the foundation. Anchor these baskets well into the stream banks and into each other.

(f) If necessary, **protect the banks** above the second layer with additional lateral layers of **thin gabions**. Fill in the gaps with compacted clayey soil.

Gabion barrier in a stream

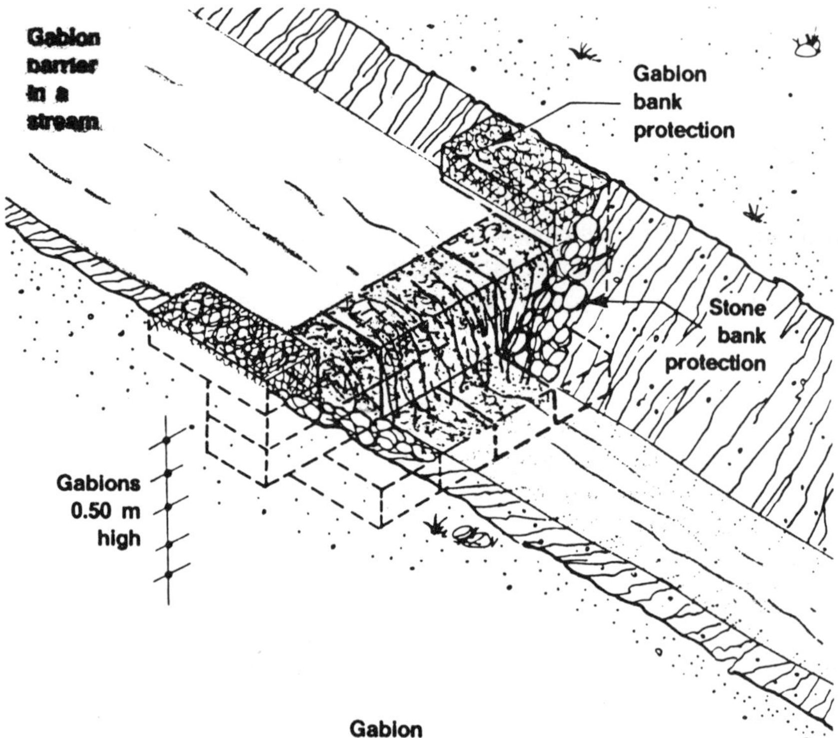

Diagram of a gabion barrier with additional bank protection

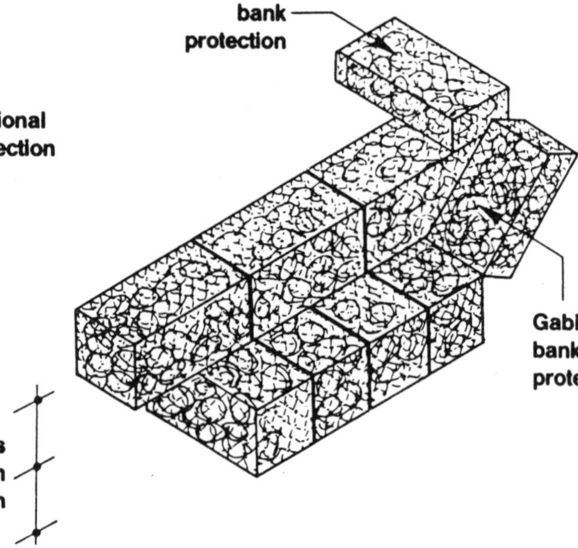

75 Adjustable diversion structures

1. Adjustable diversion structures are more expensive and more complicated to build, but they provide an easier and more precise control of the water level in the stream channel. They are permanent structures made of reinforced concrete and removable planks. In the next paragraphs, you will learn about two simple designs for adjustable diversion structures. They can be changed to suit local conditions.

Two-pillar barrage

2. You can build a narrow adjustable barrage 2.5 to 3 m long and 1 to 1.5 m high, using **reinforced concrete** and **strong planks** 5 cm thick.

3. For a barrage made of 1 m planks and consisting of two columns 1 m high, you will need the following materials:

- concrete for the foundation: 2.8 × 0.8 × 0.25 m = 0.56 m^3
- concrete for two pillars: 0.36 m^3 × 2 = 0.72 m^3
- reinforcement of pillar: steel bars 6 mm in diameter:
  for verticals:
  (14 × 1.10 m) × 2 = 30.8 m
  for cross-ties:
  (4 × 1.90 m) × 2 = 15.2 m
  (4 × 1.35 m) × 2 = 10.8 m
- reinforcement of foundations: steel bar 8 mm in diameter, 4 × 2.70 m = 10.8 m; and steel bar 6 mm diameter, 14 × 0.60 m = 8.4 m.

Alternatively, reinforcement mesh, such as 10 cm square, 6 mm thickness can be used.

Plan of two-pillar barrage

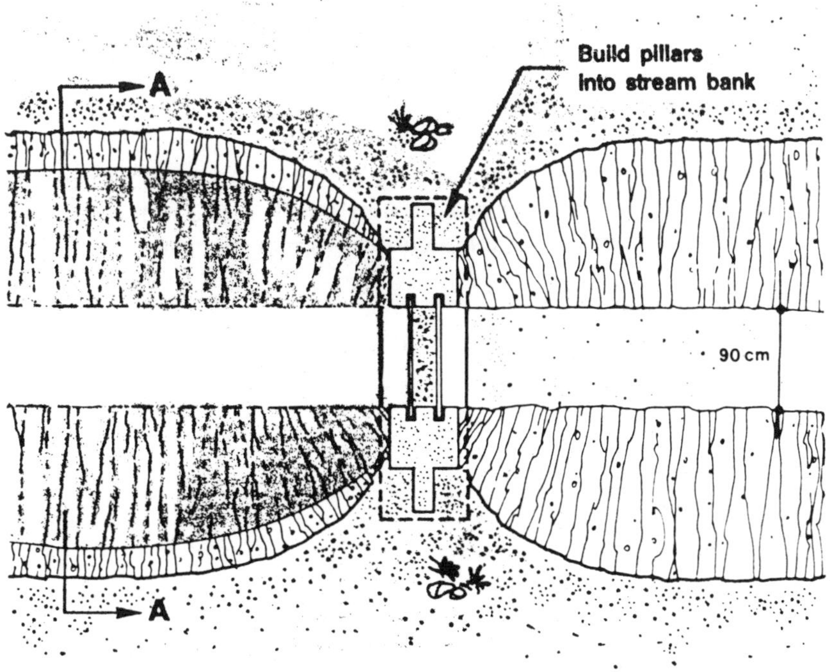

Section AA

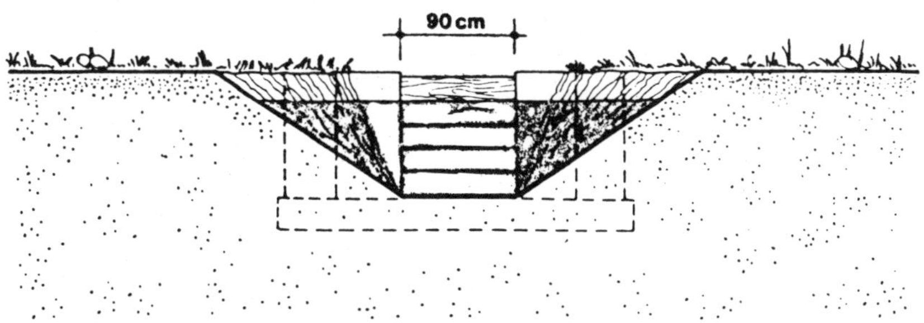

Three-pillar barrage

4. You can build **a wider adjustable diversion structure barrage** 4 to 7 m long and 1 to 1.5 m high using two lateral concrete pillars and one or more central ones, connected by two series of strong planks 5 cm thick.

5. For a barrage 1 m high made out of 1-m-long planks and with one central pillar, you will need the following **materials:**

- concrete for the foundation: $4.2 \times 0.8 \times 0.3$ m = 1.01 m^3
- concrete for three pillars: $(0.36\ m^3 \times 2) + (0.3\ m^3) = 1.02\ m^3$
- reinforcement of pillars, steel bars 6 mm diameter:

  for verticals:
  $(10 \times 1.10\ m) + 30.8\ m = 41.8\ m$
  for cross-ties:
  $4 \times 1.05\ m = 4.2\ m$
  $8 \times 0.50\ m = 4.0\ m$
  $4 \times 0.60\ m = 2.4\ m$
  plus far end pillars:
  $15.2 + 10.8\ m = 26.0\ m$

- reinforcement of foundations: steel bar 8 mm diameter, $4 \times 4.20\ m = 16.8\ m$; and steel bar 6 mm diameter, $21 \times 0.60\ m = 12.6\ m$; or use reinforcing mesh as mentioned in previous example.

Plan of three-pillar barrage

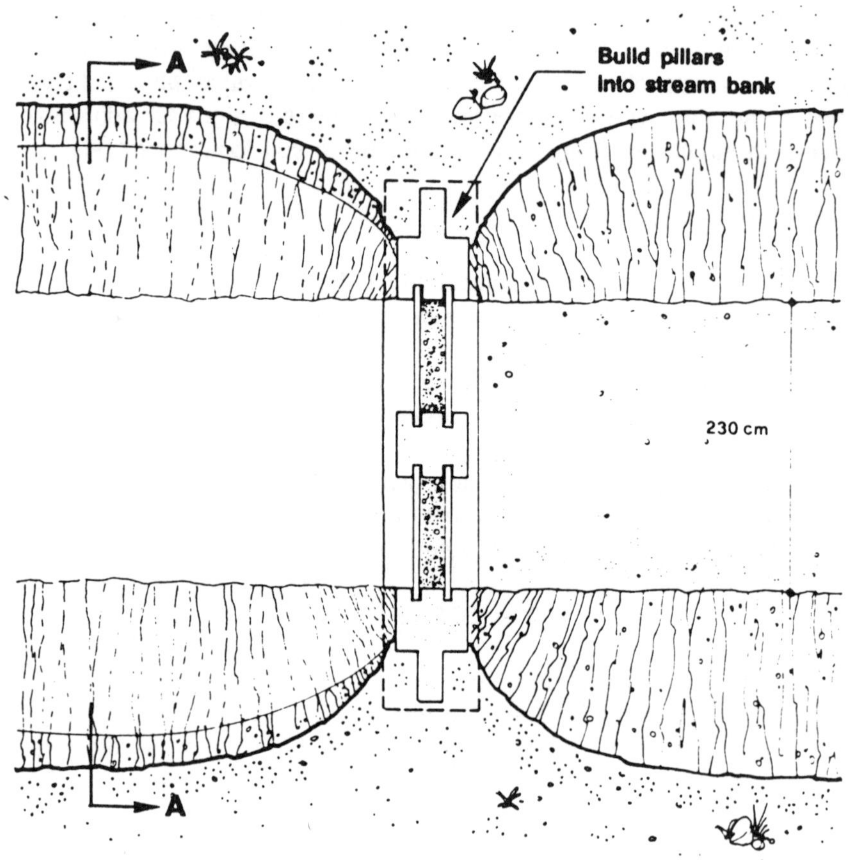

Section AA

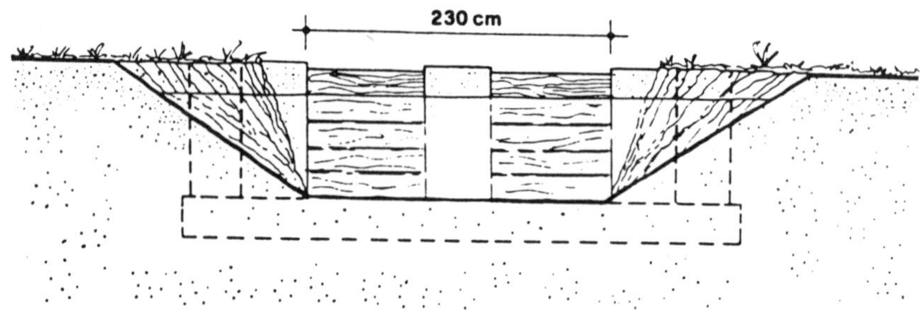

6. Bury the foundations of the barrage in the dry streambed, anchoring them as far as possible into solid footing. The top level of the foundation should be about 5 cm below the level of the streambed.

7. Build each pillar into the banks of the stream. If necessary, build lateral wings from stones or concrete. You may use additional planks and fill the space between them with well-compacted clay soil. To avoid erosion, reinforce the stream bank next to each pillar with stones.

8. To make the concrete forms and fix the reinforcement well, you may need the assistance of a good mason.

Note: if you are unsure about the stability of the streambed, it may be safer to join the foundations to form a single foundation spanning the stream. This will require more material but will retain a fixed shape if the bed should erode.

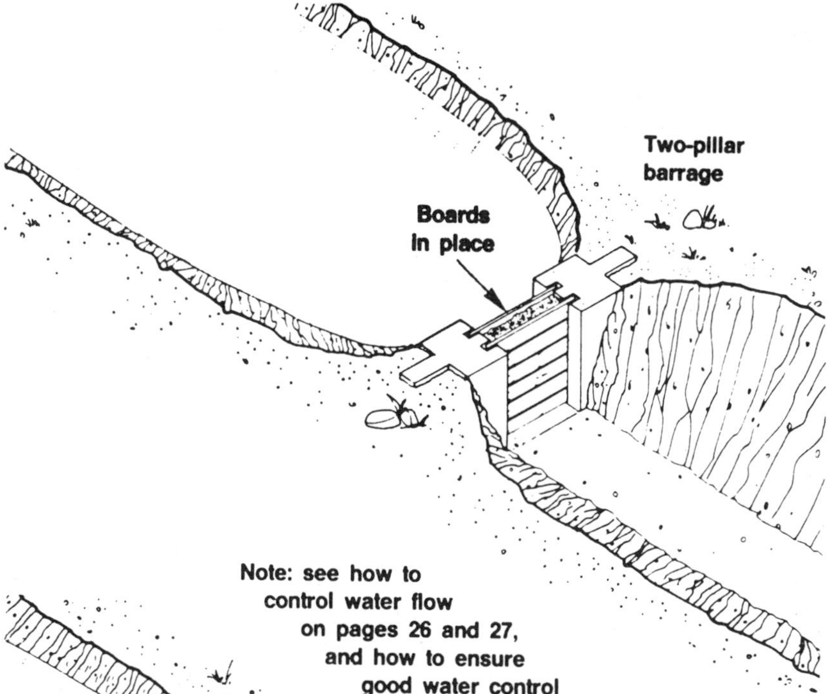

Note: see how to control water flow on pages 26 and 27, and how to ensure good water control on page 28

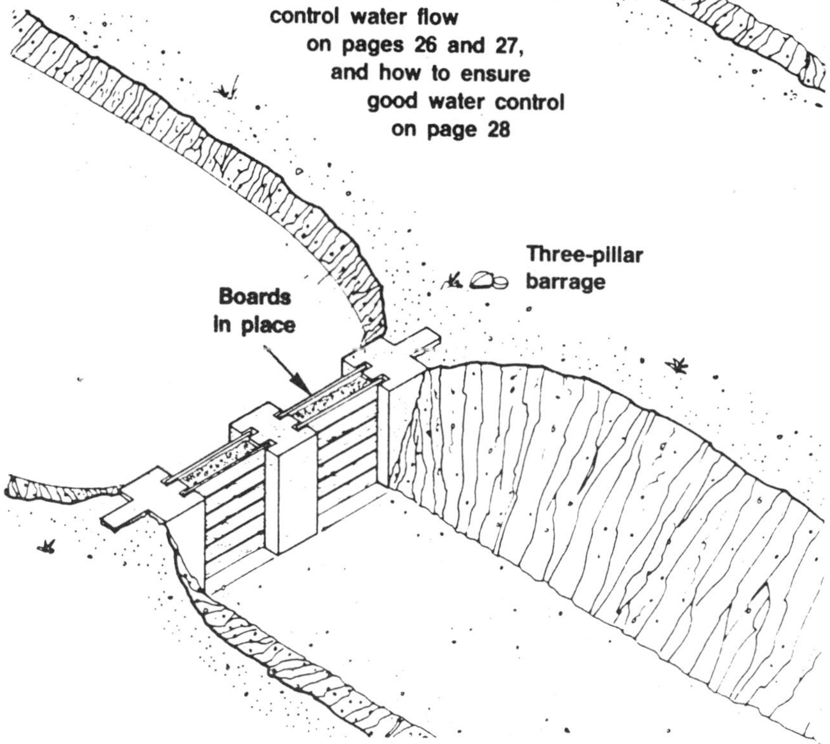

Building a two-pillar barrage

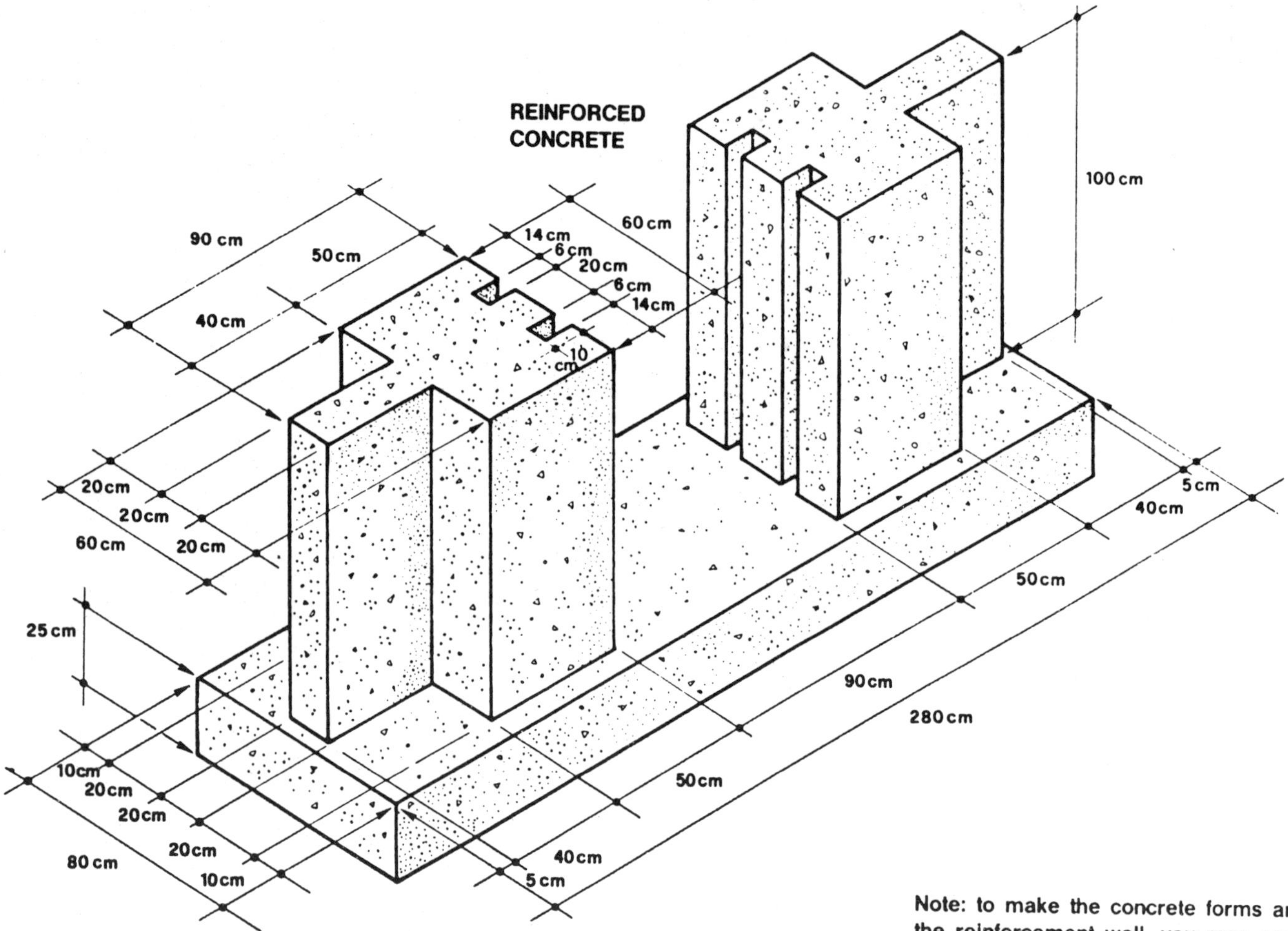

Note: to make the concrete forms and fix the reinforcement well, you may need the help of a good mason

Placement of steel bar reinforcement for two-pillar barrage

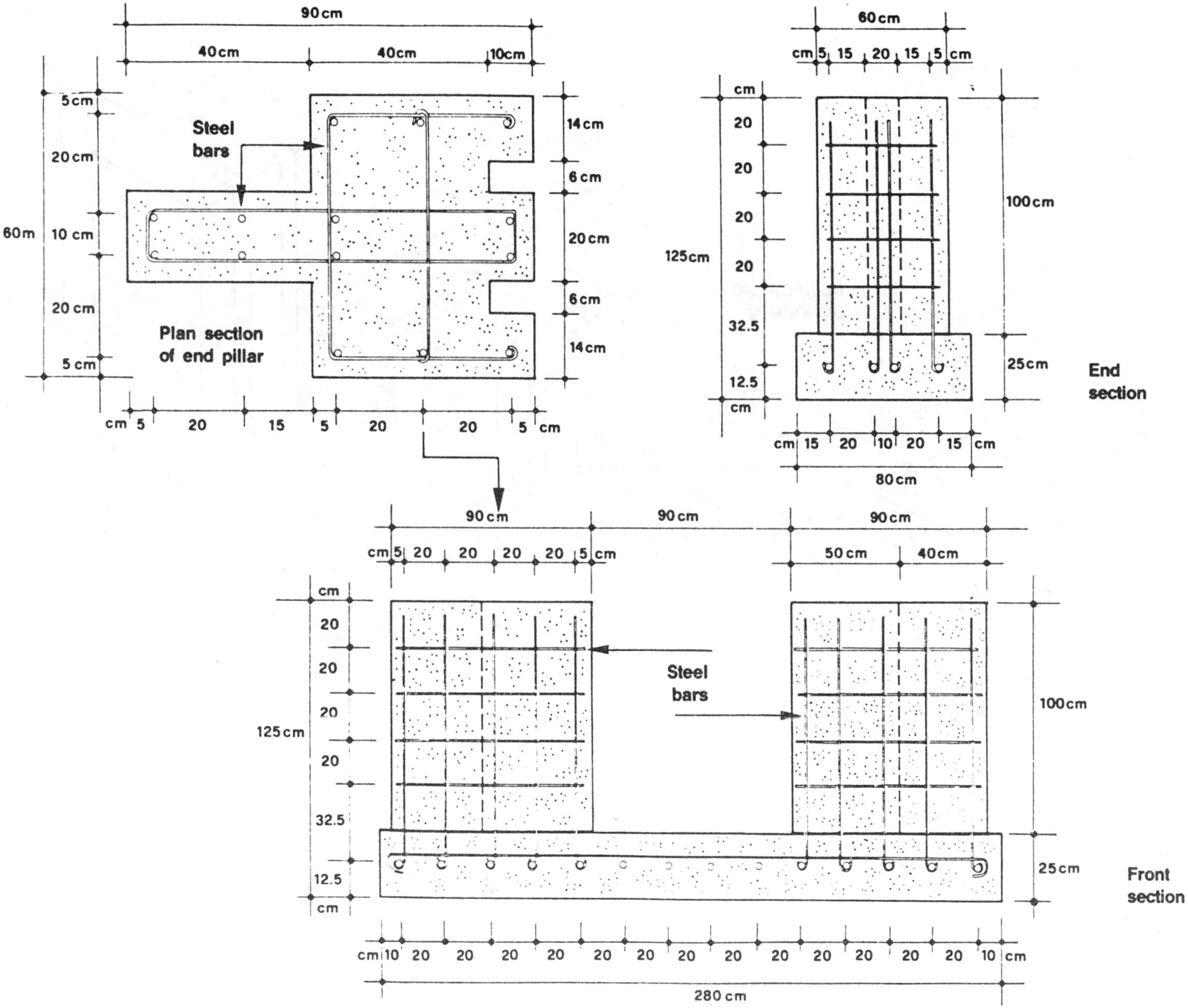

Building a three-pillar barrage

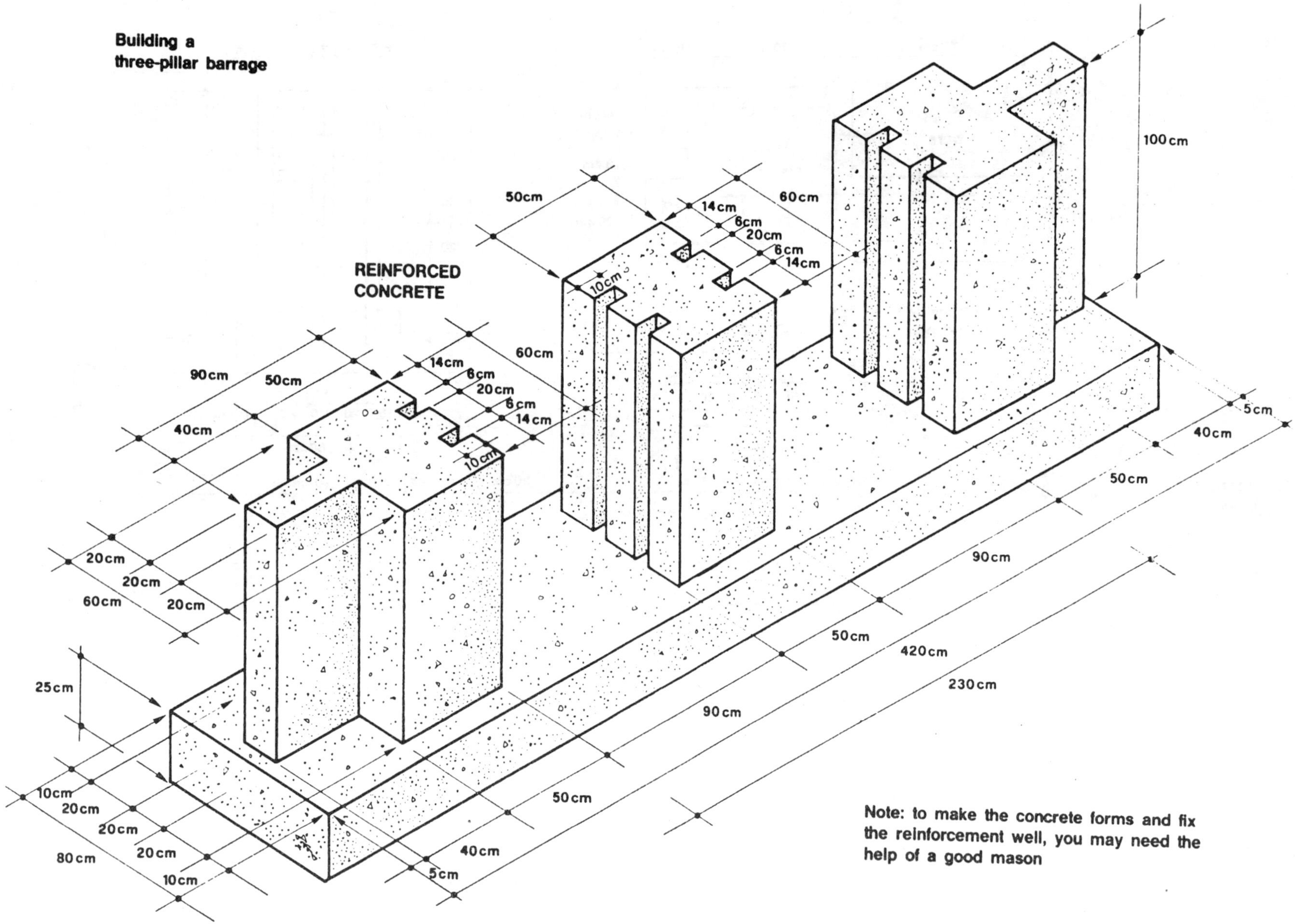

Note: to make the concrete forms and fix the reinforcement well, you may need the help of a good mason

Placement of steel bar reinforcement for three-pillar barrage

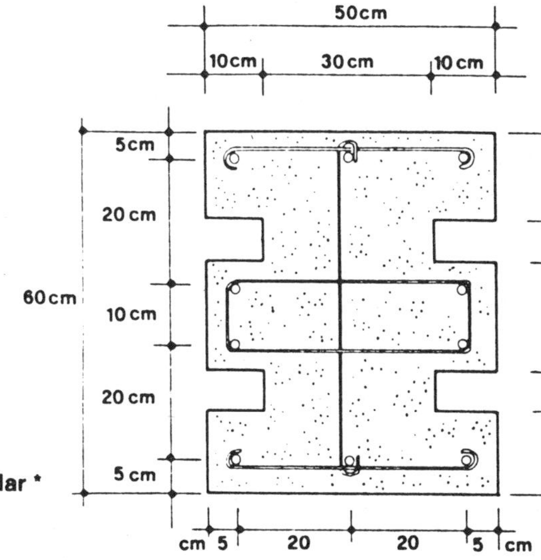

Plan section of centre pillar *

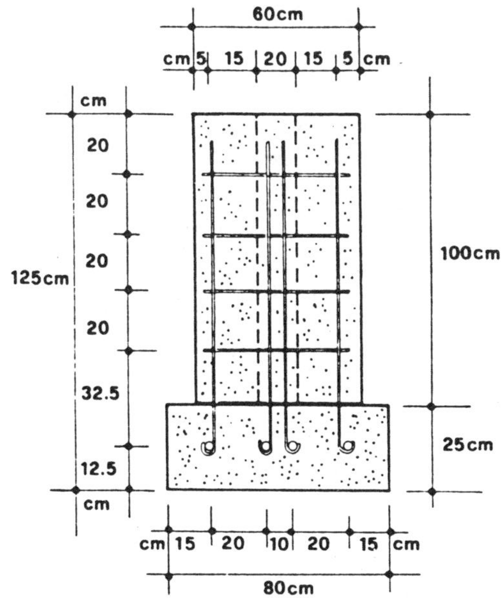

End section

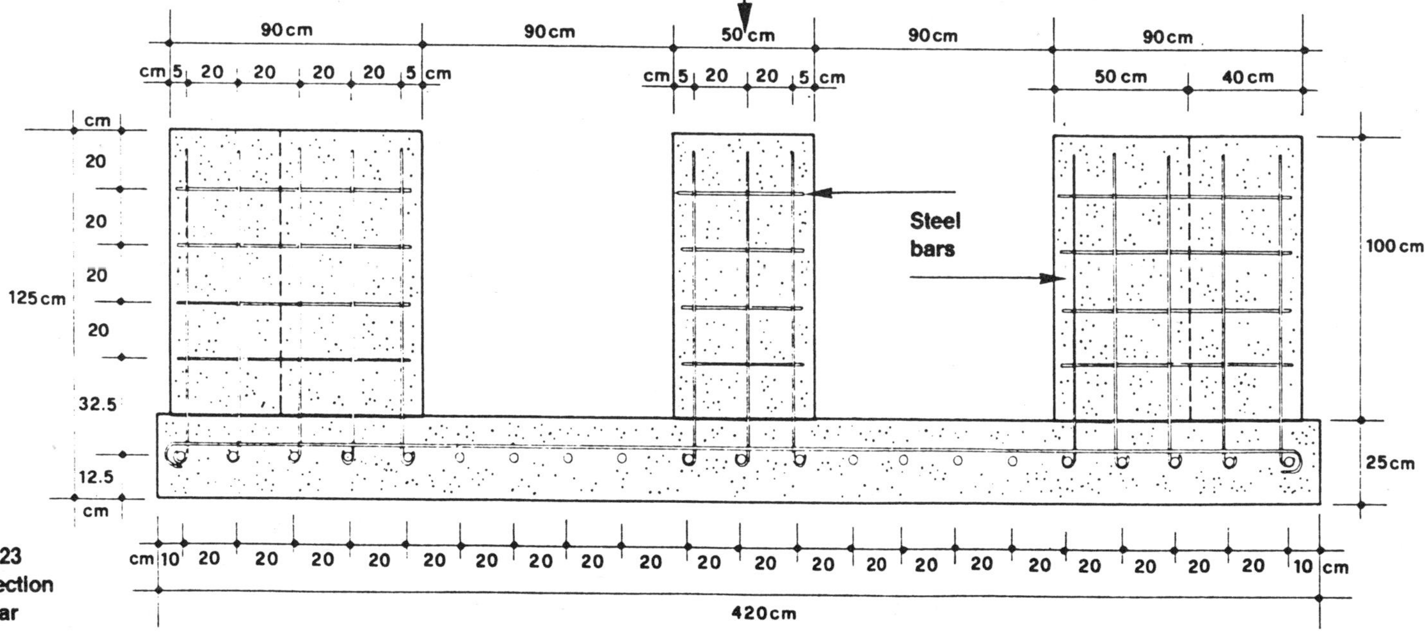

Front section

* see page 23 for plan section of end pillar

Two major types of structure

1. The previous sections described how to define the level and size of major water intake structures. We now consider the types of structures to be used. There are **two basic types:**

- an **underflow intake,** in which water flows below the control structure, which is raised or lowered to adjust flow;
- more commonly, an **overflow intake,** in which water flows over the control structure, which can also be raised or lowered.

Controlling water flow

2. You can chiefly control water flow in two ways:

- with **sluice boards,** used both for overflow and underflow intakes;
- with **a penstock,** or sliding metal door, controlled by setting with pegs or bolts, or with an adjustable handle. It is used for underflow intakes; it is usually more expensive than sluice boards but offers more precise control.

3. Both of these systems are set in a holding structure, which can be built of wood, bricks or blocks, concrete or steel like the adjustable diversion structures described in Section 75. The structures are built with one or more sets of anchoring slots or grooves in each side of the control structure, as illustrated on this and the opposite page.

Note: an intake can also be made with a **swinging arm** or **flexible stand-pipe** (see pages 130 to 132 in Section 103). This alternative is less common as a main intake, but can be convenient for controlling smaller water flows. Typical capacities of such pipes are given in Table 13 in Section 38, **Construction, 20/1**.

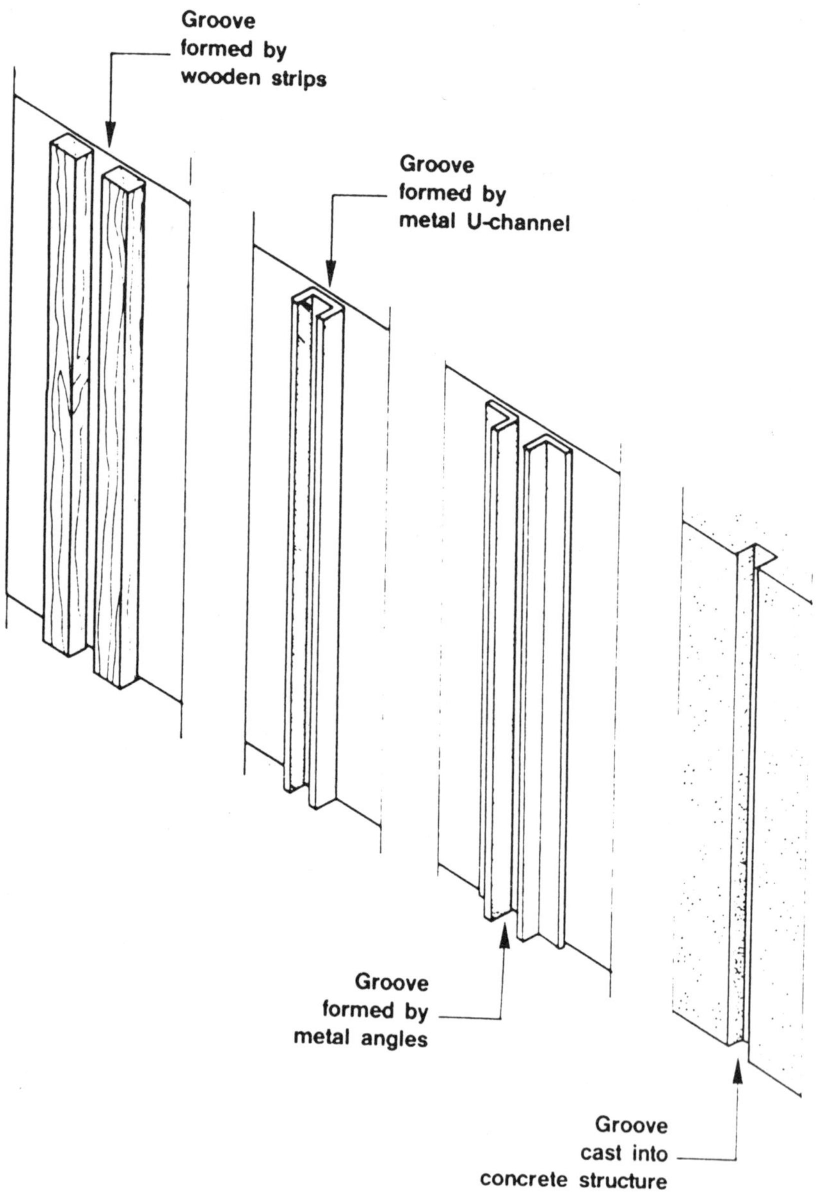

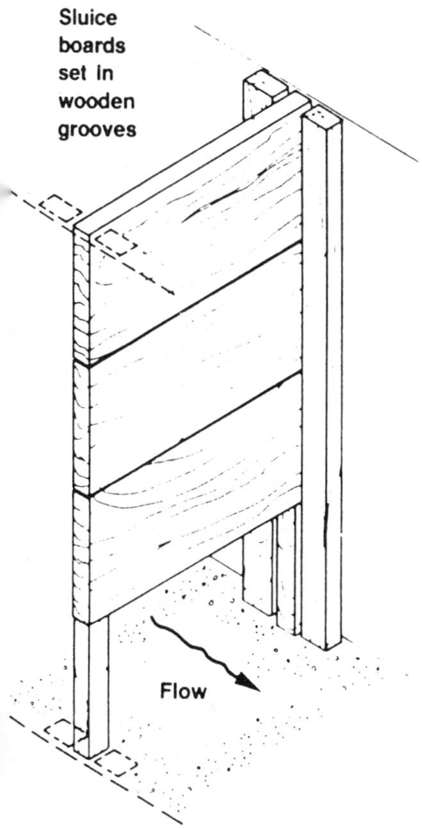
Sluice
boards
set in
wooden
grooves
Flow

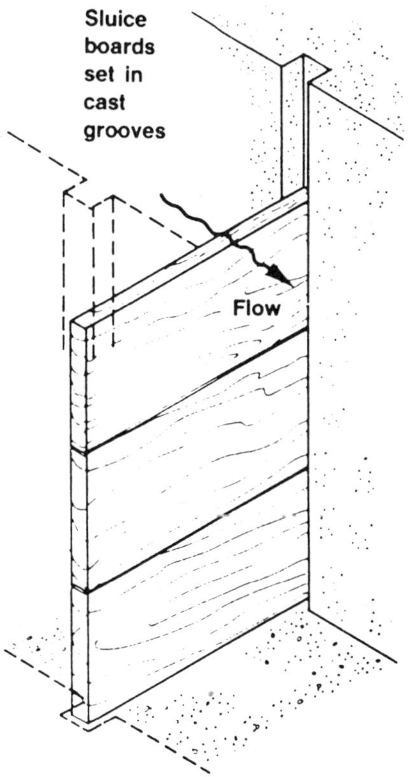
Sluice
boards
set in
cast
grooves
Flow

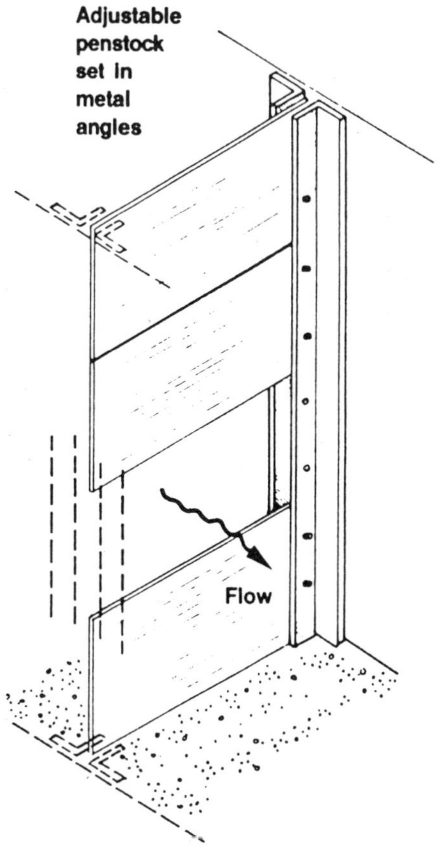
Adjustable
penstock
set in
metal
angles
Flow

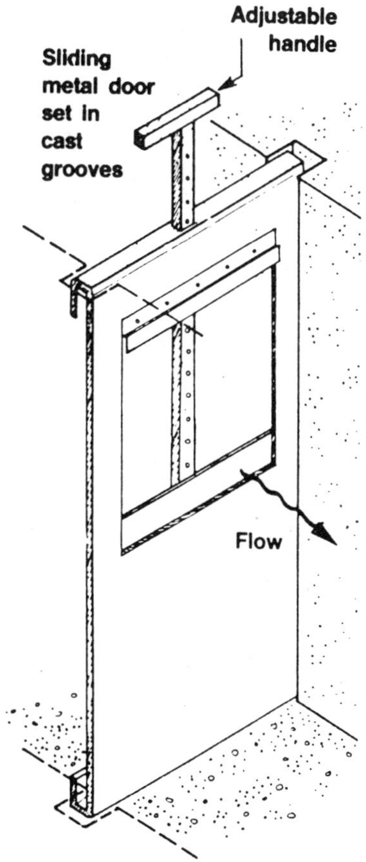
Adjustable
handle
Sliding
metal door
set in
cast
grooves
Flow

Ensuring good water control

4. While penstocks are usually designed to seal tightly in a range of conditions, **sluice boards are difficult to seal properly,** especially for wider gates, where boards are more likely to twist and warp. One useful improvement is to use sealing flaps of heavy polythene sheet or old inner tube. Usually, however, **three parallel sets of grooves** are used, two for slipping one screen and one series of boards in or out and one for adding a second series of boards when the need arises to stop the water flow completely within the feeder canal.

Estimating the flow rate through the intake

5. The flow rate through these structures when open can be estimated using **Graph 6. For overflow intakes with boards,** the control acts like a small weir (see Section 36, **Water, 4**) where water flow depends on the width of the board and the depth of the water flowing over it. **Table 32** shows typical values. **For underflow intakes such as penstocks,** the flow depends on the difference in head from one side of the sluice to the other, and on the size of the opening. **Table 33** shows typical values.

Protecting the intake from erosion

6. Care must be taken in all cases to **minimize erosion,** as the speed of flowing water may substantially increase around the gates. As a general rule, unless special designs are used (consult a hydraulics specialist), you should **limit the drop across the intake to 80 cm**.

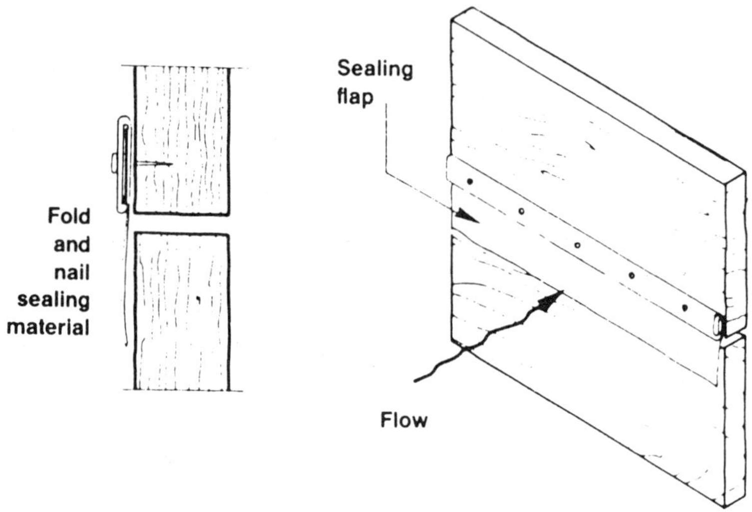

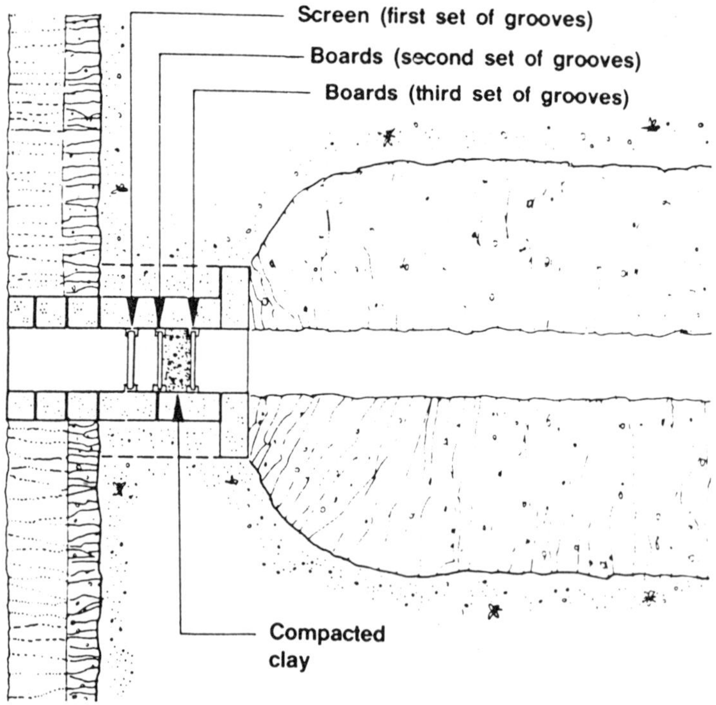

TABLE 32

Water flow over sluice boards (m^3/s)

Weir structure width (m)		0.3	0.7	1.0	1.3	1.7	2.0
Depth over weir (cm)	1	0.001	0.001	0.002	0.002	0.003	0.004
	2	0.002	0.004	0.005	0.007	0.009	0.010
	5	0.006	0.014	0.020	0.027	0.035	0.041
	10	0.016	0.040	0.057	0.074	0.098	0.115
	15	0.029	0.072	0.104	0.136	0.179	0.211
	20	0.043	0.109	0.158	0.207	0.273	0.323

TABLE 33

Water flow through penstock sluice (m^3/s)

Sluice opening area (m^2)		0.1	0.2	0.3	0.5	1.0	1.5
Head loss across sluice (cm)	1	0.027	0.055	0.082	0.137	0.274	0.412
	2	0.039	0.078	0.116	0.194	0.388	0.582
	5	0.061	0.123	0.184	0.307	0.614	0.921
	10	0.087	0.174	0.260	0.434	0.868	1.302
	15	0.106	0.213	0.319	0.532	1.063	1.595
	20	0.123	0.246	0.368	0.614	1.228	1.841

77 Screens and intake protection

1. Where conditions are likely to be turbulent, the sides and the outflow end of the structure may be reinforced using wood, light reinforced concrete, brick or boulders set in cement.

2. Intake structures can be protected from debris such as leaves or branches and from erosion by flowing water in several ways. Screens or guards can be used against debris in most cases, while gabions, wooden or bamboo piling, or rock reinforcement can be used for protection against erosion.

Using screens

3. **Screens** can be set up in a number of ways, the most common being a simple side screen. They can also be set up horizontally, as inclined screens or even in the base of the supply stream.

4. In many cases, **a single screen** is used, usually made from steel bars 6 to 8 mm in diameter spaced 20 to 35 mm apart. This screen is sufficient for clearing larger objects. If smaller particles need to be removed, an additional **screen of finer bar** (e.g. 4 to 6 mm diameter) at closer spacing (5 to 10 mm), or **steel mesh**, can be used. The additional screen may be set up inside the main screen or may be incorporated into the intake structure itself.

5. For simple structures, the screen has about the same **cross-sectional area** as the main intake. To improve flow and to ensure the screen will operate even when partly blocked, it is frequently made larger than the intake (e.g. by using inclined "V" screens or horizontal screens — see the next manual in this series, **Management for fresh-water fish culture,** *FAO Training Series*, **21/1,** Section 29).

6. Remember that if the screen starts to become blocked, it may direct water to diversion canals and so reduce the flow to the pond supply.

7. **Screens can be cleaned** by lifting the screen from its slots and brushing it, or by raising the hinged portion of a horizontal or inclined screen, or by arranging the screen so the passing water current will keep it clean. Mechanized automatic screens are also available, but these specialized installations are outside the scope of this manual.

8. You can learn more about screens in **Management for freshwater fish culture** (see above).

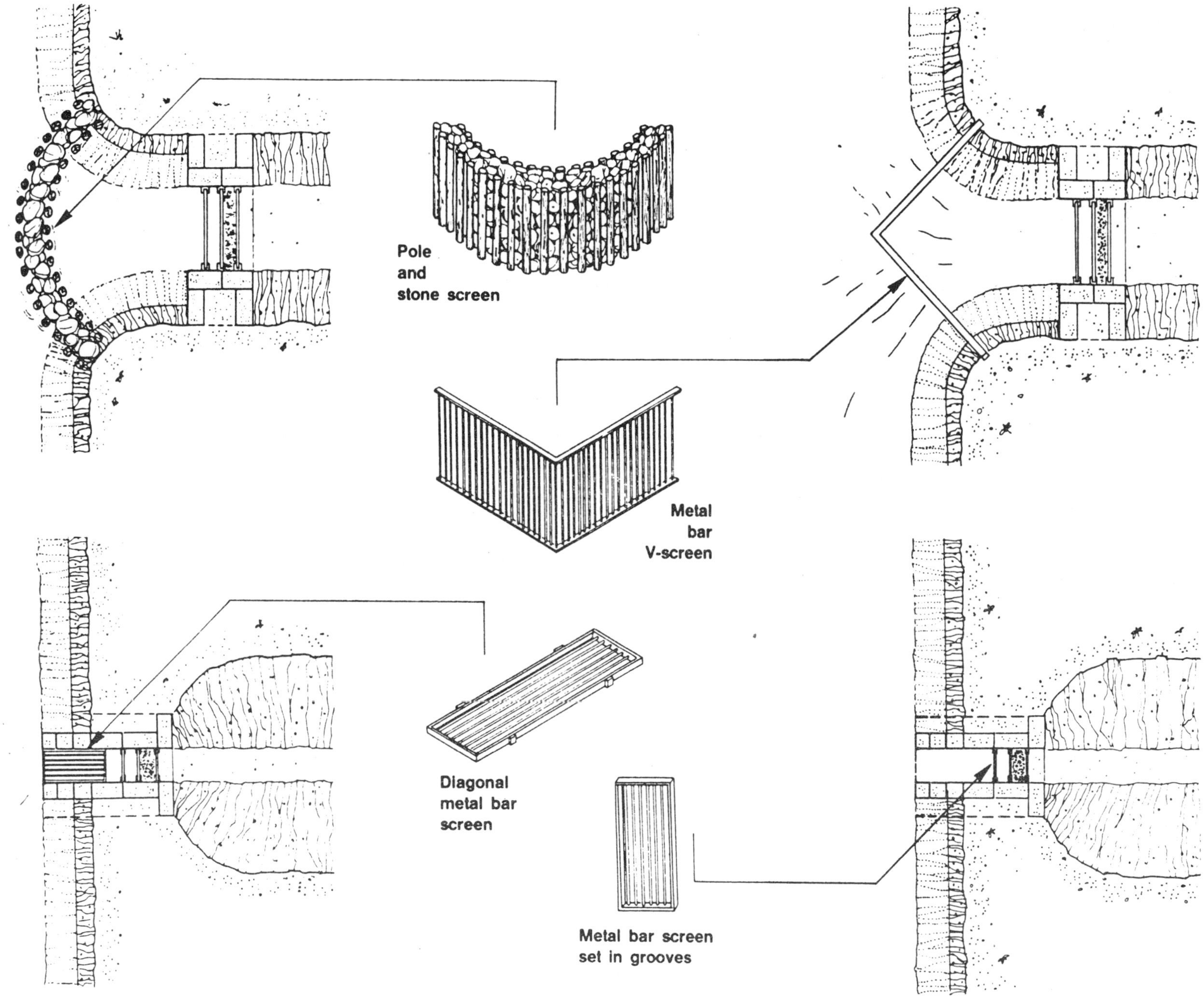
Pole
and
stone screen
Metal
bar
V-screen
Diagonal
metal bar
screen
Metal bar screen
set in grooves

Protecting the intake structures

9. Intake structures can be protected in several ways and the principles of construction are given elsewhere in the manual.

10. A light **framework of tied bamboo, woven netting, or posts and boards** can be used for wall protection. Make sure the framework is well anchored, and do not let the water work its way behind the structure. If it does, erosion can be rapid, and the structure will weaken and lose its effectiveness.

11. **Posts, tied planks or pickets** can be embedded into the stream or along the sides. If well placed, they reduce erosion. If placed across an intake area, they can also act as a coarse screen, protecting the area from large and heavy debris.

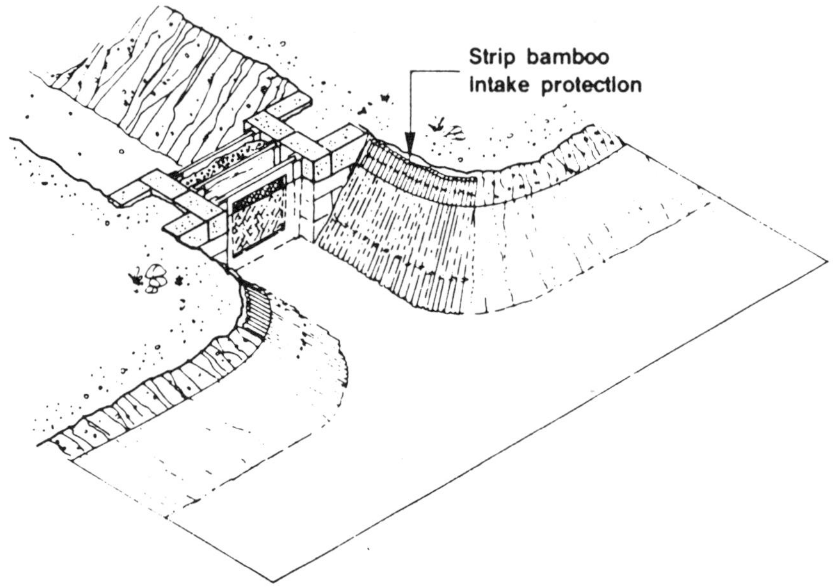

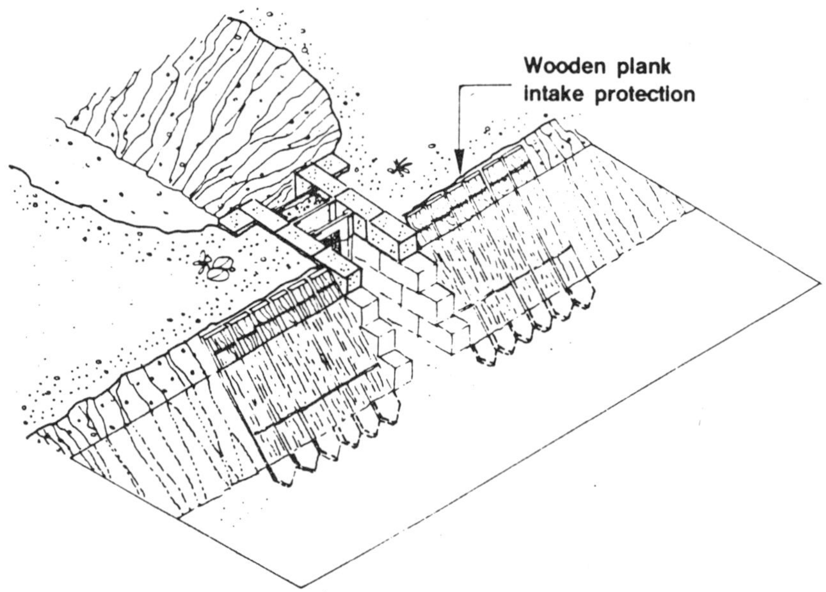

12. **Gabions** can be used around the intake and to deflect water, if for example it flows strongly against a stream bank.

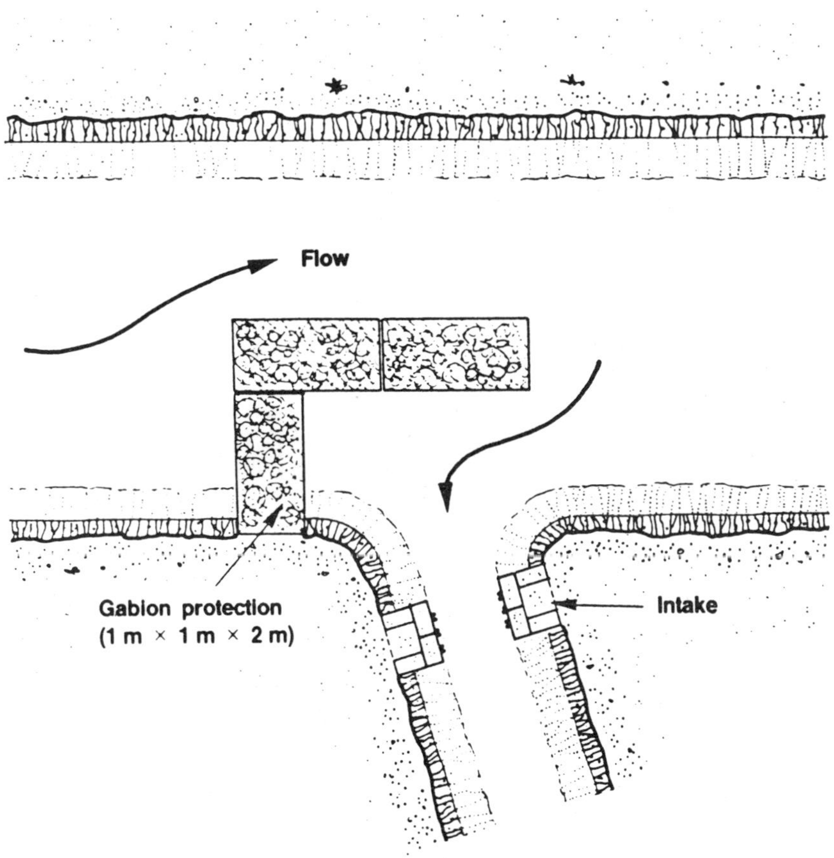

13. If **large stones or rocks** are available, they can also be used. Generally, the larger the stones, the better protection they provide.

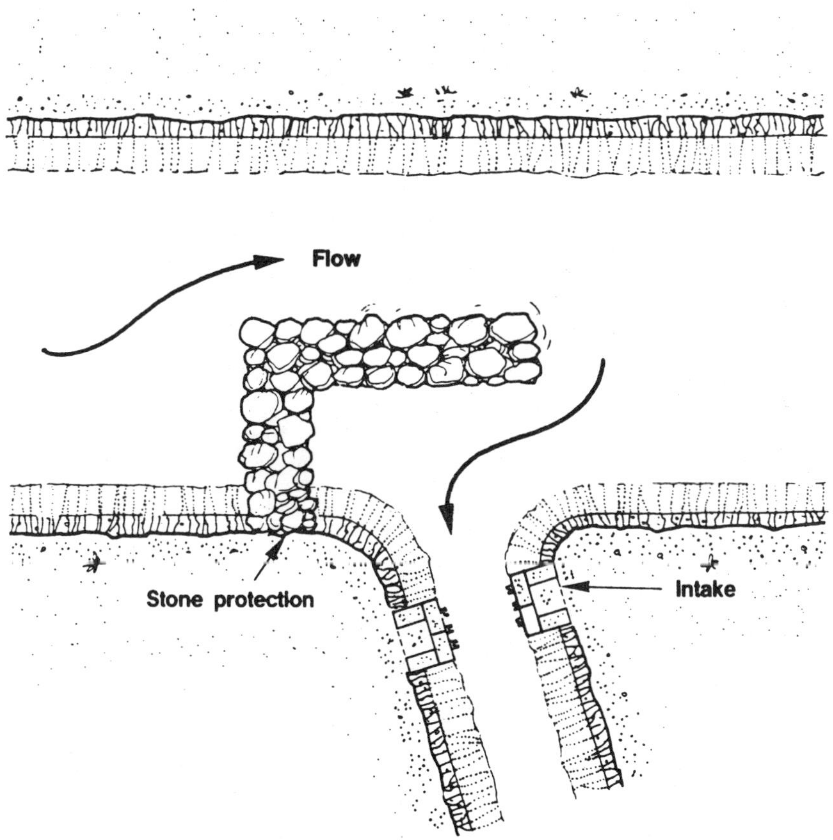

8 WATER TRANSPORT STRUCTURES

80 Introduction

1. Several different kinds of structure may be used to transport water on a fish farm. The most common one is the **open canal,** which we will consider in detail first (Sections 81 to 86). Then we will look at other common structures, including:

- simple **aqueducts** to transport water above ground level (Section 88);
- short **pipelines** to transport water above or under another structure such as a water canal or an access road (Section 89);
- simple **siphons** to transport water over an obstacle such as a pond dike (Section 89).

81 Types of open water canals

1. Different types of open water canals are used on fish farms to transport water, usually by **gravity*,** for four main purposes:

- **feeder canals** to supply water from the main water intake to the fish ponds. In a large farm with several diversion pond units, there is usually a main feeder canal branching into secondary and even tertiary feeder canals;
- **drainage canals** to evacuate water from the fish ponds, for example toward an existing valley;
- **diversion canals** to divert excess water away from barrage ponds;
- **protection canals** to divert water runoff away from the fish ponds.

2. In this chapter you will learn about feeder, drainage and diversion canals. You will learn more about protection canals later (see Section 115).

Open water canals for a small fish farm

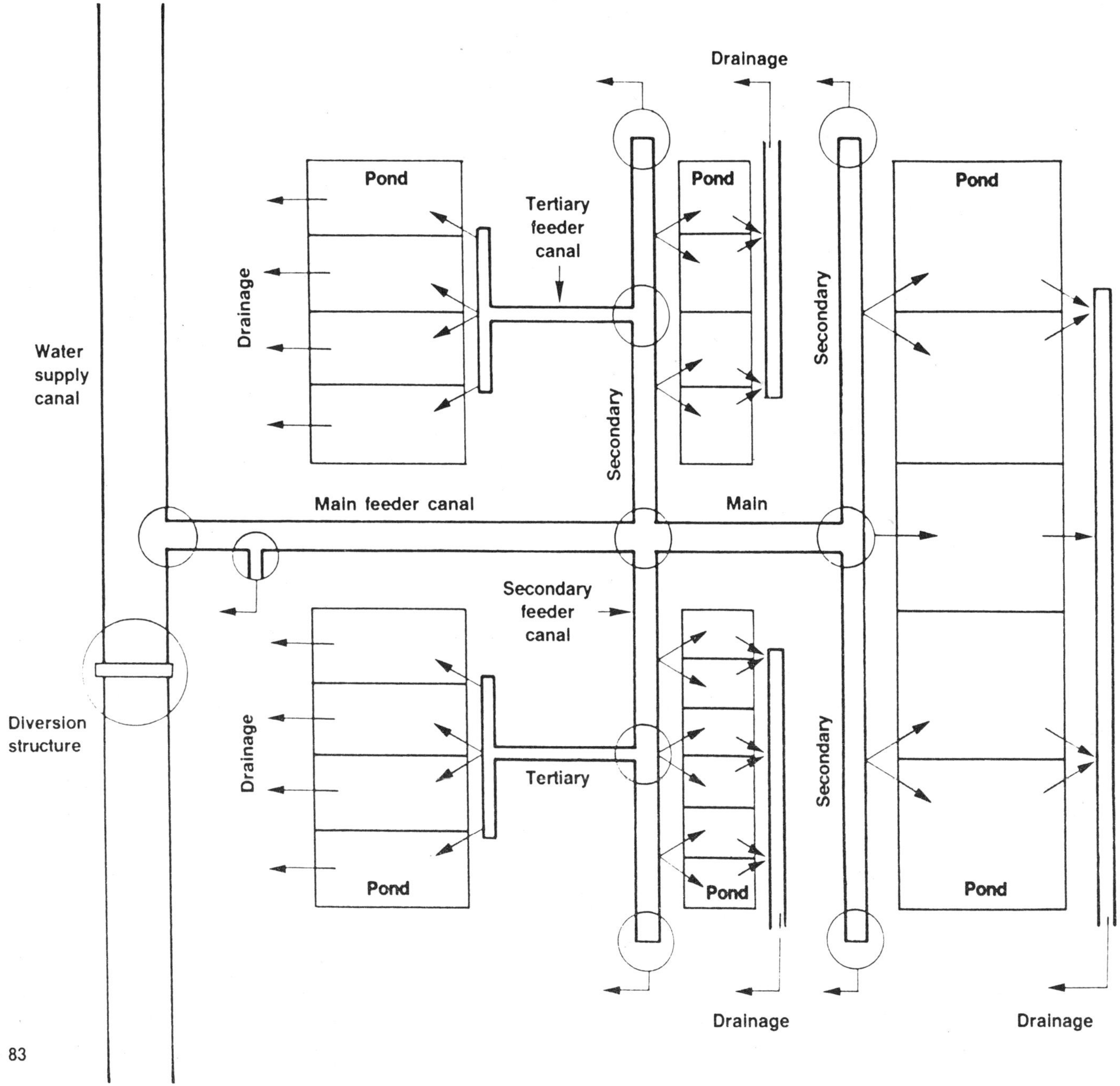

Note: see also page 83

1. All canals should be well designed to have the required **water carrying capacity**. The canals are designed using formulas that relate the carrying capacity of the canal to its **shape,** its effective gradient or **head loss,** and **the roughness** of the canal sides. The most commonly used formula incorporating all these factors is **the Manning equation:**

$$v = (1 \div n)\,(R^{2/3})\,(S^{1/2})$$

where **v** = water velocity in the canal;
n = roughness coefficient of the canal sides;
R = hydraulic radius of the canal;
S = effective slope.

2. You will learn more about these terms below. First we will consider some basic design factors.

Planning the shape of the canal

3. Water canals can have various shapes. In theory **the most efficient shape is a semi-circle,** but this is impractical for earthen canals. It is therefore generally used for precast **concrete flumes* or plastic flumes** only (see Section 87).

4. It is very common for **unlined fish farm canals** to have **a trapezoidal cross-section,** defined by:

- the width **(b)** of its horizontal bottom;
- the slope ratio **(z:1)** of its angled sides;
- the maximum water depth **(h)**; and
- the freeboard* **(f)** to protect against overflowing.

5. When water canals are **lined with bricks or concrete,** they may also have **a rectangular shape** (see Section 83).

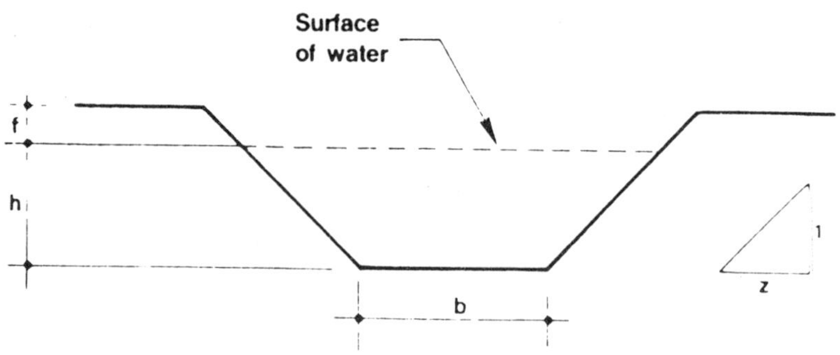

Selecting the side slope for a trapezoidal canal

6. As you learned for pond dikes, the slope of the sides of a trapezoidal canal is usually expressed as a ratio, for example 1.5:1. This ratio is defined as **the change in horizontal distance** (here 1.5 m) **per metre of vertical distance,** see Section 40, **Topography, 16/1**. The side slope can also be expressed in terms of the angle it makes with the vertical, in degrees and minutes.

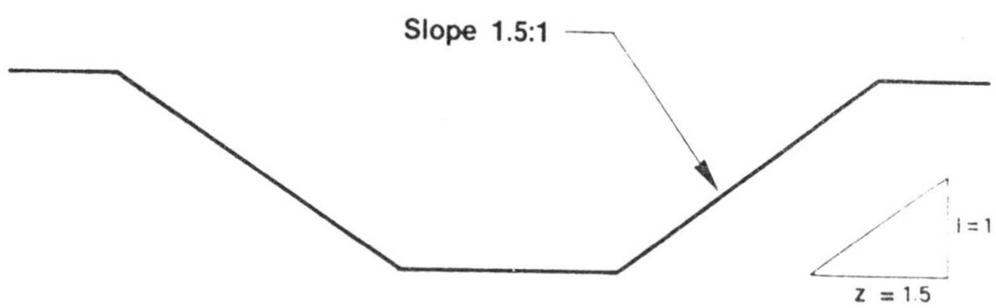

7. **The slope of the sides** to select for an earthen trapezoidal canal depends on the **type of soil** in which the sides are cut (see **Table 34**). The more stable the soil material, the steeper the slope of the sides can be. If the canal is lined, the slope of the sides also varies according to the **type of lining** used.

TABLE 34

Side slopes for trapezoidal canals in various soils

Type of soil or lining material	*Side slopes not steeper than*	
Light sand, wet clay	3:1	18° 20'
Loose earth, silt, silty sand, sandy loam	2:1	26° 30'
Ordinary earth, soft clay, loam, gravelly loam, clay loam, gravel	1.5:1	33° 40'
Stiff earth or clay	1:1	45°
Tough hardpan, alluvial soil, firm gravel, hard compact earth	0.5:1	63° 30'
Stone lining, cast-in-place concrete, cement blocks	1:1	45°
Buried plastic membrane	2.5:1	22° 30'

Selecting the slope for the bottom of a canal

8. The longitudinal **bottom slope of earthen canals** is determined according to topographical conditions:

- **in very flat areas,** the bottom slope can be nil (horizontal canal) or at the most kept to a minimum value of 0.05 percent, or 5 cm per 100 m;
- **in steeper areas,** the bottom slope should be limited to 0.1 to 0.2 percent (10 to 20 cm per 100 m) to avoid the water flowing too fast along the canals and eroding them.

9. The bottom level can be lowered whenever necessary by building **drop structures** in the canal (see Section 87).

10. **In lined canals,** such as those built of bricks or concrete, the bottom slope may be greater as there is less risk of damage by erosion.

Determining the maximum velocity of water flow in canals

11. In open canals, water velocity varies with depth and with the distance from the sides of the canal. Close to the bottom and next to the edges, the water flows less rapidly. When designing canals, you are normally concerned with **the average velocity of the water** across the entire canal cross-section.

12. **The maximum average velocity** that can be safely allowed in a canal to **avoid erosion** depends on the soil (see also Section 123, **Soil and freshwater fish culture,** *FAO Training Series,* **6**) or the lining material. Maximum permissible water velocities in canals and **flumes*** for various soils and linings are defined in **Table 35**.

Calculating the geometry of the canal and its hydraulic radius, R

13. Knowing the bottom width **b** (in m) of the canal, the maximum water depth **h** (in m) and the side slope ratio **(z:1),** you can easily calculate the following characteristics of the canal:

- **wet cross-section area A** (in m^2);
- **wet perimeter P** (in m), which is the length of the perimeter of the cross-section actually **in contact** with water, not including the water top width **B** (**Table 36,** column 5);
- **the hydraulic radius R** (in m), which is the ratio of the wet cross-section area **A** to the wet perimeter **P.** It is often used to define the shape of the canal;
- **the water top width B** (in m), the distance across the water surface.

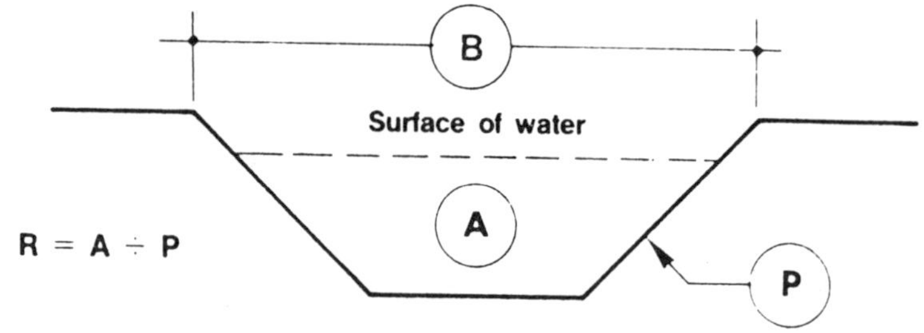

14. The geometry of the wet cross-section of canals is summarized in **Table 36** for the three most common shapes: rectangular, trapezoidal and triangular.

Note: the greater the value of **R**, the greater the flow in the canal.

TABLE 35

Maximum permissible average velocities of water in canals and flumes

Type of soil or lining	*Maximum permissible average velocity* (m/s)
UNLINED CANALS	
Soft clay or very fine clay	0.2
Very fine or very light pure sand	0.3
Very light loose sand or silt	0.4
Coarse sand or light sandy soil	0.5
Average sandy soil and good loam	0.7
Sandy loam, small gravel	0.8
Average loam or alluvial soil	0.9
Firm loam, clay loam	1.0
Firm gravel or clay	1.1
Stiff clay soil, ordinary gravel soil, or clay and gravel	1.4
Broken stone and clay	1.5
Coarse gravel, cobbles, shale	1.8
Conglomerates, cemented gravel, soft slate	2.0
Soft rock, rocks in layers, tough hardpan	2.4
Hard rock	4.0
LINED CANALS	
Cast-in-place cement concrete	2.5
Precast cement concrete	2.0
Stones	1.6 - 1.8
Cement blocks	1.6
Bricks	1.4 - 1.6
Buried plastic membrane	0.6 - 0.9
FLUMES	
Concrete or smooth metal	1.5 - 2.0
Corrugated metal	1.2 - 1.8
Wood	0.9 - 1.5

TABLE 36

Geometry of the canal cross-section below water level

Canal cross-section	*Cross-section area* A (m^2)	*Wetted perimeter* P (m)	*Hydraulic radius* R = (2) ÷ (3) (m)	*Water top width* B (m)
(1)	(2)	(3)	(4)	(5)
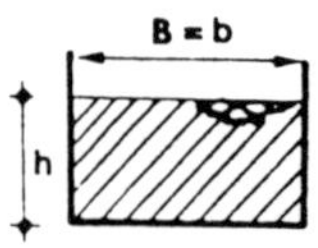	bh	b + 2h	$\frac{bh}{b + 2h}$	b
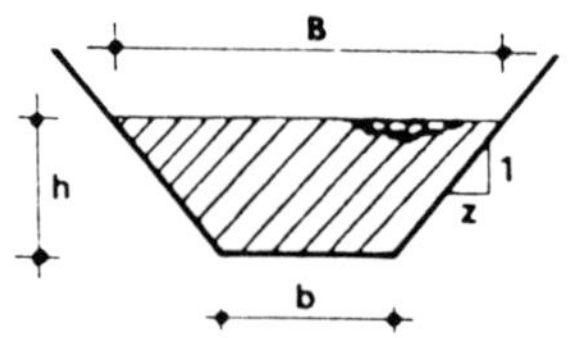	(b + zh) h	$b + 2h\sqrt{1 + z^2}$	$\frac{(b + zh)\, h}{b + 2h\sqrt{1 + z^2}}$	b + 2zh
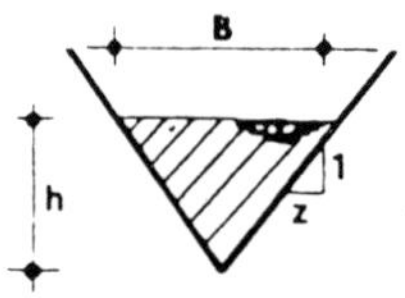	zh^2	$2h\sqrt{1 + z^2}$	$\frac{zh}{2\sqrt{1 + z^2}}$	2zh

Abbreviations: **b** = bottom width (in m)
h = maximum water depth at centre of canal (in m)
z = side slope, horizontal change per unit vertical change

TABLE 37

Coefficient of roughness (Manning) for open canals and flumes

Conditions of water flow	n	1/n
EARTHEN CANALS, UNLINED		
Clean and smooth earth, recently completed	0.017	58.82
Little curving, in solid loam or clay, with silt deposits, free from growth, in average condition	0.025	40.00
Short grass, few weeds	0.024	41.67
Dense weeds in deep water	0.032	31.25
Irregular soil with stones	0.035	28.57
Badly maintained, dense weeds through flow depth	0.040	25.00
Clean bottom, brush on sides	0.070	14.29
LINED CANALS		
Concrete bricks	0.020	50.00
Concrete, cast, unfinished, rough	0.015	66.67
Concrete, trowel finished, smooth	0.013	76.92
Bricks, rough walls	0.015	66.67
Bricks, carefully built walls	0.013	76.92
Boards, with algae/moss growth	0.015	66.67
Boards, quite straight, no growth	0.013	76.92
Boards, well planed and fitted	0.011	90.91
Plastic buried membrane	0.027	37.04
FLUMES/GUTTERS/AQUEDUCTS		
Concrete	0.012	83.33
Metal, smooth	0.015	66.67
Metal, corrugated	0.021	47.62
Wood and bamboo (smooth)	0.014	71.43

TABLE 38

Carrying capacity (litres/second) of earthen trapezoidal canals
(side slope 1.5:1; coefficient of roughness 0.20 - 0.25)

Longitudinal slope canal	*Water depth* (m)	*Bottom width of canal* (m)						
		0.10	0.15	0.20	0.30	0.40	0.50	0.75
0.05 percent (S = 0.0005)	0.05	—	—	—	—	—	1.40	2.10
	0.10	—	—	—	—	5.05	6.24	9.15
	0.12	—	—	—	5.21	7.47	9.07	13.33
	0.14	4.17	5.22	6.58	8.18	10.31	12.33	17.48
	0.16	5.85	7.03	8.33	10.84	13.59	16.14	23.08
	0.18	7.83	9.38	10.97	13.90	17.38	19.87	29.78
	0.20	10.20	11.53	13.94	17.11	21.81	25.65	36.54
	0.22	13.10	15.12	17.52	22.11	26.76	31.84	43.88
	0.24	16.15	19.74	21.93	28.20	32.10	38.12	52.84
	0.30	29.28	32.80	37.24	44.86	53.72	61.61	81.50
	0.40	62.72	69.46	76.94	88.30	104.80	116.16	153.45
	0.50	116.14	124.19	134.65	152.25	175.12	192.00	316.30
0.1 percent (S = 0.001)	0.05	—	—	—	—	1.67	2.09	3.18
	0.10	2.60	—	4.23	5.7	7.10	8.80	12.96
	0.12	3.99	5.04	6.08	8.29	10.50	12.87	18.85
	0.14	5.91	7.23	8.44	11.50	14.52	17.13	24.81
	0.16	8.18	10.17	11.79	14.86	19.03	22.40	32.15
	0.18	11.27	13.10	15.56	19.97	24.36	28.48	41.27
	0.20	14.96	16.72	19.91	24.91	30.45	37.09	52.50
	0.22	18.05	21.09	25.05	31.93	37.80	45.65	63.38
	0.24	23.56	26.62	30.96	40.02	46.50	55.06	75.88
	0.30	41.53	47.87	52.63	61.72	74.77	88.02	118.18
	0.40	91.60	98.75	109.92	126.14	147.68	163.68	220.16
	0.50	164.55	179.28	187.34	216.56	248.29	271.99	355.86
0.2 percent (S = 0.002)	0.05	0.60	0.98	1.16	1.79	2.37	2.94	4.51
	0.10	3.74	4.60	5.73	8.04	9.90	12.37	18.31
	0.12	5.78	7.11	8.64	11.72	14.52	18.28	26.65
	0.14	8.63	10.22	11.68	16.59	20.32	24.73	35.44
	0.16	11.75	14.15	16.67	21.55	27.18	33.03	46.76
	0.18	15.95	18.96	22.23	28.16	34.22	41.42	59.20
	0.20	21.37	23.88	28.05	34.66	43.89	52.55	73.50
	0.22	26.40	29.72	34.98	43.90	53.51	63.91	87.76
	0.24	32.80	38.20	42.58	56.07	63.84	76.24	103.99
	0.30	59.82	66.30	72.87	90.20	105.23	125.74	167.08
	0.40	126.00	139.60	153.89	178.28	209.61	232.32	306.90
	0.50	231.41	257.72	269.30	308.44	350.24	392.00	514.00

The coefficient of roughness of a canal

15. The **coefficient of roughness (n)** expresses the resistance to water flow created by the sides and the bottom of a canal. The greater the value of **n**, the greater the roughness of the canal walls, and the more difficult it is for the water to flow through the canal.

16. The values of **the coefficient of roughness** under various conditions are summarized in **Table 37**. Their reciprocal value **(1 ÷ n)** is also tabulated, to be used in later calculations.

The significance of slope or gradient

17. In simple cases, you can assume that **the bottom of the canal will slope downstream**. In fact, water will flow in canals as long as **the water level** is higher at the upstream end than the downstream end. If a canal has **a horizontal floor**, the gradient can be taken as the difference in head between upstream and downstream. The slope **S** of the canal bottom is expressed as metres of head per metre of canal length, for example **S** = 0.01 or 1 percent. **The greater the value of S, the greater the flow**.

18. Note that for steady, even flow and to minimize the risk of sedimentation, the canal should be built so that its **bottom slope follows the overall gradient**, i.e. the depth remains constant. Because they are easier to construct, however, canal bases are often made horizontal.

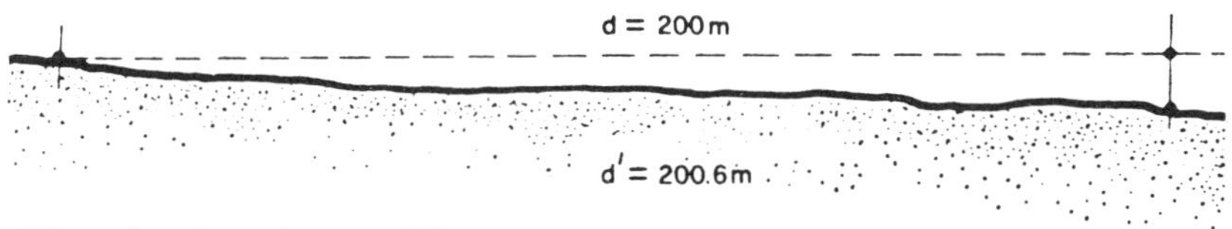

When the slope is very little you can either measure distance (d) on the horizontal or distance (d') on the ground, with very little difference in the measurements

Predicting the water capacity of earthen canals

19. The Manning equation can be applied directly (see paragraph 25 in this section), or it can be used in a number of simplified forms.

20. If you plan to build **a standard trapezoidal canal** with a bottom width **b** = 1 m, sides slope **z:1** = 1.5:1, and a flat longitudinal slope **S** = 0.0001 - 0.0002 (0.01 - 0.02 percent), you can predict **the approximate water capacity Q** (in m^3/s) of such a canal by assuming that the average water velocity will be **v** = 0.3 - 0.5 m/s, as follows:

Q = wet cross-section area x v

Example

If we select **v** = **0.3 m/s** because of the relative roughness of the walls, the water carrying capacity of such a canal is estimated as follows:

Water depth h (m)	*Wet cross-section** A (m^2)	*Water carrying capacity* Q (m^3/s)**	Q (m^3/day)
0.1	0.115	0.0345	2 981
0.2	0.260	0.0780	6 739
0.3	0.435	0.1305	11 275
0.4	0.640	0.1920	16 589
0.5	0.875	0.2625	22 207

* **A** = (b + zh) h with **b** = 1 m and **z** = 1.5; **h** from column 1
** **Q** = **A** x 0.3; to obtain litres per second (l/s), multiply by 1 000

21. Another simple method is to use **a table providing estimates of the water carrying capacity** for a number of canal sizes, depths of water and longitudinal slopes. **Table 38** provides such data for a trapezoidal canal cut in ordinary soil with side slope 1.5:1.

Predicting the water carrying capacity of lined canals

22. If you plan to build **a rectangular canal lined with brick, block or concrete** (see Section 83), you can estimate its water carrying capacity (in l/s) as follows:

Bottom width (m)	*Water depth* (m)	*Longitudinal slope* (percent)			
		0.02	0.05	0.10	0.15
0.30	0.30	20-30*	30-40	40-60	40-70
0.50	0.40	40-70	70-120	100-160	120-200
0.80	0.60	140-240	230-370	320-530	400-650

* Use first number for canals with rough walls and second number for those with smooth walls

Examples of rectangular lined canals

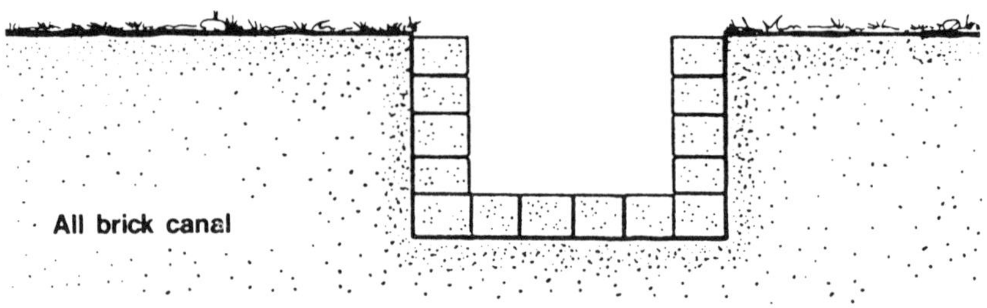

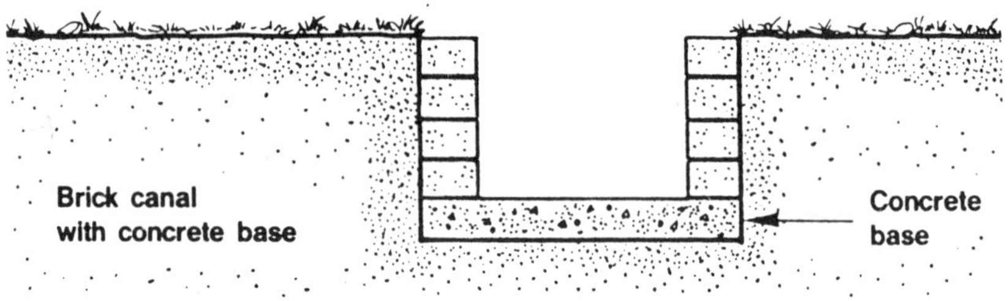

Using graphs to design a water canal

23. Graphs can also be easily used to design a water canal. See for example:

- **Graph 7,** which gives the water carrying capacity of trapezoidal earthen canals with **smooth walls,** side slope 1:1 and a bottom slope **S** = 0.1 percent;
- **Graph 8,** which gives the water carrying capacity of similar canals with rough **walls**.

24. You can use such a graph in two ways:

(a) **You fix the characteristics** of your water canal and you determine from the graph its carrying capacity.

Example

The canal has the following characteristics:

- bottom width = 1.20 m
- water depth = 0.40 m
- side slope = 1:1
- bottom slope = 0.1 percent
- **n** = 0.020 (regular soil)

Using **Graph 7**, you determine point **A**. It corresponds to a carrying capacity **Q** = 620 m^3/h.

(b) **You fix the water carrying capacity** of the canal, and you determine from the graph the characteristics you require.

Example

If the canal has to have a carrying capacity of **Q** = 425 m^3/h, being dug in a **stony soil** (**n** = 0.035) with side slope 1:1 and slope **S** = 0.1 percent, use **Graph 8**. Following line **Q** = 400 m^3/h, select a relatively wide bottom value (for example, 1.50 m) and determine point **A** at **Q** = 425 m^3/h. From this point, determine water depth = 0.30 m on the left scale.

GRAPH 7

Carrying capacity of earthen trapezoidal canals with smooth walls (side slope 1:1, coefficient of roughness n = 0.020; slope S = 0.1 percent)

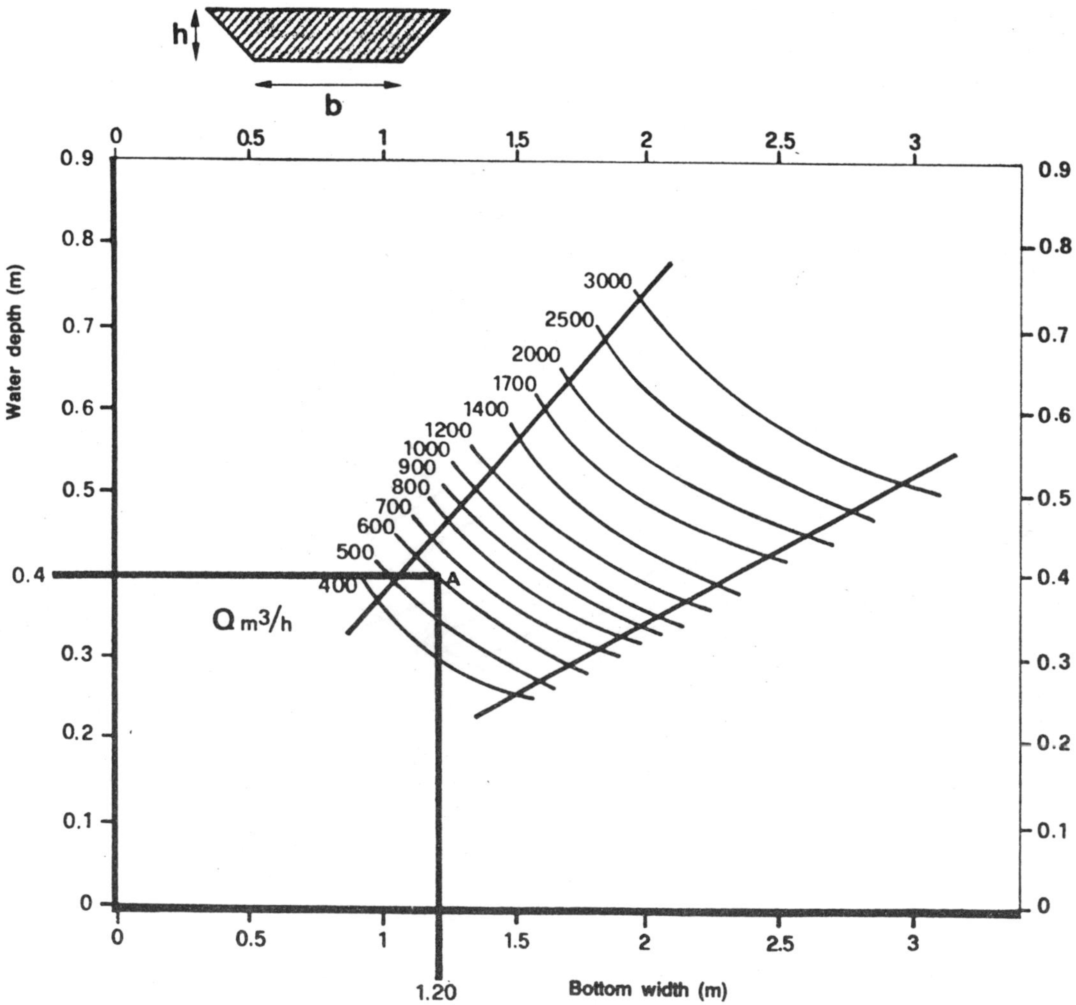

Remember: **Q m³/h = 86.4 Q l/s**

GRAPH 8

Carrying capacity of earthen trapezoidal canals with rough walls (side slope 1:1; coefficient of roughness n = 0.035; slope S = 0.1 percent)

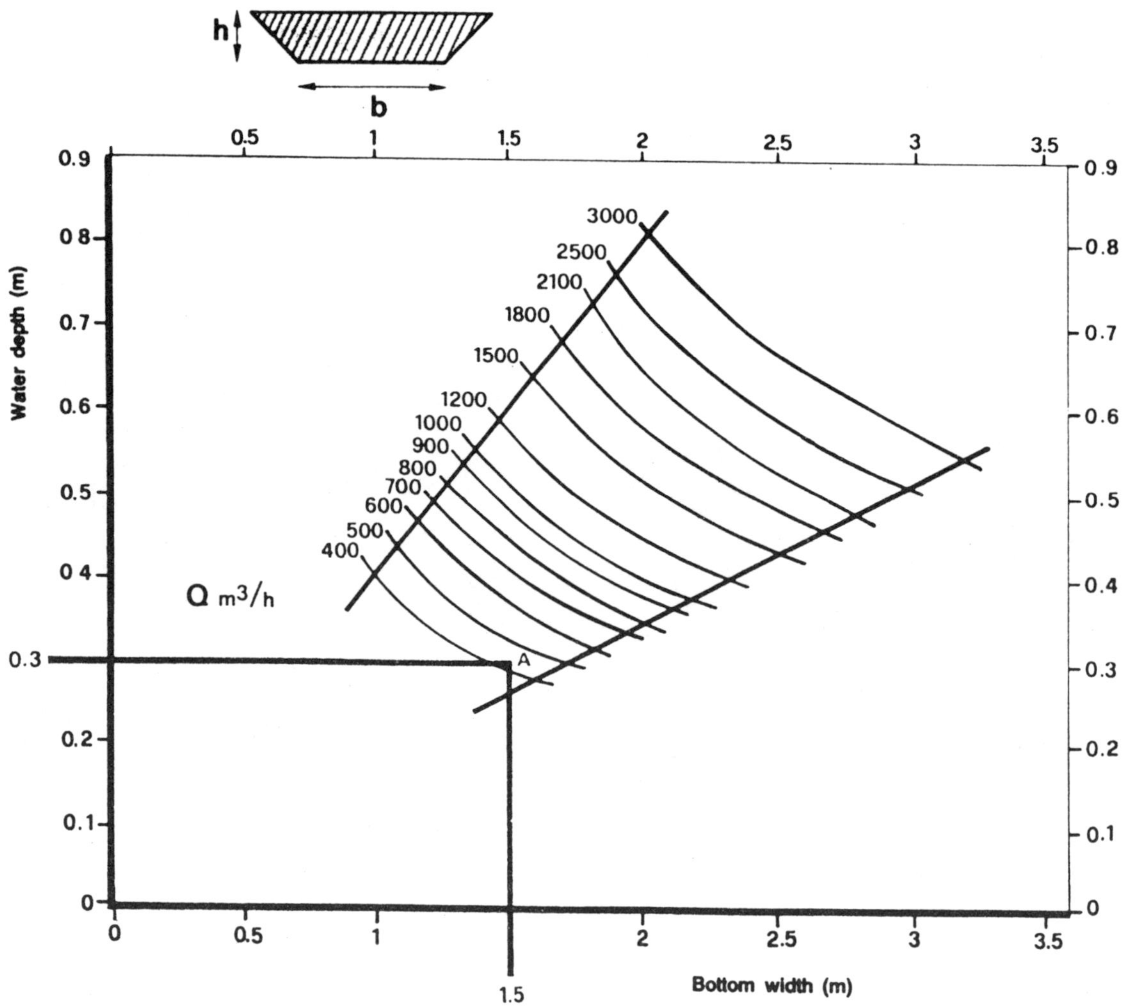

Remember: Q m³/h = 86.4 Q l/s

Using the Manning equation to calculate the carrying capacity of a water canal

25. It is relatively easy to calculate directly **the carrying capacity** (in m^3/s) of **any open water canal with a uniform steady flow** by using **the Manning equation** in the form

$$Q = A\ (1 \div n)\ R^{2/3}\ S^{1/2}$$

where, as you learned previously,

- **A** = wet cross-section area in m^2 **(Table 36)**;
- **R** = hydraulic radius, in m **(Table 36)**;
- **S** = longitudinal slope of the canal bottom;
- **n** = coefficient of roughness **(Table 37)**.

26. To assist you with the calculations, you can also consult:

- **Table 37**, which gives some values of **(1 ÷ n)**;
- **Table 39**, which gives $\sqrt{1 + z^2}$, for common values of **z**;
- **Table 40**, which gives the values of one-half powers **($S^{1/2}$)**;
- **Table 41**, which gives the values of two-third powers **($R^{2/3}$)**.

Example

A trapezoidal canal has the following characteristics:

- bottom width **b** = 0.50 m
- water depth **h** = 0.40 m
- **n** = 0.030
- **S** = 0.003
- side slope **z:1** = 1.5:1

Determine its carrying capacity as follows:

- **A** = (b + zh) h = [0.50 m + (1.5 × 0.40 m)] × 0.40 m = **0.44 m²**
- **(1 ÷ n)** = 1 ÷ 0.030 = 33.33
- **R** = A ÷ b + 2h ($\sqrt{1 + z^2}$) = 0.44 m² ÷ [0.50 m + (2 × 0.40 m) (1.80)] = 0.44 m² ÷ 1.94 m = **0.227 m (Table 39)**
- $R^{2/3}$ = $(0.227\ m)^{2/3}$ = **0.372 m (Table 41)**
- $S^{1/2}$ = $(0.003)^{1/2}$ – **0.055 (Table 40)**
- **Q** = (0.44 m²) (33.33) (0.372 m) (0.055) 0.300 m³/s = **300 l/s**

Calculating and checking the average water velocity in the canal

27. There are several ways by which you can calculate the average water velocity in an open canal. For example, you can use one of the following three simple methods:

(a) **Knowing the water flow Q** (m^3/s) being carried by the canal with a **wet cross-section area A** (m^2), determine the **average water velocity v** (in m/s) as

$$v = Q \div A$$

Example

If for the above canal **Q** = 0.300 m³/s and **A** = 0.44 m², then **v** = 0.300 m³/s ÷ 0.44 m² = **0.68 m/s**

(b) The **average water velocity v** (in m/s) can also be calculated directly using the standard Manning formula together with **Tables 37, 40 and 41**:

$$v = (1 \div n)\ R^{2/3}\ S^{1/2}$$

Example

If for the above canal, **n** = 0.030, **R** = 0.227 m and **S** = 0.003, then **v** = (33.33) (0.372 m) (0.055) = **0.68 m/s**

TABLE 39

Common values of $\sqrt{1+z^2}$

z	1	1.5	2	2.5	3
$\sqrt{1+z^2}$	1.41	1.80	2.24	2.69	3.16

Remember: **z** from the side slope ratio expressed as **z:1**

TABLE 40

Common values of $S^{1/2}$

S	$S^{1/2}$	S	$S^{1/2}$	S	$S^{1/2}$	S	$S^{1/2}$
0.0001	0.0100	0.0010	0.0316	0.0020	0.0447	0.0030	0.0548
0.0002	0.0141	0.0011	0.0332	0.0021	0.0458	0.0032	0.0566
0.0003	0.0173	0.0012	0.0346	0.0022	0.0469	0.0034	0.0583
0.0004	0.0200	0.0013	0.0361	0.0023	0.0480	0.0036	0.0600
0.0005	0.0224	0.0014	0.0374	0.0024	0.0490	0.0038	0.0616
0.0006	0.0245	0.0015	0.0387	0.0025	0.0500	0.0040	0.0632
0.0007	0.0265	0.0016	0.0400	0.0026	0.0510	0.0042	0.0648
0.0008	0.0283	0.0017	0.0412	0.0027	0.0520	0.0044	0.0663
0.0009	0.0300	0.0018	0.0424	0.0028	0.0529	0.0046	0.0678
		0.0019	0.0436	0.0029	0.0539	0.0048	0.0693
						0.0050	0.0707

Remember: **S** = bottom slope expressed in units of vertical fall (m) per unit of horizontal distance (m)

$S^{1/2} = \sqrt{S}$

TABLE 41

Common values of $R^{2/3}$
R = hydraulic radius (in m)*

R	Second decimal									
	0	1	2 ↓	3 ↓	4	5	6	7	8	9
0.0	0.000	0.046	0.074	0.097	0.117	0.136	0.153	0.170	0.186	0.201
0.1	0.215	0.220	0.243	0.256	0.269	0.282	0.295	0.307	0.319	0.331
0.2 →	0.342	0.353	0.364	0.375	0.386	0.397	0.407	0.418	0.428	0.438
0.3	0.448	0.458	0.468	0.477	0.487	0.497	0.506	0.515	0.525	0.534
0.4	0.543	0.552	0.561	0.570	0.578	0.587	0.596	0.604	0.613	0.622
0.5	0.630	0.638	0.647	0.655	0.663	0.671	0.679	0.687	0.695	0.703
0.6	0.711	0.719	0.727	0.735	0.743	0.750	0.758	0.765	0.773	0.781
0.7	0.788	0.796	0.803	0.811	0.818	0.825	0.832	0.840	0.847	0.855
0.8	0.862	0.869	0.876	0.883	0.890	0.897	0.904	0.911	0.918	0.929
0.9	0.932	0.939	0.946	0.953	0.960	0.966	0.973	0.980	0.987	0.993

* See **Table 36**

How to use this table: **If R = 0.227 m, determine $R^{2/3}$:**

- in the first column, locate the **R** value down to its first decimal (0.2)
- follow this line to the right until the column giving the second decimal (2)
- mark this number down = 0.364, the answer for **R** = 0.220 m
- follow the line until the next column to the right (3)
- mark this number down = 0.375, the answer for **R** = 0.230 m
- as **R** = 0.227 m is intermediate between these two values, you have to interpolate
- calculate the difference between the two previous numbers; 0.375 − 0.364 = 0.011
- divide this difference by 10 : 0.011 ÷ 10 = 0.0011
- multiply the result by the third decimal of the **R**-value 0.227 m: 0.0011 × 7 = 0.0077
- add this to the smallest number read from the table earlier: 0.364 + 0.0077 = 0.3717 = 0.372
- **$R^{2/3}$ = (0.227 m)$^{2/3}$ = 0.372 m**

(c) You can use a graphic method to determine **the average water velocity v** (in m/s) as

$$v = C\sqrt{RS}$$

where

- **C** is obtained from **Graph 9** in function of the roughness coefficient (1 ÷ n, see **Table 37**) and the hydraulic radius **R** (see **Table 36**); and
- $\sqrt{RS}$ is obtained from **Graph 10** in function of **R**, the hydraulic radius and **S**, the longitudinal slope of the canal bottom.

Example

For the same data as in the above example, find:

- from **Graph 9,** for **R** = 0.227 m and **(1 ÷ n)** = 33.33, **C** = 26
- from **Graph 10,** for **R** = 0.227 m and **S** = 0.003, $\sqrt{RS}$ = 0.0262
- $v = C\sqrt{RS}$ = 26 × 0.0262 = 0.6812 = **0.68 m/s**

28. Once you know the average water velocity **v** (in m/s), you can compare its value with **the maximum permissible average velocity** in your particular canal (see **Table 35**). The velocity **v** calculated by design should be smaller than the maximum permissible value, to avoid erosion of the canal.

Example

If the canal is dug in sandy loam, the maximum permissible average velocity is 0.8 m/s and your designed value **v** = 0.68 m/s is acceptable.

Determining the characteristic dimensions of the optimum trapezoidal canal

29. If the **water carrying capacity Q** (in m/s) of an earthen trapezoidal canal is known (once you have planned the fish farm, for example), it is easy to determine the characteristic dimensions for the best canal. Proceed as follows:

(a) According to soil quality, determine **the maximum permissible average velocity v max** (m/s) from **Table 35** and the side slope of the canal **(z:1)** from **Table 34**.

(b) Define **the coefficient of roughness n** from **Table 37**.

(c) Calculate the optimum **wet cross-section area** (in m^2) as: **A = Q ÷ v max**.

(d) Obtain the square root of **A** as $\sqrt{A}$.

(e) From **Table 42**, calculate **the characteristic dimensions of the optimum canal** by multiplying this square root by the numbers indicated on the line corresponding to the selected side slope **z:1**.

Example

To design a trapezoidal canal to be dug in firm loam for a water carrying capacity of 1.5 m^3/s, proceed as follows:

(a) From **Table 35,** maximum permissible average velocity **v max** = 1 m/s.

(b) From **Table 34,** assume side slope 1.5:1.

(c) From **Table 37,** assume **n** = 0.025.

(d) Calculate **A** = 1.5 m^3/s ÷ 1 m/s = 1.5 m^2.

(e) Calculate $\sqrt{A} = \sqrt{1.5\ m^2}$ = 1.225 m.

(f) Using **Table 42** for side slope 1.5:1, calculate the canal characteristics:

- water depth **h** = 0.689 $\sqrt{A}$ = 0.689 × 1.225 m = 0.84 m
- bottom width **b** = 0.417 $\sqrt{A}$ = 0.417 × 1.225 m = 0.51 m
- water top width **B** = 2.483 $\sqrt{A}$ = 2.483 × 1.225 m = 3.04 m
- wet perimeter **P** = 2.905 $\sqrt{A}$ = 2.905 × 1.225 m = 3.559 m
- hydraulic radius **R** = 0.344 $\sqrt{A}$ = 0.344 × 1.225 m = 0.421 m

GRAPH 9

Values of the coefficient C

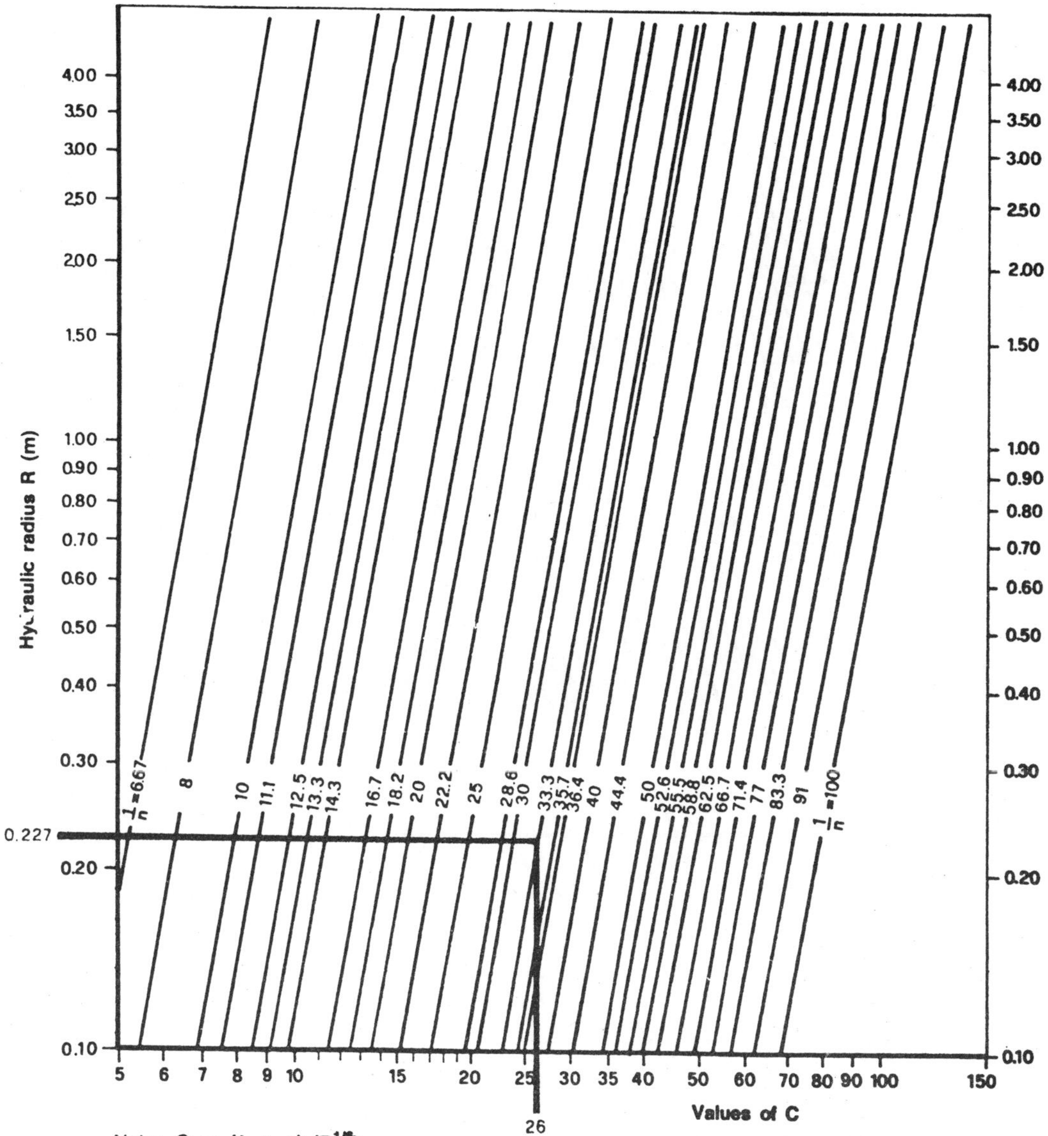

Note: $C = (1 \div n)(R^{1/6})$

GRAPH 10

Values of the coefficient $\sqrt{RS}$

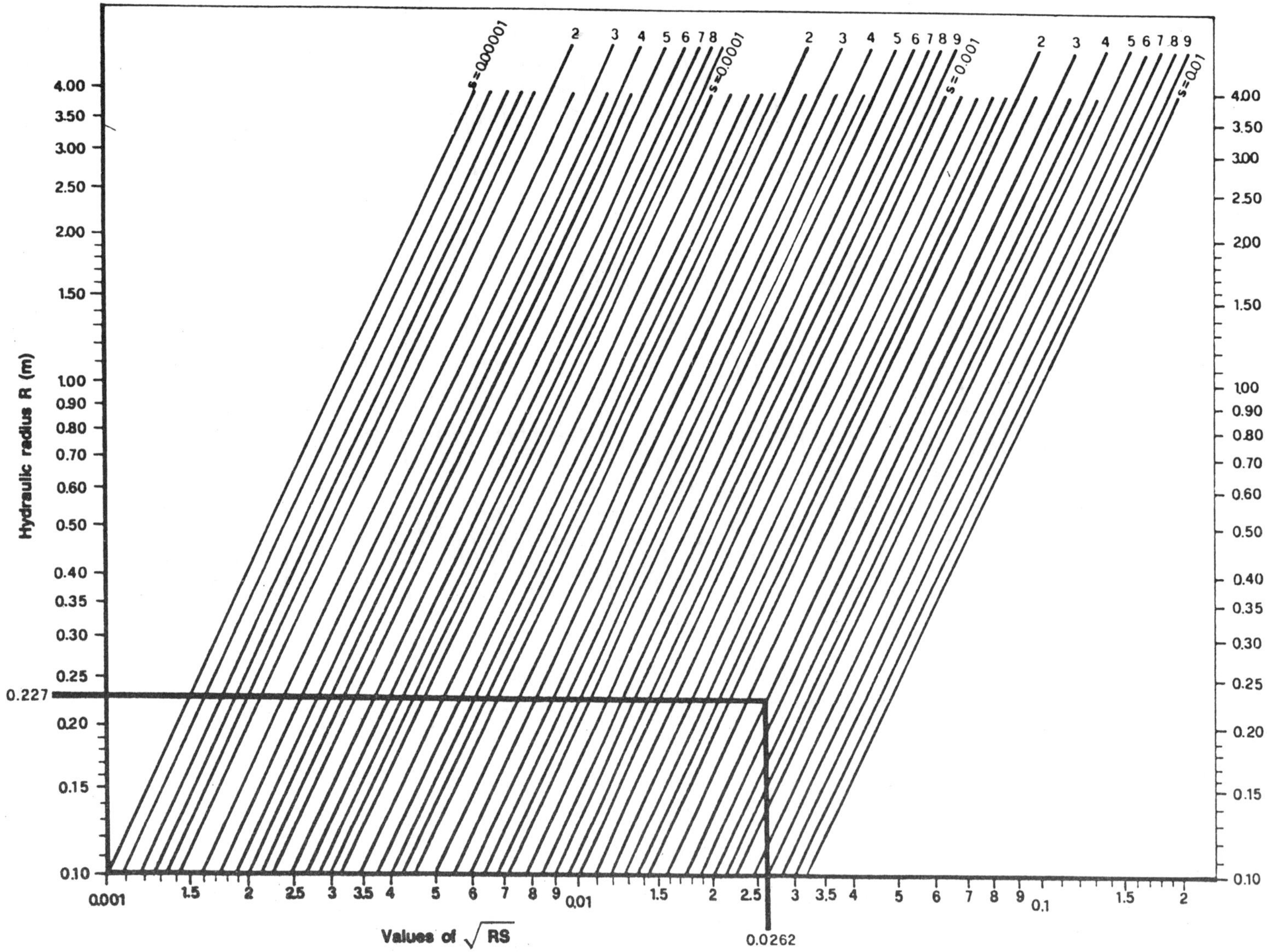

TABLE 42

Factors for determination of characteristic dimensions of the optimum trapezoidal canal (Multiply $\sqrt{A}$ by the given factor to obtain the dimension)

Canal side slope	*Characteristic dimensions*				
	h	b	B	P	R
1:1	0.739	0.612	2.092	2.705	0.370
1.5:1	0.689	0.417	2.483	2.905	0.344
2:1	0.636	0.300	2.844	3.145	0.318
2.5:1	0.589	0.227	3.169	3.395	0.295
3:1	0.549	0.174	3.502	3.645	0.275

where **h** = water depth (m)
b = bottom width (m)
B = water surface width (m)
P = wetted perimeter (m)
R = hydraulic radius (m)
A = cross-section area (m^2)

TABLE 43

Allowable curves for canals

Capacity of canal (m^3/s)	*Minimum radius* (Rm)	
	Unlined canal	Lined canal
< 0.3	100 m	40 m
0.3-3	150 m	60 m
3-15	300 m	100 m

Determining the slope of the canal bottom

30. For a given canal, **the longitudinal slope S** can be calculated as

$$\mathbf{S = (nv \div R^{2/3})^2}$$

where **n** is the coefficient of roughness **(Table 37)**;
v is the average water velocity, in m/s;
R is the hydraulic radius, in m **(Table 41)**.

Example

For the above designed canal

$S = [(0.025)\ (1\ m/s) \div (0.421^{2/3})]^2$
$S = [0.025 \div 0.562]^2$
$S = 0.002 =$ **0.2 percent**

Losing water from an earthen canal

31. Water losses in earthen canals result from **evaporation** (1 to 2 percent) and **seepage** (5 to 40 percent). Seepage losses, which are by far the most important, vary according to the **type of soil** in which the canal is dug.

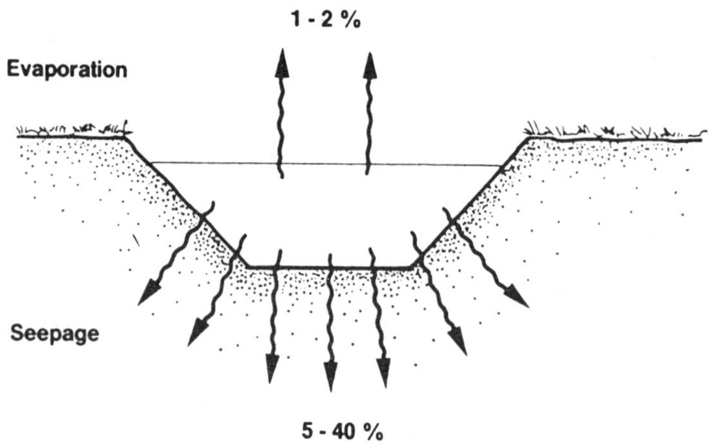

Water losses by type of soil

*Type of soil**	*Average daily water losses per m^2 of wet perimeter*	
	(m^3/day)	(l/day)
Impervious clay	0.09	90
Clayey loam	0.18	180
Sandy clay or conglomerates	0.25	250
Sandy clay loam or loam	0.40	400
Sandy loam	0.50	500
Sand and gravel or alluvium	0.70	700
Gravelly and porous	1.00	1 000
Very porous gravels	1.80	1 800

* See **Soil, 6.**

Example

An earthen canal built in sandy loam has a wet perimeter **P** = 3.559 m. If its total length is 78 m, the wet soil area is 3.559 m × 78 m = 277.6 m^2. Total seepage losses will average 277.6 × 0.50 m^3/day = 138.8 m^3/day.

32. **When designing feeder canals,** it is advisable to include **water losses averaging 10 to 20 percent** according to the type of soil present.

33. If your water canal is very long, you can also adopt the rule of thumb saying that you will lose **10 percent of your water for each kilometre of canal**.

Example

If at the main water intake you have 100 l/s of water available, after 1 km only 90 l/s will remain, and after 2 km only 81 l/s will remain.

Determining the freeboard for the canal*

34. Until now you have learned a lot about the **wet cross-section** of canals. But, as already briefly mentioned at the beginning, the sides of the canal should be built a little higher than designed for a certain water carrying capacity to avoid overflow. This additional height of the walls, above normal water level, is called **the freeboard**.

35. The freeboard varies according to the kind of canal:

- **in earthen canals,** it varies from 20 to 50 cm
- **in lined canals,** it varies from 10 to 20 cm.

36. You will learn more about freeboard in the next sections.

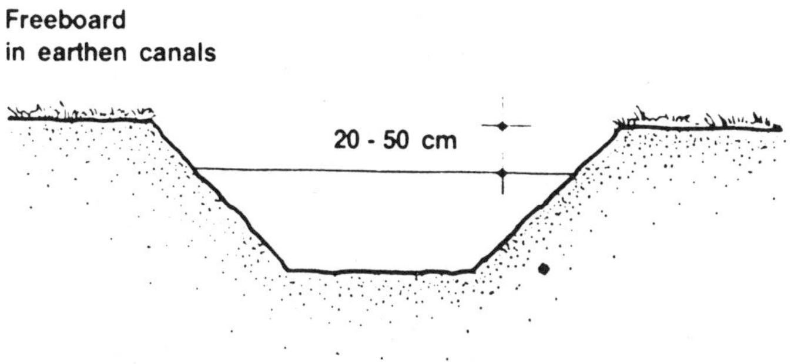

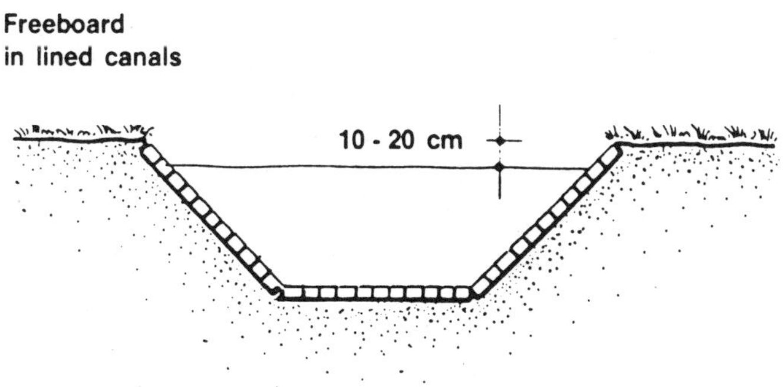

Making curves in canals

37. In some locations you may need to make the canal curve, for example to avoid a specific feature, or to take advantage of topographical conditions. **Table 43** shows **the minimum radius (Rm)** of the curve allowable. As a general rule:

- in firm soils, **Rm** = 20 × bed width in m;
- in loose soils, **Rm** = 30 to 50 × bed width in m.

38. If necessary, make **the freeboard higher on the outside of the curve**, and line it to prevent erosion. For tighter bends, it is better to use stilling basins (see Section 117) or junction boxes (see Section 87).

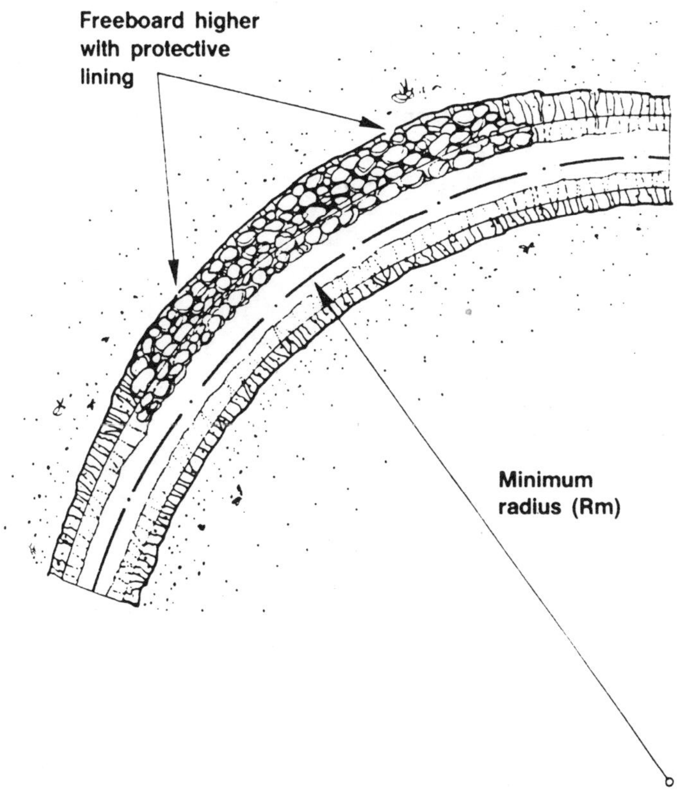

Some other points to remember

39. In many cases you may have several choices of width, depth, gradient, side slope, etc. Some other practical factors may help you define your choice:

(a) **If the water carries silt**, too low a velocity will encourage the silt to settle out. You may wish to design an area specifically for this purpose.

(b) If you need **to cross the canal**, it may be better to make it narrower at this point, possibly by lining the walls.

(c) If there are **difficult or permeable soils** at lower levels, you may wish to keep canals wide and shallow.

(d) If there are **standard tools** available for construction and maintenance, such as a bulldozer blade or a backhoe shovel, you may prefer to base canal sizes on these. Similarly, if you are using sheet polythene or **concrete slabs*** to line the canal, you may wish to size the canals to suit standard dimensions.

(e) Remember to allow **sufficient capacity for floodwater** likely to reach the canal.

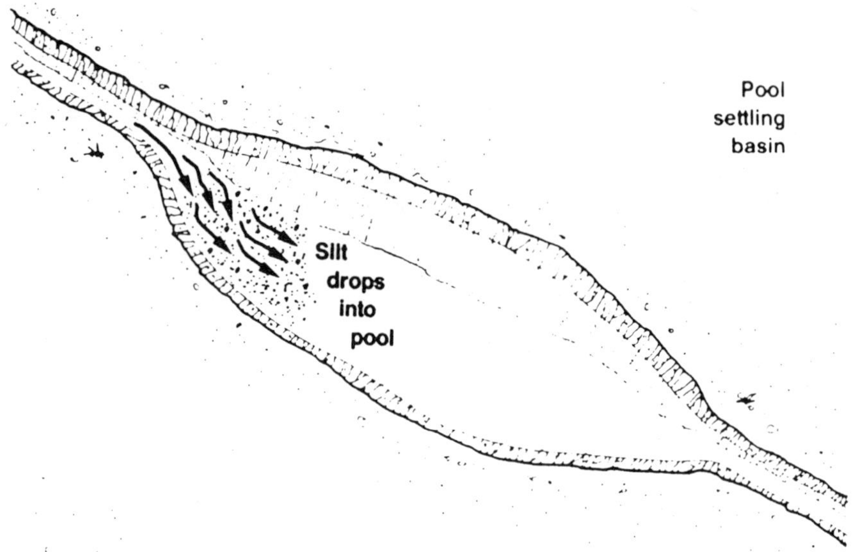

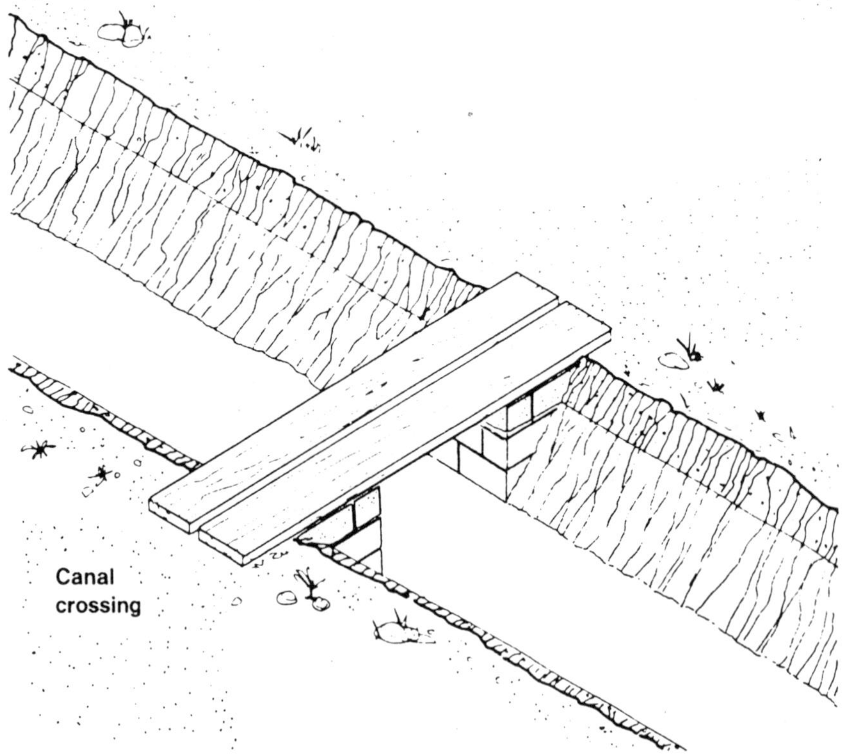

Locate canals above permeable soil

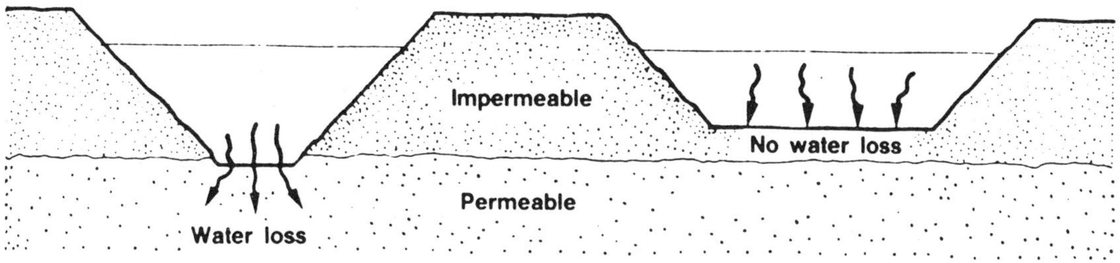

Canal bottom width equal to width of backhoe shovel

Barrage pond with diversion structure

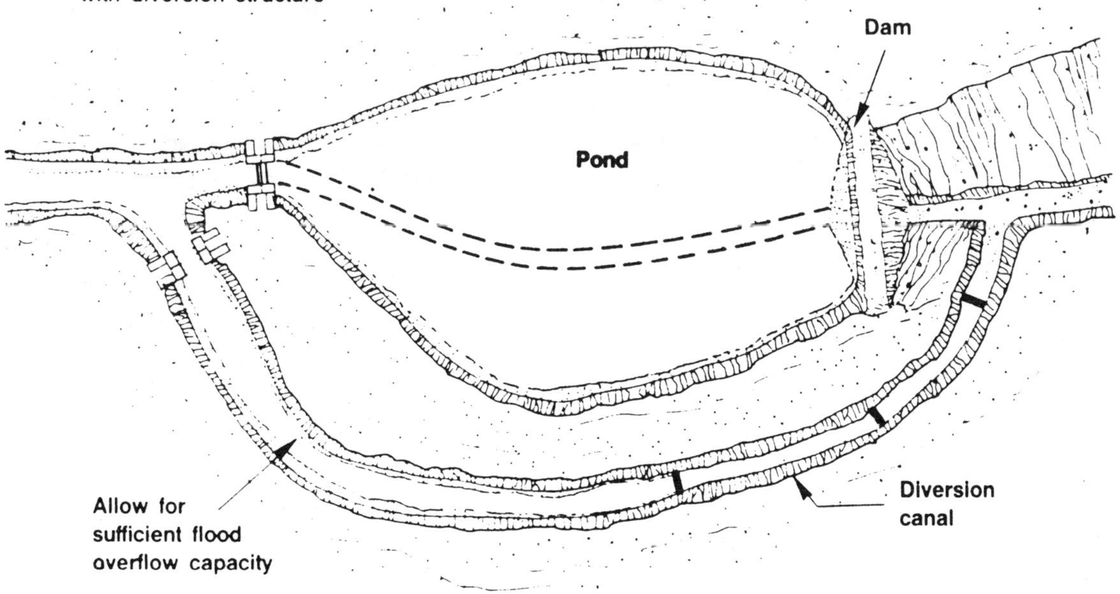

83 Water feeder canals

1. Water feeder canals, or supply canals, connect the **main water intake** to the various fish rearing facilities of the fish farm, and, in particular, to the **diversion ponds**. They can be ranked as primary (main), secondary or tertiary feeder canals according to their function on the farm.

Special points about feeder canals

2. When planning and designing your feeder canals, you should remember the following:

(a) **The main feeder canal** should bring the water to the farm site by **gravity* at the highest possible level**.

(b) On the farm sites, the canals should **bring water to each facility by gravity**.

(c) At each facility, **the water level should be high enough** to enable its drainage by **gravity** at any time.

(d) **If pumping is needed**, it is usually simpler **to pump into** a gravity supply canal, rather than pump out of each pond. Avoid having to do both.

(e) The level of **the canal bottom** should ideally be **at least 10 cm higher** than the normal water level in the pond it supplies. If the site slope is very gentle, however, the canal upper surface can be as little as 5 cm above the normal pond water level.

(f) **The main feeder canal** should be **as short as possible**. If, for example, the longitudinal slope of the stream is less than 2 percent, it is best to raise its water level using a diversion structure to avoid digging too long a feeder canal (see Sections 82 to 89).

(g) **The longitudinal bottom slope** should be **as small as possible**. To decrease the canal level, it is best to use **drop structures** (see Section 95).

(h) The main canal should be designed so that **all the ponds of the farm** can be filled within five (total water area 5 ha) to 30 days (total water area 25 ha).

Example

A 4-ha fish farm contains 40 000 m^3 of water in its fish ponds. These ponds should be filled within five days and therefore the water flow requirement **Q** 40 000 m^3 5 days 8 000 m^3/d 0·093 m^3/s 93 l/s. So the main feeder canal should be designed for a carrying capacity of 93 l/s 12 l/s (water losses) **105 l/s**

(i) Each pond should be **filled within a minimum period of time** according to its size, varying from a few hours for small ponds to a few days for large ponds. Design its feeder canal accordingly.

(j) If possible, it is best to be able **to fill two ponds simultaneously**. Design the feeder canal accordingly.

(k) As far as possible, you should **standardize the size** of the feeder canals.

(l) Feeder canals should be dug **past the last pond** they supply to a drainage point, to act as an overflow and automatically evacuate any excess water away from the farm.

(m) **If there is a risk of excessive water runoff**, you should build a **protection canal** (see Section 115).

3. For economic reasons, **the size of the feeder canals should be kept to a minimum**. When deciding on their dimensions, remember that:

(a) If you select the **minimum wet cross-section A**, you may have to **increase the water velocity v** to reach the necessary water carrying capacity **Q** because **Q = vA**. Be careful not to exceed the maximum permissible average velocity (see **Table 35**).

(b) If you need **to increase the carrying capacity** of a canal, it is **better to widen** than to deepen it.

(c) It is easier to maintain **a shallow canal** than a deep one.

(d) However, **seepage losses** are usually higher in surface soil. If your water supply is limited, a deeper canal might be better.

Example

Common characteristics of trapezoidal earthen feeder canals

Canal dimensions	*Small farm* (Q = a few l/s)	*Medium farm* (Q = 20-50 l/s)
Bottom width (m)	0.25	0.50
Water depth (m)	0.15-0.20	0.15-0.25
Freeboard (m)	0.10-0.20	0.20-0.30
Bottom slope (%)	0	0.1
Side slope	1.5:1	1.5:1
Top width (m)	1-1.45	1.55-2.15

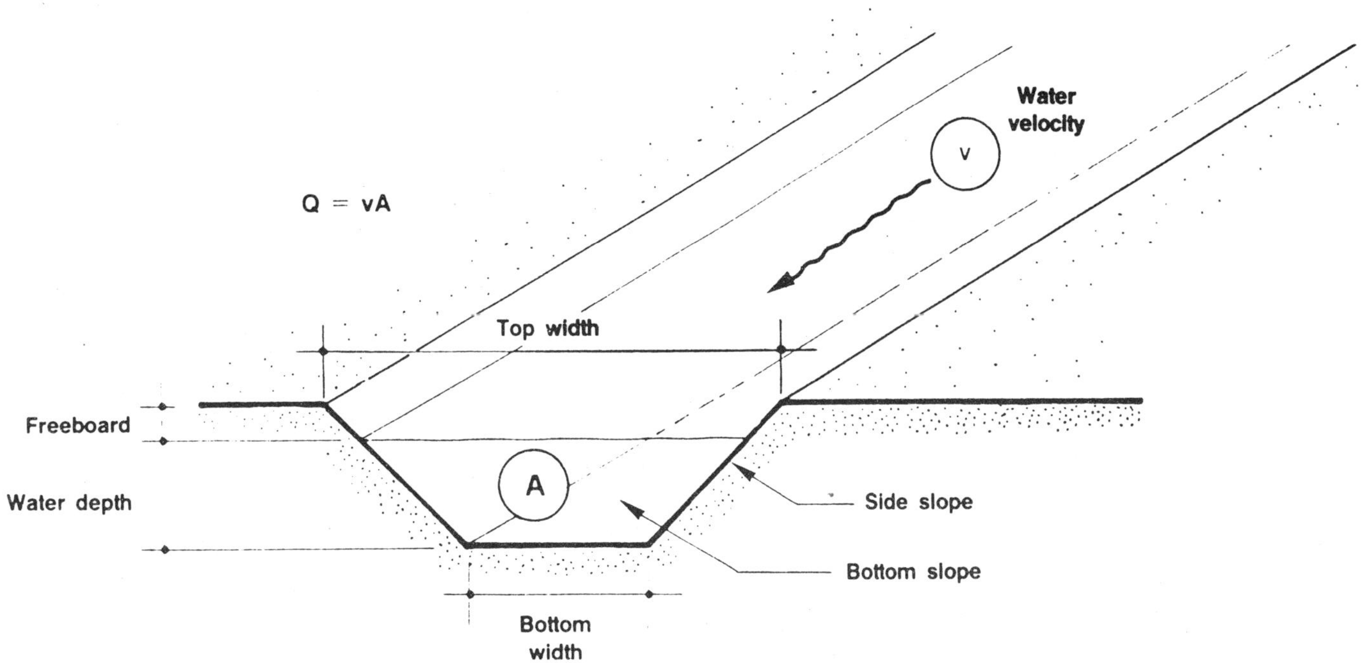

When to use lined feeder canals

4. You may need to use lined feeder canals when:

- **the water supply** is limited and **soil seepage** is high;
- **the lining material** is locally available at reasonable cost;
- feeder canals have to be built **on the top of pond dikes**;
- feeder canals have to be built in **soils very susceptible to erosion**.

5. In certain situations, it may be more advantageous to build **lined feeder canals** using, for example, clay, fired bricks, concrete blocks or concrete. Butyl rubber or heavy-gauge polythene **sheeting** can also be used, although care has to be taken to avoid damage when laying the material. The sheets are usually laid singly in a series anchored at the top edge by bedding in trenches. Semi-circular, square or rectangular precast **concrete sections** can also be used, as is commonly done in irrigation schemes.

Brick or block feeder canal

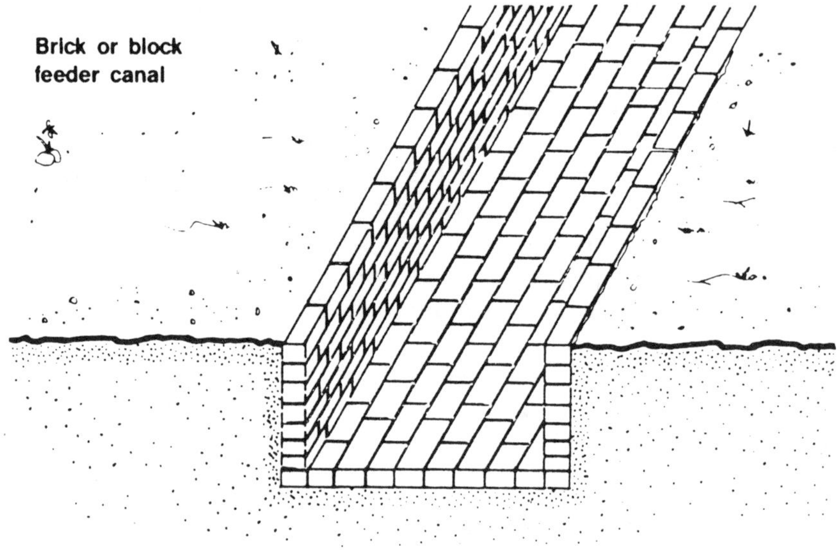

Concrete feeder canal

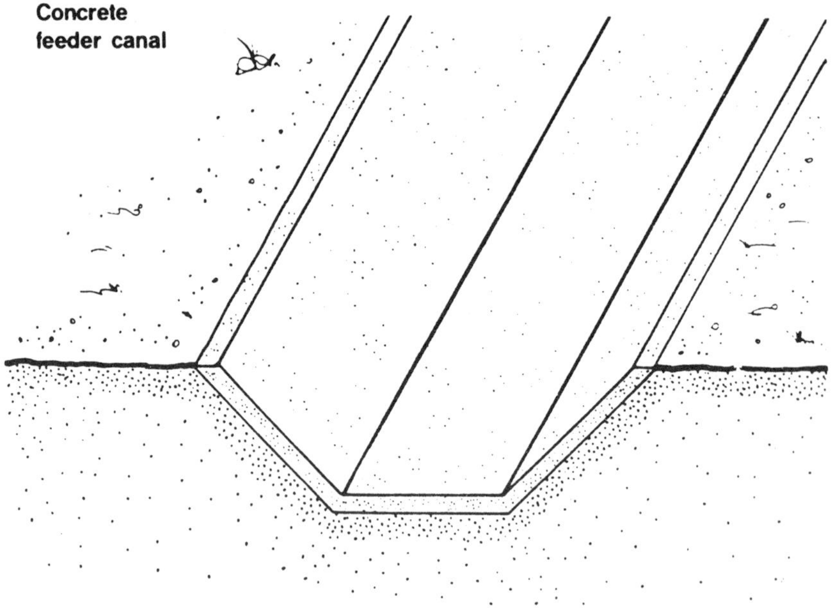

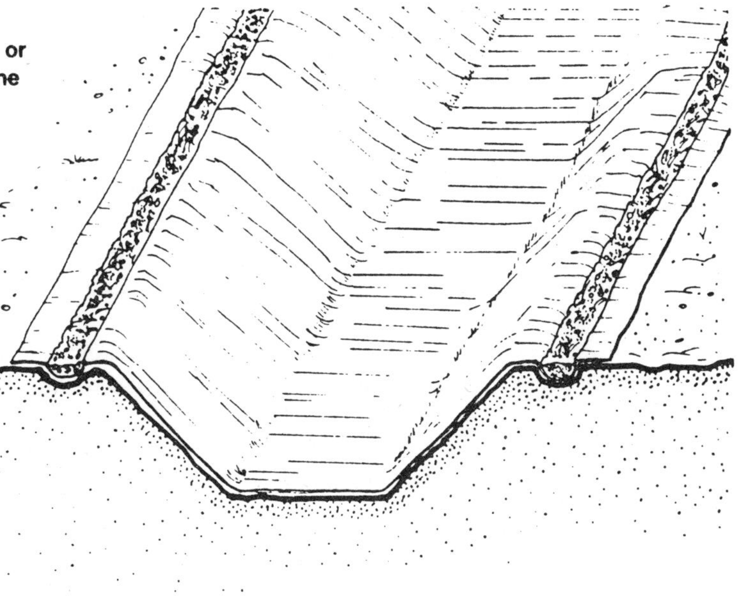

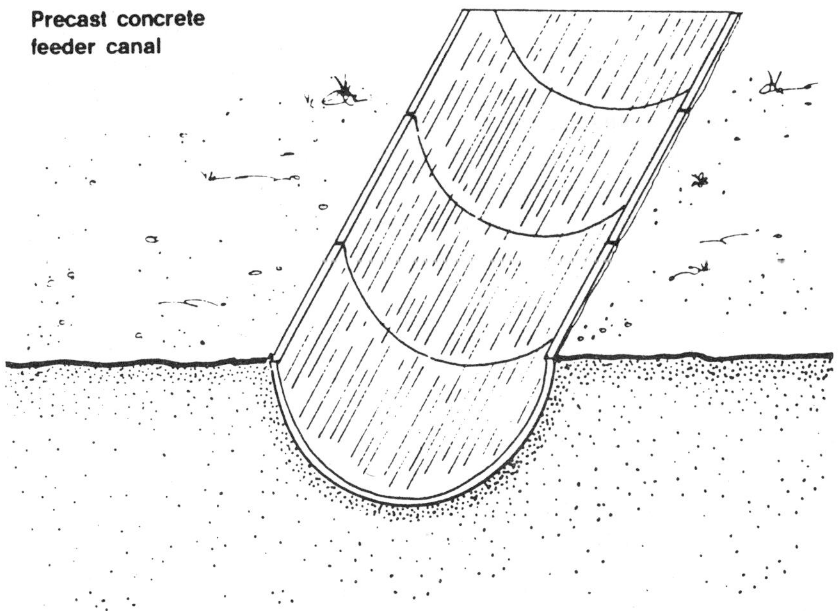

Important facts about lined feeder canals

6. Lined feeder canals have **several advantages**:

- **water losses** are greatly **reduced**, normally averaging not more than 30 l/m^2 of wet perimeter per day;
- feeder canals **can be relatively smaller**, for example, when built in a rectangular or semi-circular shape;
- **the bottom slope** can be increased, because the maximum permissible velocity of water is greater (see **Table 35**), which can contribute to size reduction;
- they can be **built above ground level** or partially buried, which may considerably reduce the earthwork;
- they are easier and cheaper **to maintain**;
- they are not damaged by burrowing animals; and
- they do not deteriorate when kept dry.

7. **Major problems** with canal linings may include:

- **deterioration of joints**, especially in concrete linings;
- **cracking of the lining** due to settlement of fill, erosion of material, swelling of clay soil or very poor quality of concrete;
- cracks in joints and penetration of cracks by **weeds** are a frequent cause of progressive deterioration of canal linings;
- higher **initial investment**;
- **flexible liners** may tear, harden in sunlight or may pull out of their bedding-in trenches.

8. **Common dimensions of lined feeder canals** vary according to their shape:

- **for rectangular cross-sections**, see Section 82;
- **for trapezoidal cross-sections**, the bottom width of the canal usually varies from 0.5 h to 1 h, **h** being the water depth.

84 How to prepare for the construction of a canal

1. You have already learned in the previous volume in this series, **Topography, 16/2** (Section 114), how you should first **survey the possible route** the canal can take between the main water intake and the fish ponds by **contouring and profiling** (longitudinal profile and cross-sections).

2. You have also learned how **to stake out the centre line*** of the canal, once its route has been defined, taking into account the possible **presence of rocks** underground by spot checks with a soil auger (see **Soil, 6**). If necessary, identify the location of the **drop structures** you will have to build to avoid bottom slopes that are too steep (see Section 87).

3. Now you should carefully **stake out the canal cross-section**.

4. **If the canal is all in cut*** and requires no artificial banks:

- on each side of the centre stakes, set **bottom stakes** to show the width of the canal bottom;
- on each side of the centre stakes, set **slope stakes** to show the intersections of the canal side slopes with the actual ground surface. On sloping ground, determine the distance as explained in the next subsection, paragraphs 3 and 9.

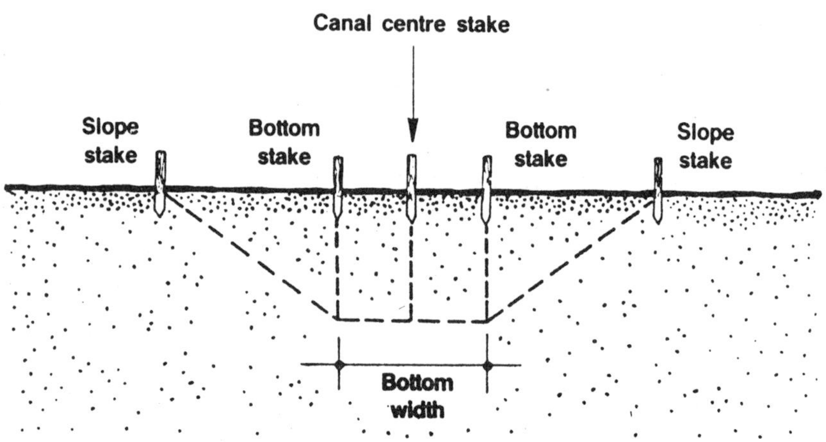

5. If **the canal has two artificial banks**, its cross-section being built **partly in cut and partly in fill***·

- on each side of the centre stakes set **bottom stakes** and **slope stakes** as above, showing the limits of the cross-section in cut;
- on each side of the canal centre stakes, set **bank centre stakes** at distances which remain constant as long as the canal section does not change;
- on the external side of the bank centre stakes, set the **bank slope stakes**, showing the limits of the cross-section in fill.

6. **If the canal is on the side of a hill and has one artificial bank** built on the downhill side of the canal, the canal bottom should always be built in cut for its full length:

- on each side of the centre stakes, set **bottom stakes** and **slope stakes** as above, showing the limit of the cross-section in cut;
- on the downhill side of the canal centre stakes, set the **bank centre stakes** at a distance which remains constant as long as the canal section does not change;
- on the downhill side of these bank centre stakes, set the **bank slope stakes**, showing the downhill limit of the cross-section in fill.

7. As far as possible, you should try **to balance the earthwork in cut and fill**. This is best done from cross-section profiles (see Section 96, **Topography, 16/2**).

Balance the earthwork
in cut and fill

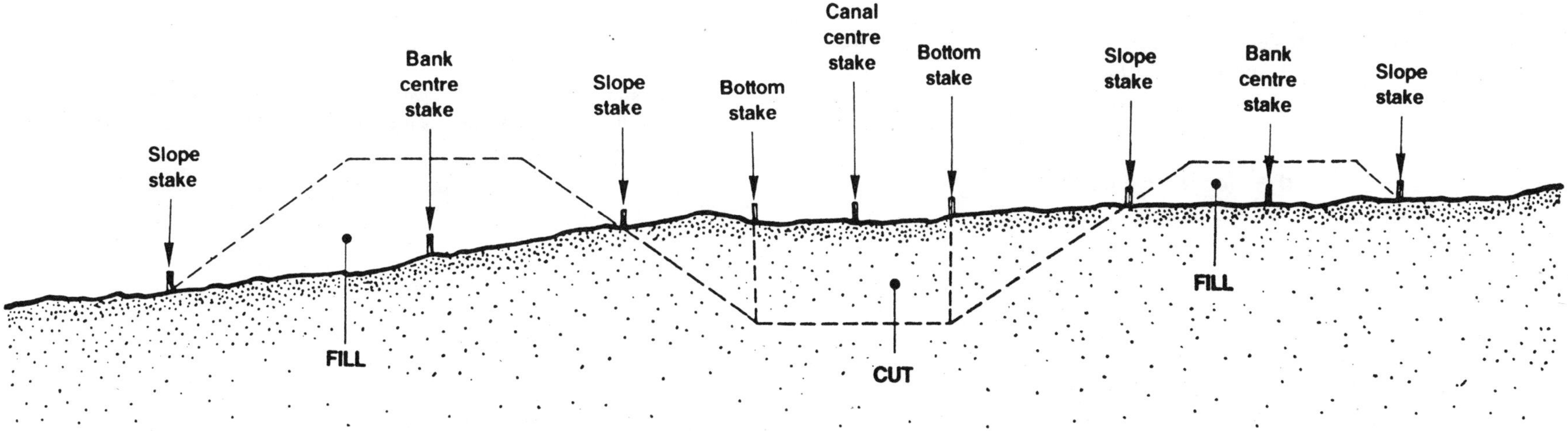

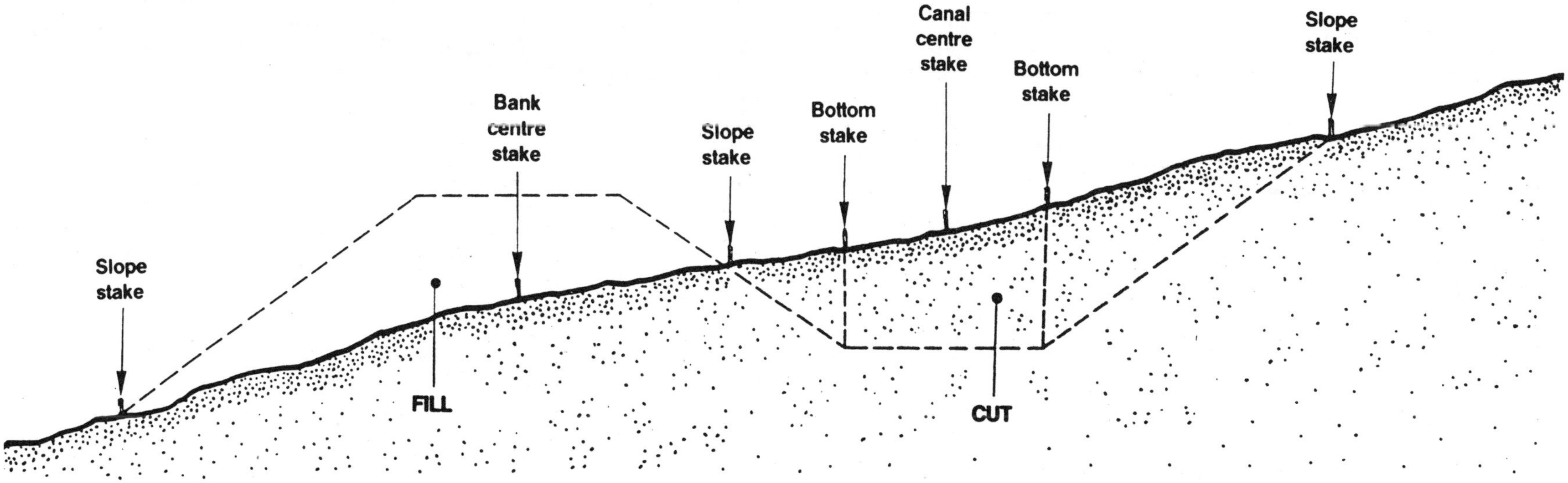

Setting slope stakes on sloping ground

8. The distances from a centre line at which to set the slope stakes vary with the slope of the ground. To establish their correct position you will need to proceed step by step, by trial and error, as described in the example.

9. The method to use is based on the fact that the **distance d** (in m) from a centre stake to the slope stake should be:

$$\mathbf{d = (b \div 2) + (xz)}$$

where **b** is the **canal bottom width**, in m;
x is the **depth** or **height of cut or fill** at the slope stake, in m; and
z is the **side slope**, or ratio of horizontal distance to drop or rise.

Example

A canal is to be cut in a hillside with the following characteristics:
b = 0.80 m and **z** = 1.5. The slope stake is to be set on the left of the canal centre stake.

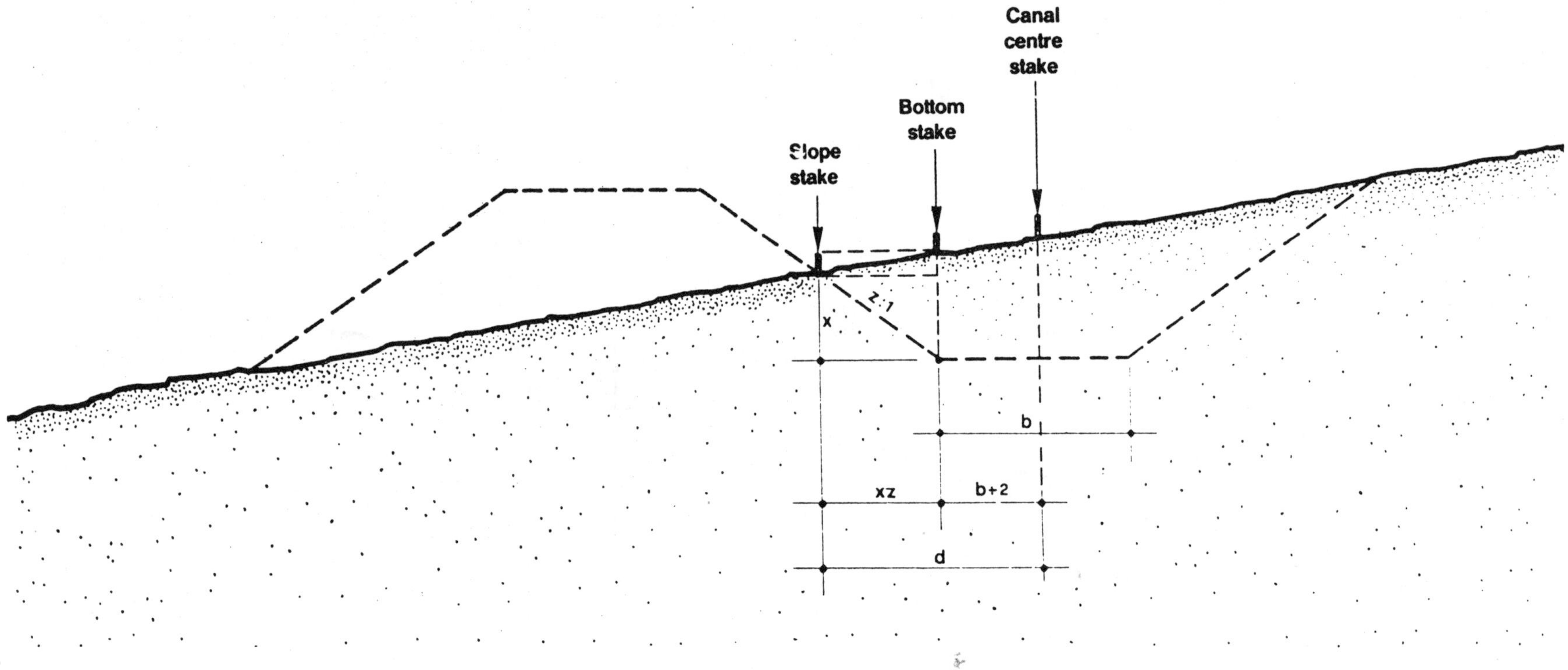

10 Using the example on page 66, proceed in the following way:

(a) From a fixed **levelling station** at **0**, establish the **horizontal line of sight WY** using a sighting level (see Sections 54 to 59, **Topography, 16/1**).

(b) Measure height **AB** = 0.30 m at the canal centre stake.

(c) From the design, you know the elevation of point **C**, on the centre line of the canal bottom as **E(C)** = 102.33 m.

(d) From the longitudinal profile of the canal centre line, you know the elevation of point **B** as **E(B)** = 102.73 m.

(e) Calculate height **BC** = E(B) − E(C) = 102.73 m − 102.33 m = 0.40 m.

(f) Calculate height **AC** = AB + BC = 0.30 m + 0.40 m = 0.70 m.

(g) **As a first trial,** hold the levelling staff at **D** and measure **DG** = 0.42 m.

(h) Calculate **x(D)** = AC − DG = 0.70 m − 0.42 m = 0.28 m.

(i) Using the formula, obtain the computed **d(D)** = (0.80 m ÷ 2) + (0.28 m × 1.5) = 0.82 m.

(j) Measure horizontal distance **AG** = 0.37 m; this distance is smaller than the computed distance **d(D)**. Therefore, the slope stake position should be further away from the centre stake.

(k) **As a second trial**, hold the levelling staff at **E** and measure **EH** = 0.53 m.

(l) Calculate **x(E)** = AC − EH = 0.70 m − 0.53 m = 0.17 m.

(m) Using the formula, obtain the computed **d(E)** = (0.80 ÷ 2) + (0.17 m × 1.5) = 0.655 m.

(n) Measure horizontal distance **AH** = 0.73 m; this distance is greater than the computed distance **d(E)**. Therefore, the slope stake position should be closer to the centre stake.

(o) **Move the levelling staff** slightly closer to the centre stake and eventually hold it at **F** where you measure **FK** = 0.49 m.

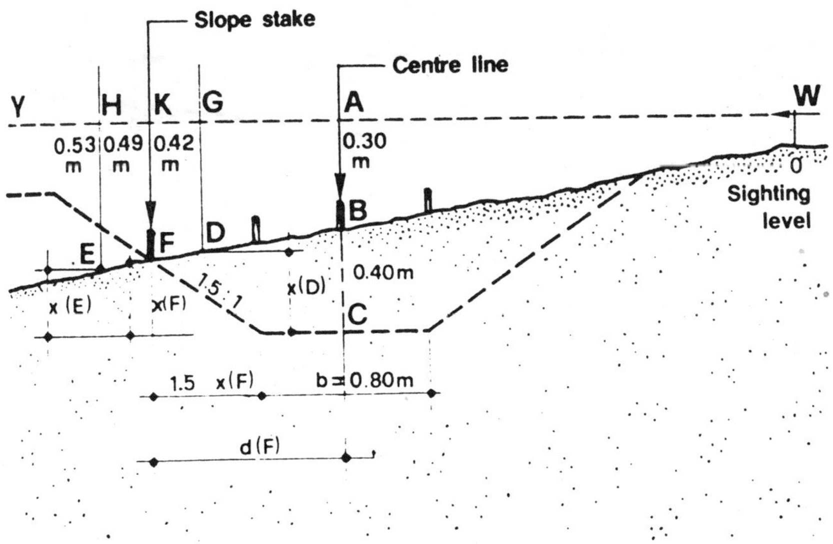

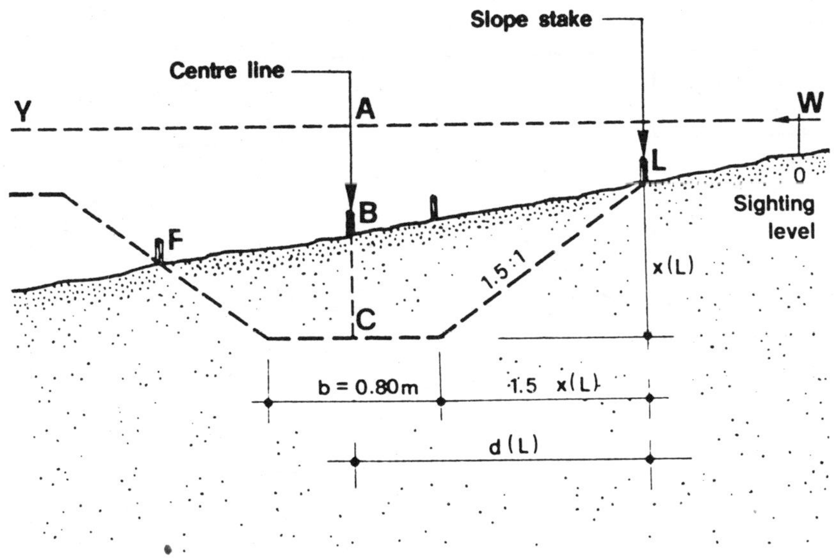

(p) Calculate **x(F)** = AC − FK = 0.70 m − 0.49 m = 0.21 m.

(q) Computed **d(F)** = (0.80 ÷ 2) + (0.21 m × 1.5) = 0.715 m.

(r) Measure horizontal distance **AK** = 0.72 m. This distance is practically equal to the computed distance **d(F)** which shows that **F is the correct location for slope stake F**.

(s) Repeat the same procedure, from **the same levelling station O**, to determine the correct location **for slope stake L**.

(t) Repeat the same procedure to locate **bank slope stakes** whenever necessary, with:

- **b** = top width of the bank;
- **z** = side slope of the bank;
- **x** = height of the bank (fill) at the slope stake.

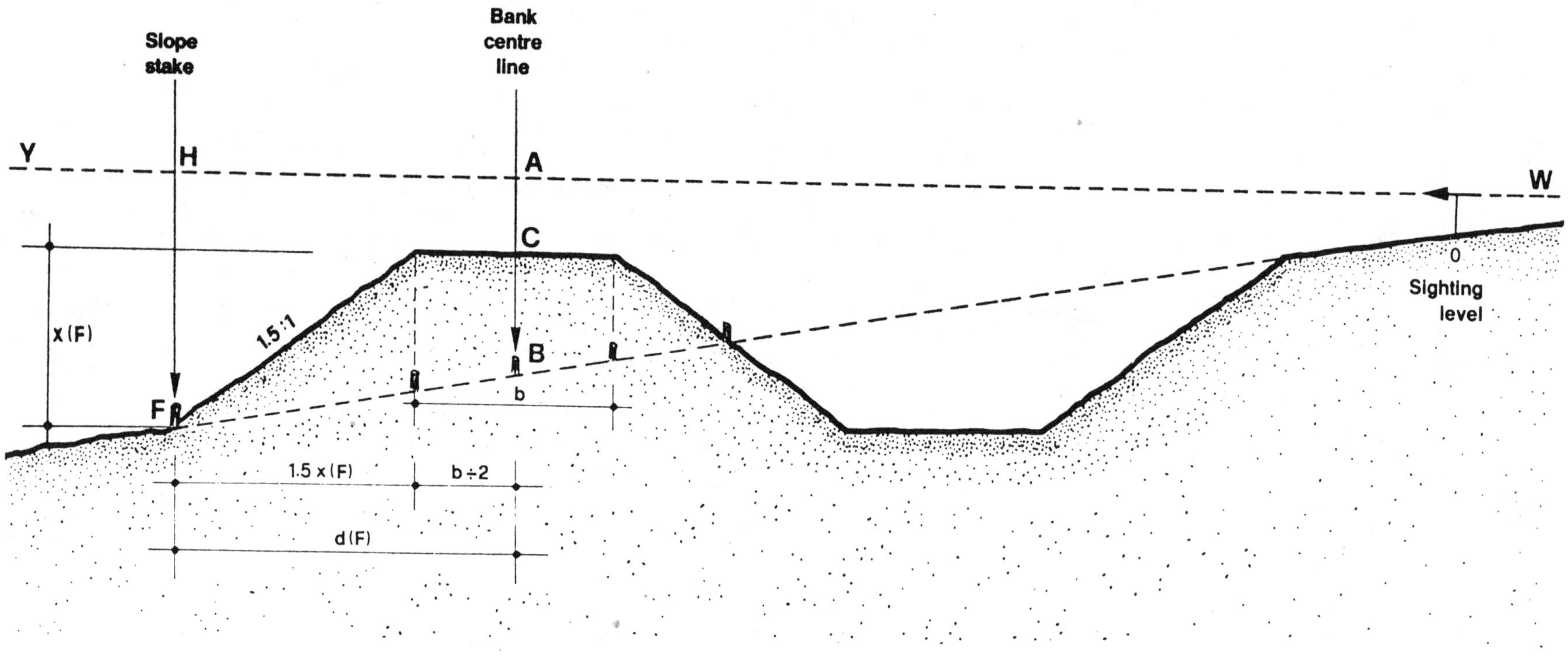

Building an earthen canal by hand

11. An earthen canal is constructed in several steps.

(a) Mark the length of the canal with centre, slope and bottom stakes as discussed above. Stretch a length of strong cord along the bottom stakes to mark the first cut. Dig out a vertical trench as wide as the canal bottom:

- dig to a depth of 10 cm higher than the final depth using the centre stakes as reference points;
- leave sections of earth to hold the stakes in place until you have finished digging the centre trench;
- carry away the earth that you dig out to build banks or throw it downhill to avoid later erosion by runoff into the finished canal.

Note: if you use the earth to build banks, make sure that it is well compacted (see Section 62, **Construction, 20/1**).

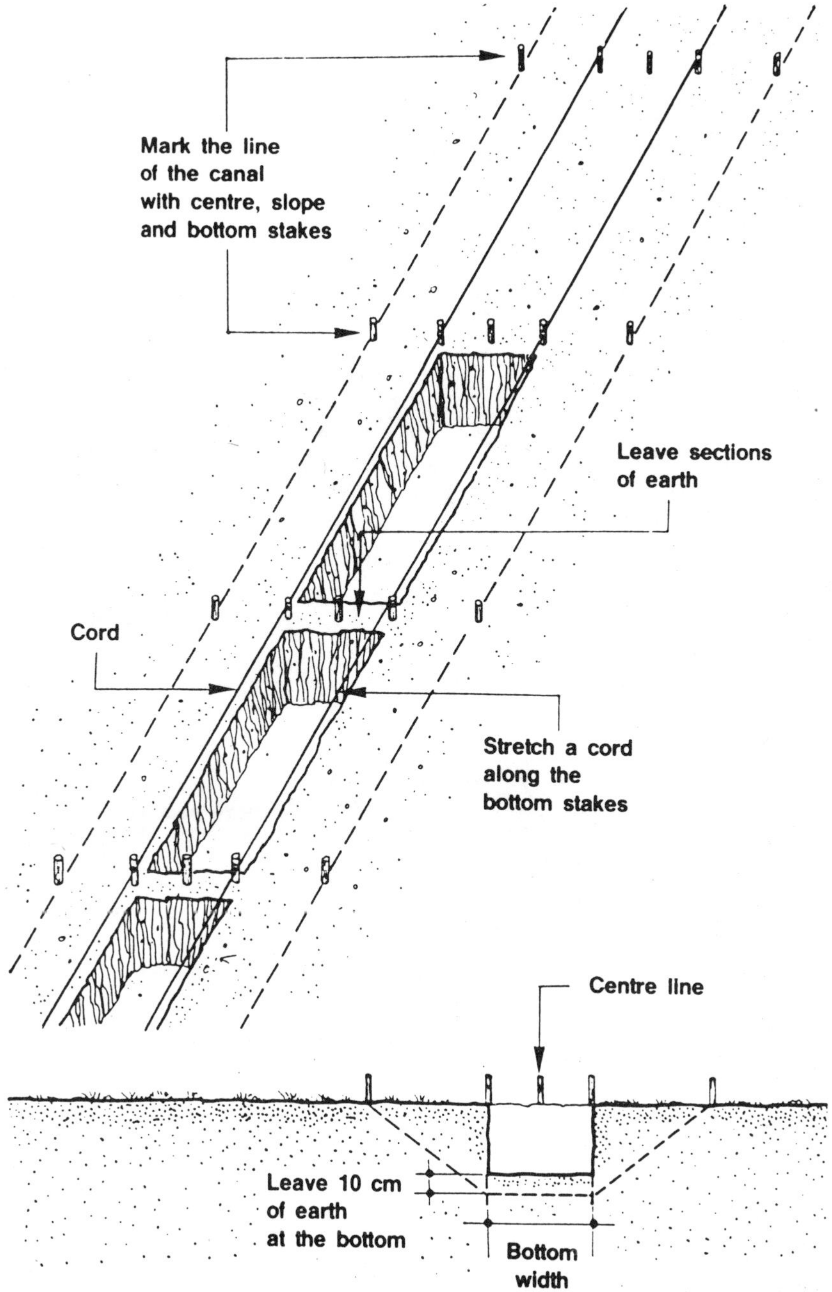

(b) Move the lengths of cord out to the slope stakes to mark the next cut. Remove the centre and bottom stakes and the sections of earth that you left to hold these stakes in place.

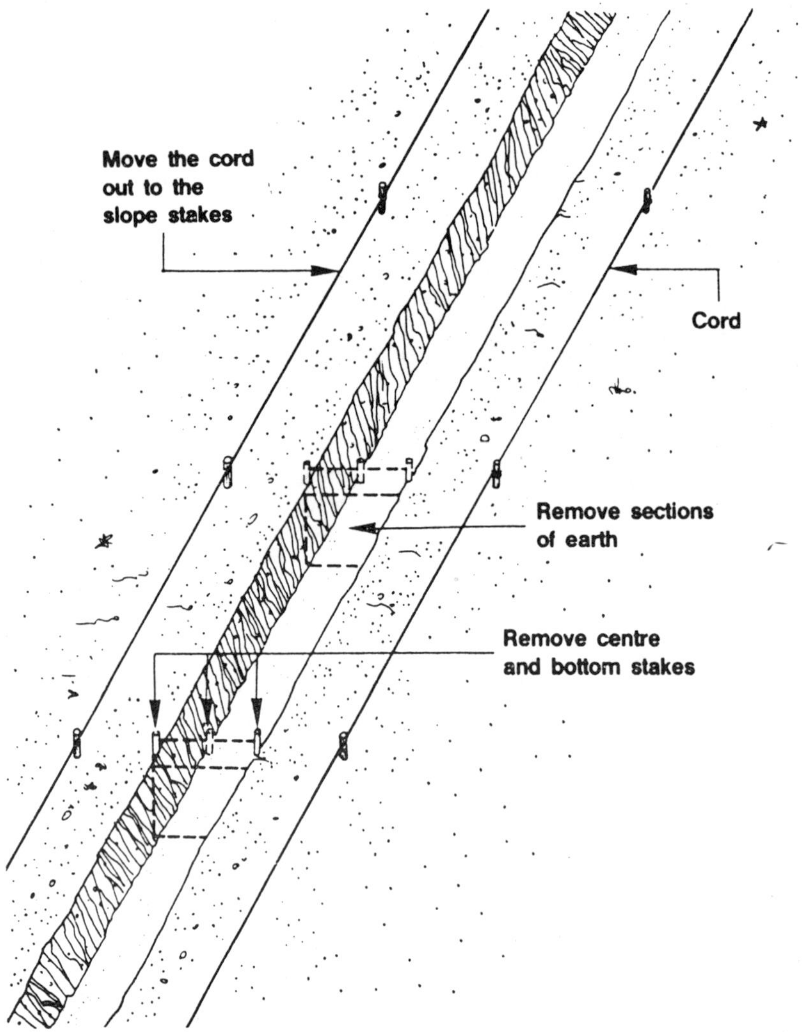

(c) Dig out the remaining 10 cm of earth in the bottom of the canal and, if necessary, adjust the slope of the bottom as you were told in the previous manual, **Topography, 16/2**, Section 114.

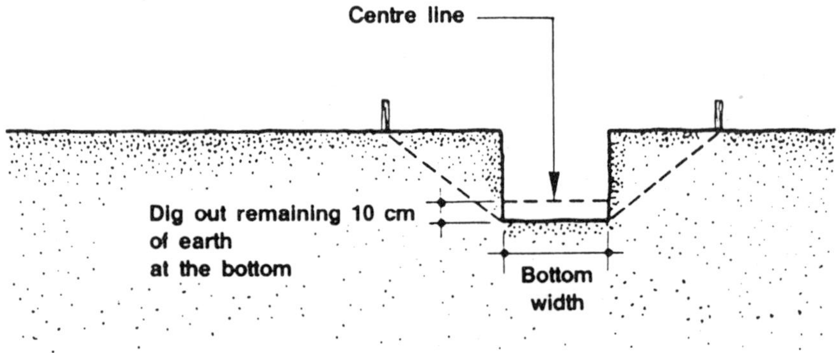

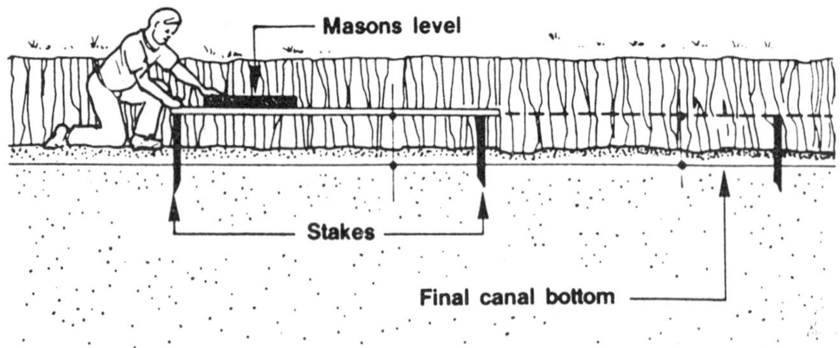

(d) Cut out the sides of the canal from the side stakes obliquely to the edges of the canal bottom:

- use a wooden gauge to check the cross-section of the canal as you dig;
- get rid of the earth as you were told before.

(e) Complete the construction of the banks if necessary, levelling the top and forming the external side slopes. Plant grass to avoid erosion (see **Construction, 20/1**, Section 69).

(f) Build the **water control structures** before letting any water flow through the canal (see Section 87).

(g) After you have finished **check that the canal functions as designed** by letting in some water before starting the construction of the diversion ponds.

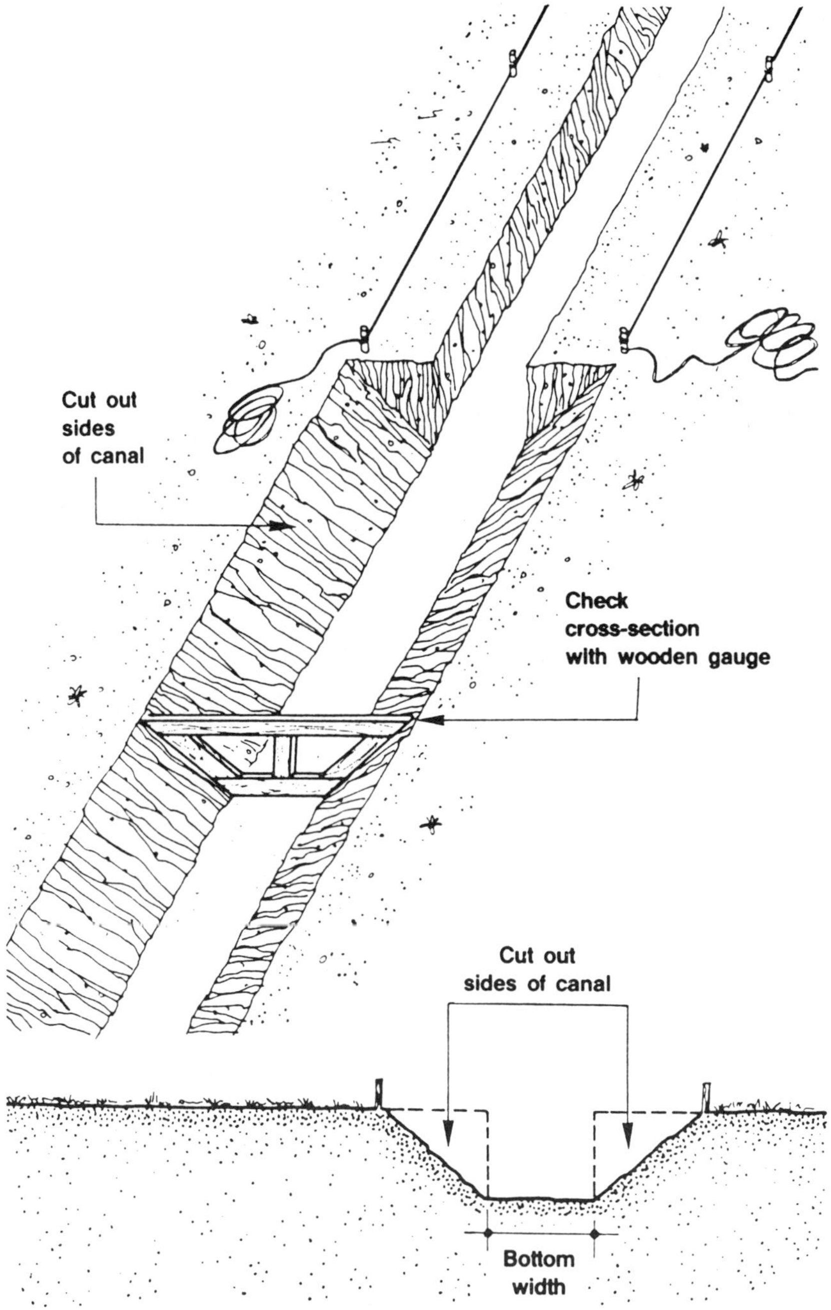

Building a clay-lined canal

12. Where good clay is available (see **Soil, 6**) and **where the canal is not dried out seasonally**, a clay lining can reduce seepage by 75 to 80 percent:

- allow an additional 40 to 45 cm around the bed and sides;

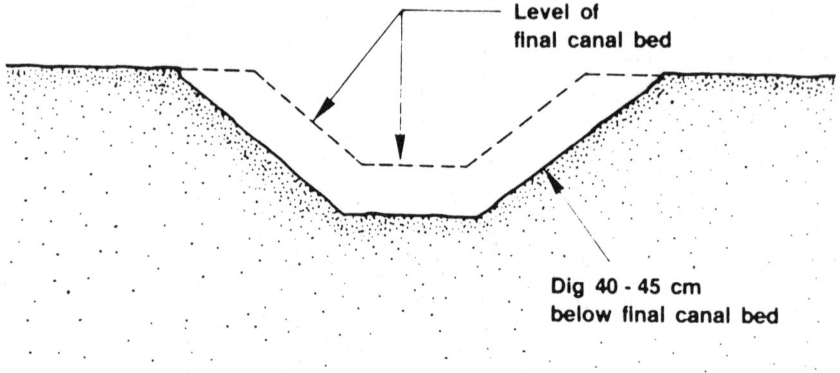

- spread a layer of 7.5 to 15 cm of clay, by itself or mixed 2:1 with sand and gravel, knead in well by foot or with a sheeps-foot roller;

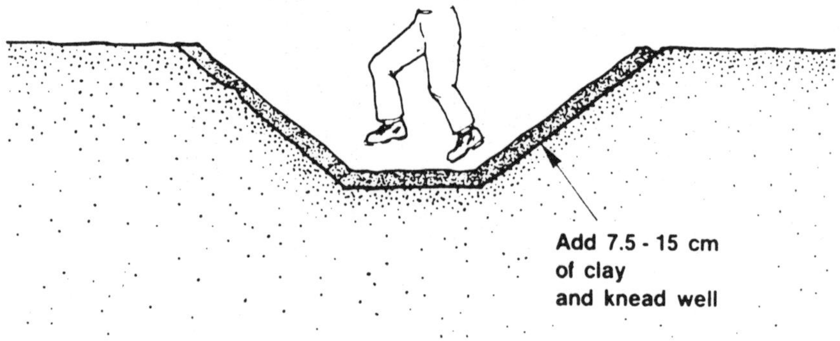

- cover this layer with 25 to 30 cm of silt, kneading and compacting well.

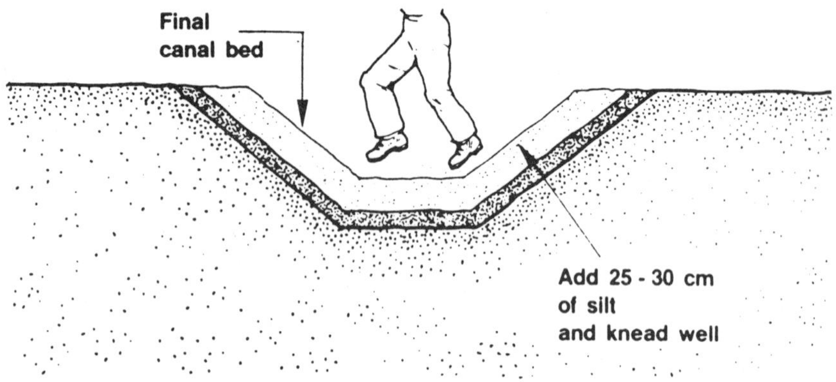

Building a concrete-lined canal

13. Where sand, gravel and Portland cement are available, **poured concrete linings** are usually preferred to precast materials or fired bricks, because of their long life. Their construction also requires less effort.

14. Small cement-concrete linings can be easily formed by **plastering the faces of the canal with mortar**, after the ditch has been properly dug to its designed dimensions (increased from 2.5 to 5 cm by the thickness of the mortar). Refer to Section 33, **Construction, 20/1**, where you learned about **cement mortars**. Another possibility is to use **soil cement** if soils are suitable.

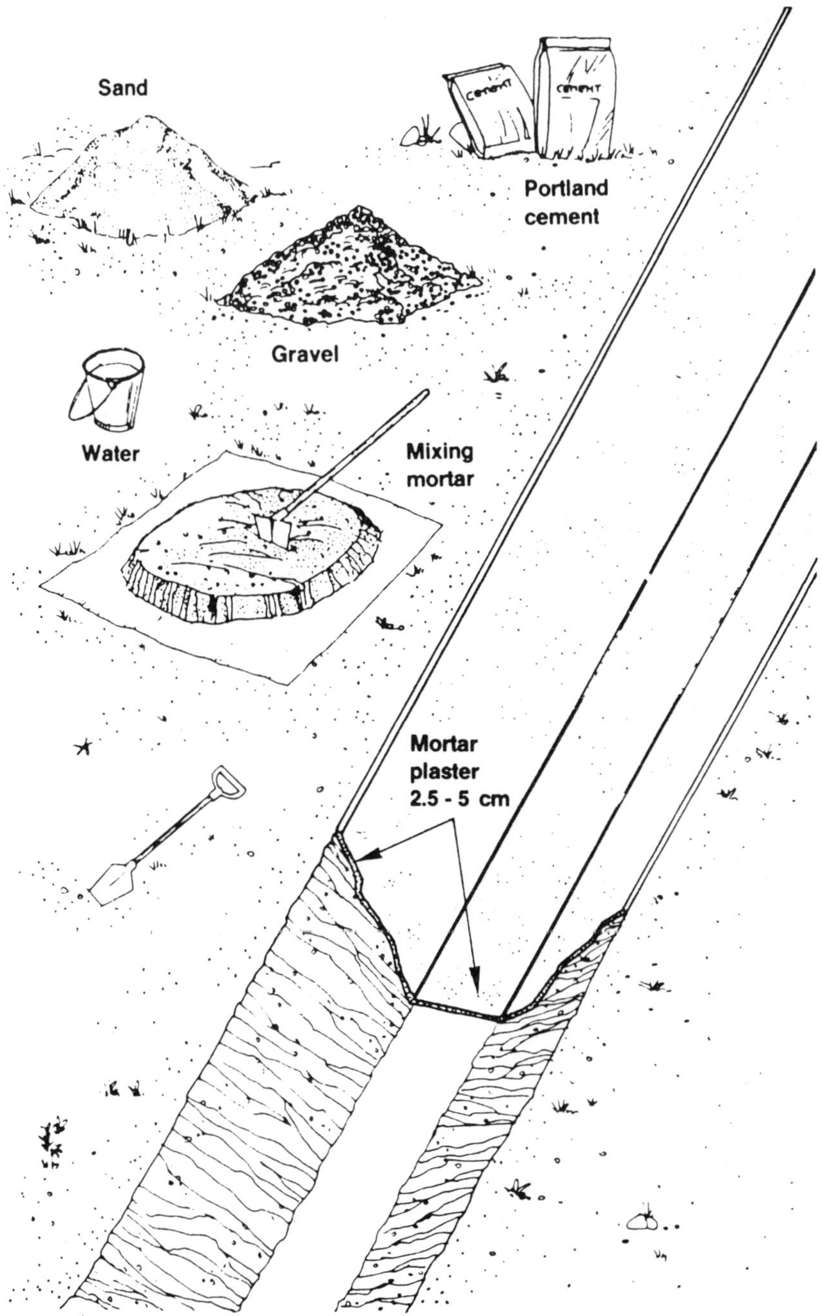

15. Larger canals are better lined by **using wooden forms. Cement concrete** (see Section 34, **Construction, 20/1**) is then placed between the forms and the earthen walls. Remember the following points:

(a) **The required thickness** of the lining varies from 5 to 7.5 cm, depending on the size of the canal.
(b) **Side slopes** may be 1:1 in rocky soils, but for most other soils a slope of 1.5:1 is preferred.
(c) Dig **the earthen canal** as explained above, increasing its designed dimensions according to the thickness of the concrete.
(d) First line **the canal bottom** by building a series of **concrete guide strips about 10 cm wide at 10 m intervals**, their top surfaces being at the **exact elevations** indicated by the design.
(e) Let these strips harden, and complete the bottom lining by **pouring concrete between them**, using them as reference points to give the canal bottom its designed height and slope.

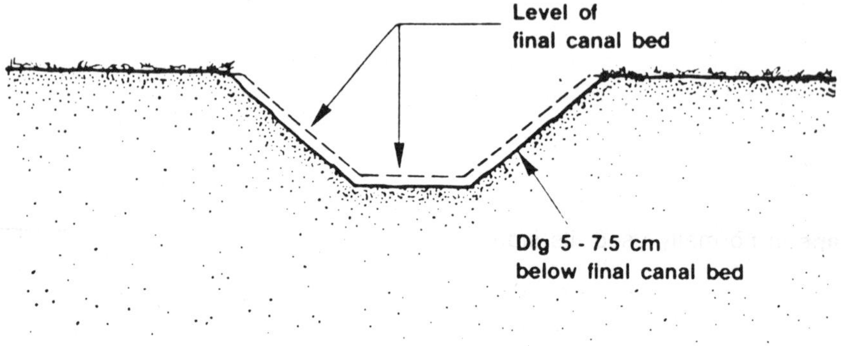

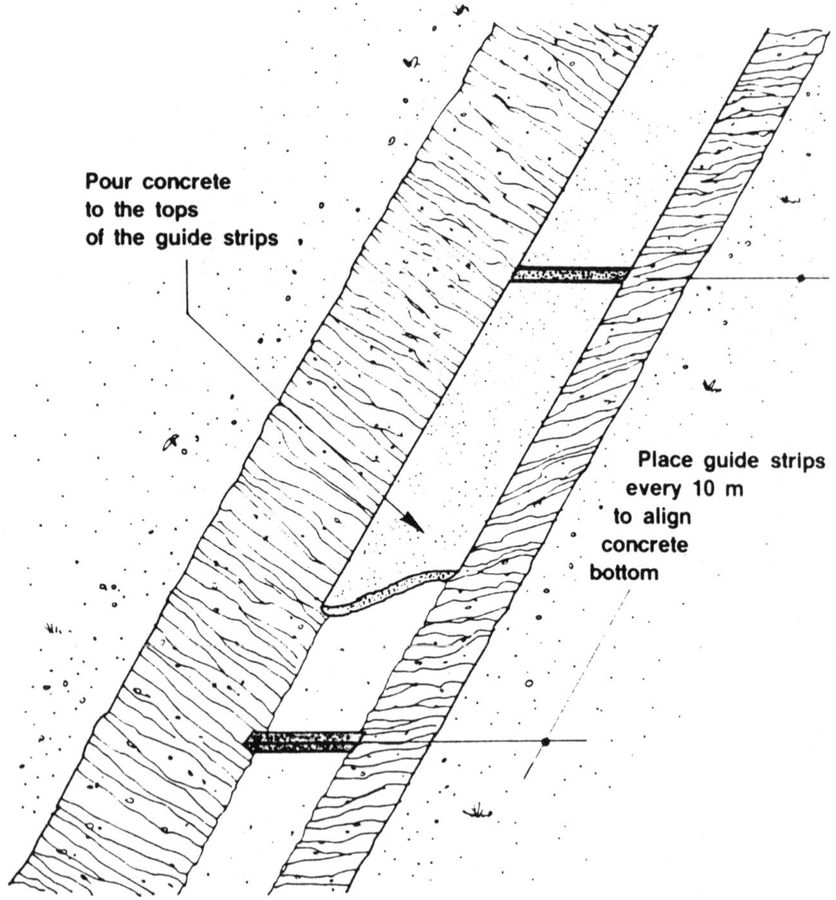

(f) Metal or wooden forms can be used when pouring the concrete.

(g) Line **the canal walls** after securing the forms well. Ensure a good connection with the bottom slab.

(h) **Cure the linings properly** for several days (see Section 34, **Construction, 20/1**).

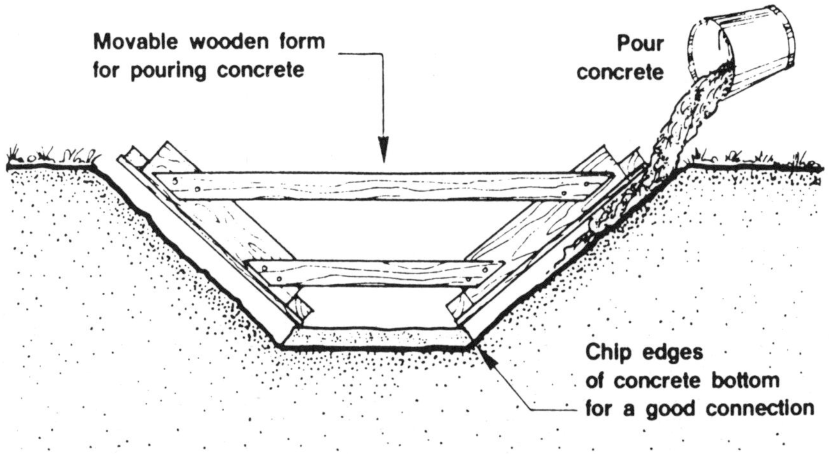

Note: joints should be placed between poured sections at 2 to 4 m intervals to prevent cracking resulting from expansion or contraction of the concrete caused by temperature and humidity changes.

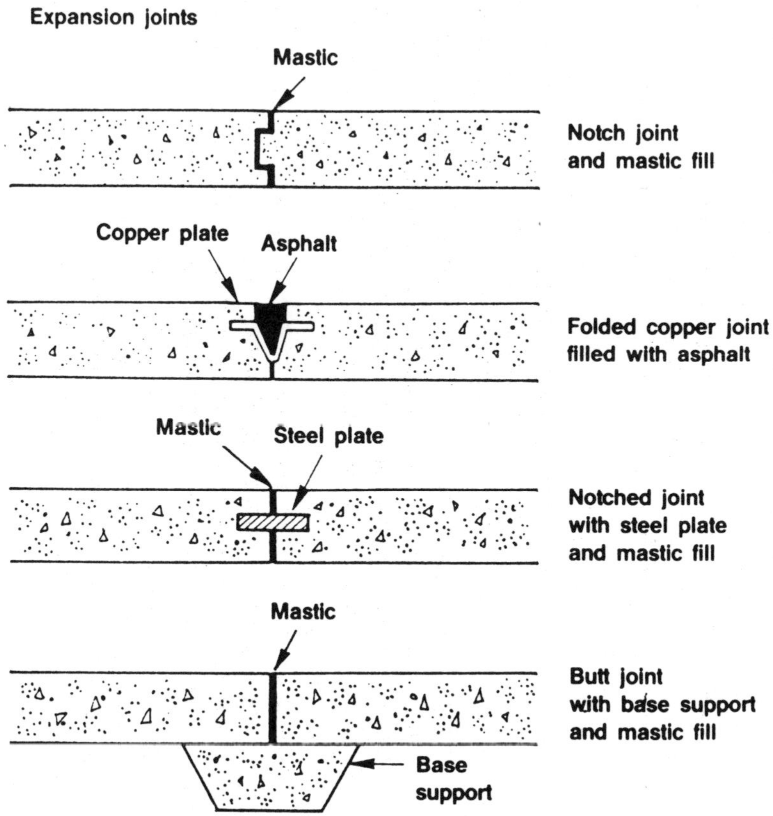

Building a brick or block-lined canal

16. Small brick or block-lined canals can be built either in a rectangular or in a trapezoidal shape. The floor may be built either of brick or concrete. For additional seepage control, a plastic lining may be set behind the walls and under the floor.

17. For larger channels **a trapezoidal shape** is normally used; its side walls are supported by the cut embankment or by lateral fill, which must be firm and well compacted to avoid settlement and cracking of the bricks.

18. Standard or **vertical-wall canals** for fish farm use are normally of single brick or block thickness, although where lateral strengthening is required, small piers or bracing points may be incorporated. If hollow or notched concrete blocks are used, simple reinforcement can easily be added.

Rectangular

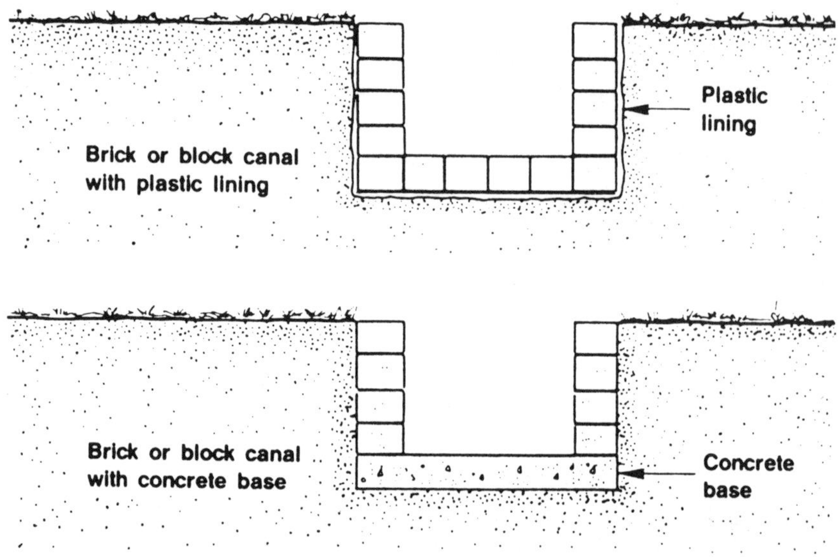

Trapezoidal

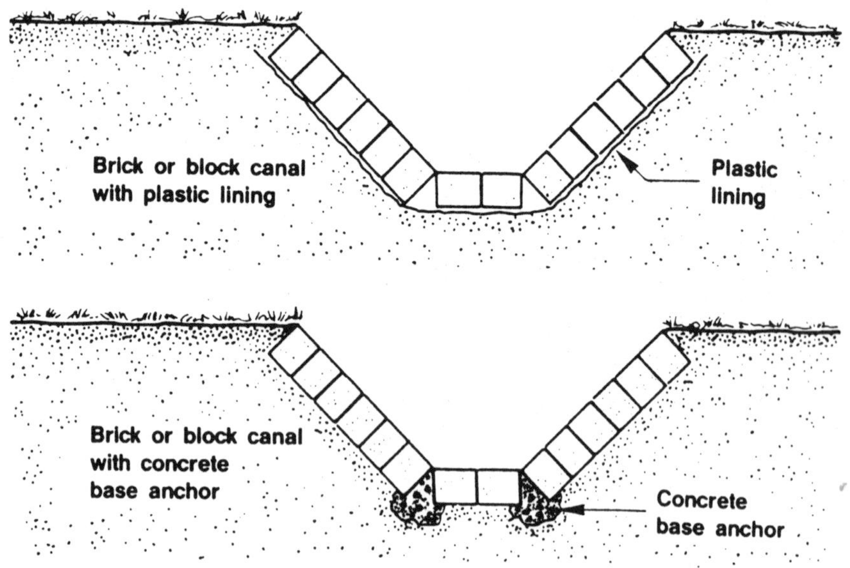

Note: brick is laid either in conventional horizontal runs, face or side down, or diagonally in "herringbone" fashion, bedded on 10 to 15 mm of 1:5 cement mortar, with 1:3 jointing mortar.

19. **Rectangular channels** can be constructed either inset in an embankment or with filled sides (with fill on each side not exceeding 30 cm), or partially or completely freestanding, in which case lateral strengthening may be needed to protect against impact damage.

20. For **brick channels**, the outside fill slope should be no steeper than 45° (1:1 ratio); thus a rectangular channel filled to 60 cm of its height will have fill at least 60 cm wide on either side. Trapezoidal canals are filled to their upper edge or higher (forming a small embankment). As the slope is based on the upper canal width, the quantity of fill material is consequently much greater and the base much wider than that of vertical-sided channels.

21. **The main problems of using brick** are the relative cost and slow speed of construction, and rigidity, which may lead to cracking if the base material is not compacted properly or if lateral soils (e.g. clays) are prone to swelling or settlement. Plastic lining may be used to limit damage from seepage, which may otherwise accelerate cracking and deterioration. Take particular care to ensure that bricks are sound and that mortar is of good quality and will not wash out.

Various examples of rectangular canals

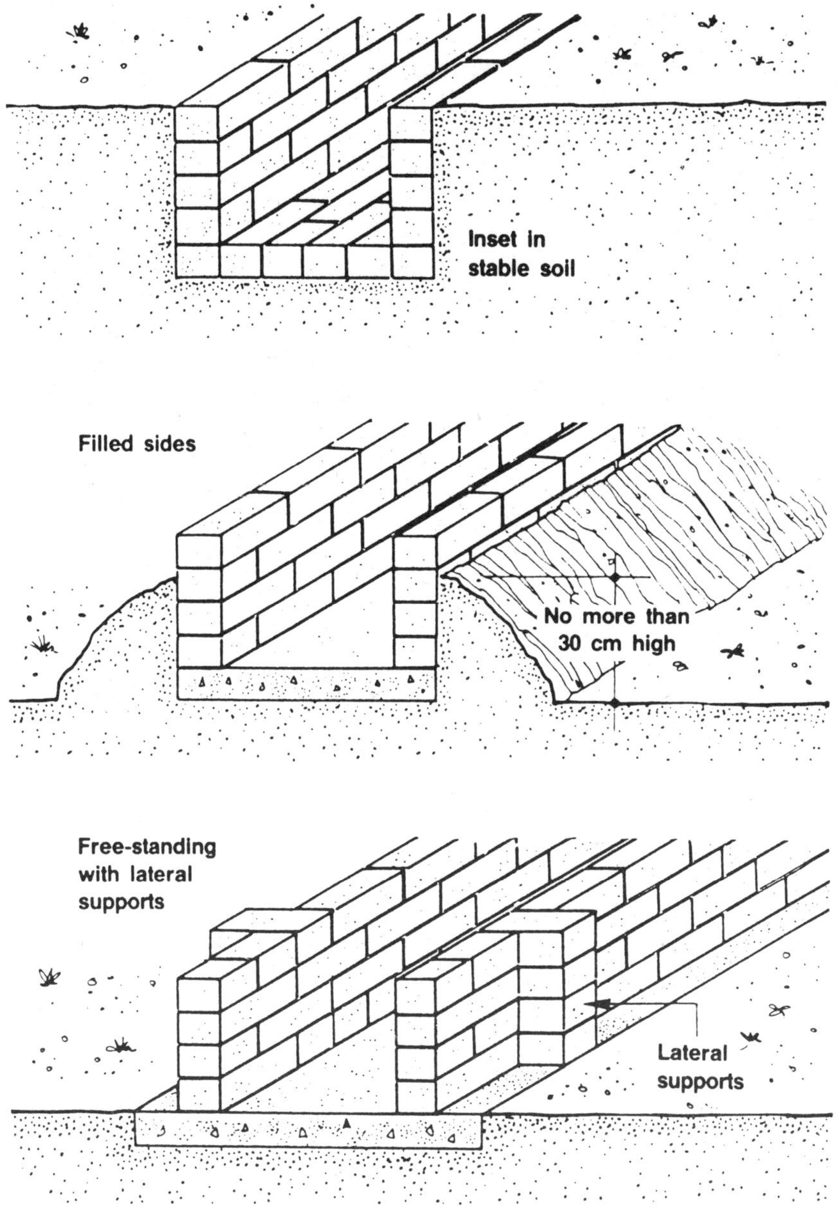

Using a cement or stone slab lining*

22. **Precast cement units or cut stone slabs** may also be used for lining, set either on a slab, brick or concrete floor. These are usually in trapezoidal canals, with lateral support at the top edge. As with bricks, poor base materials may result in cracking of the mortar and of the slabs themselves.

23. Normally **single slab units** are used on side walls. They limit effective wall height and hence canal depth. A mortar fill of 5 to 10 mm is normal, and the bottom edge of the wall slabs should be well supported on the base to avoid collapse. As with brick walls, an external plastic lining can help reduce deterioration through cracking.

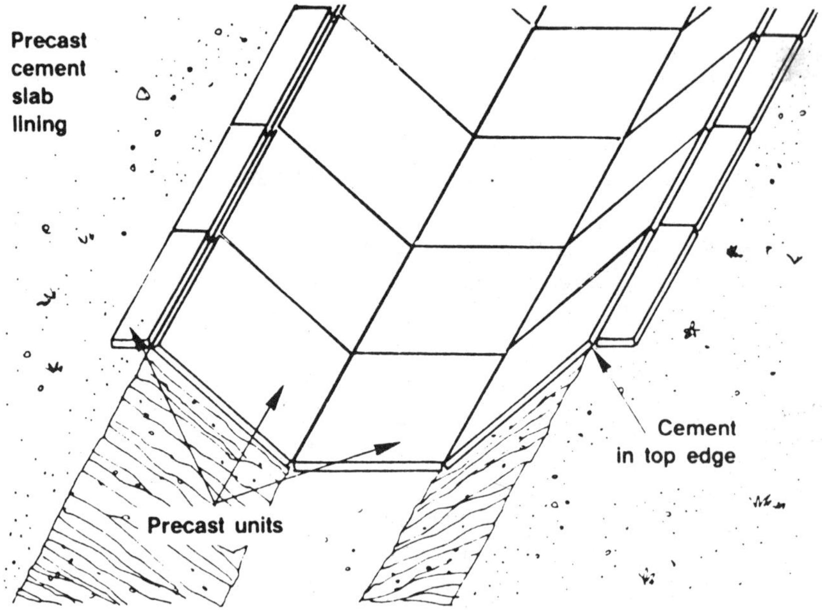

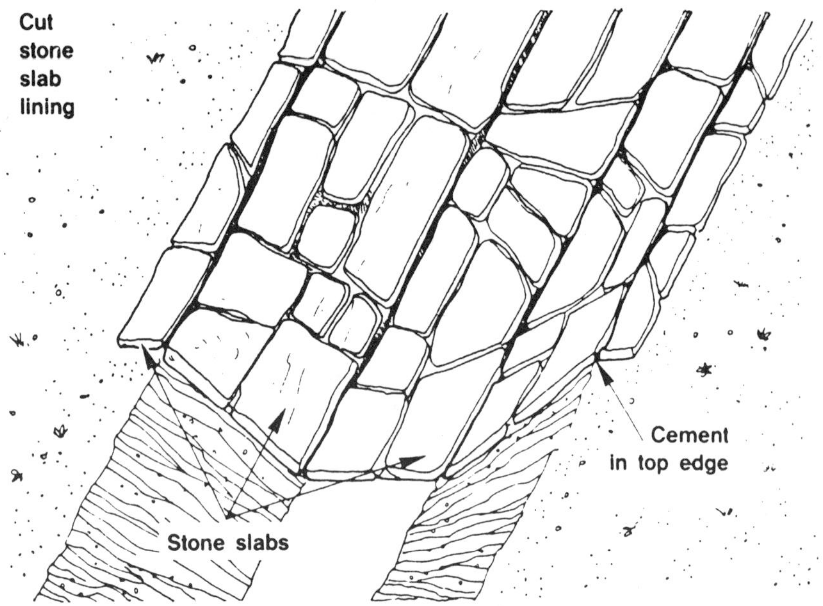

Using precast linings

24. A range of **manufactured concrete linings** is produced for agriculture, drainage and waste water use. If available locally, they may be an effective choice. They are normally integrally cast — walls and base form a single unit — and can be used backfilled, in embankments or freestanding. They are produced either lightly reinforced or unreinforced. The unreinforced linings are weaker and may be somewhat permeable if poor quality aggregates are used. In this case, internal plastering or external lining with polythene is recommended. Units are now also available in glass reinforced plastic (fibreglass) — which is light, strong and smooth-walled, but relatively expensive — and in glass reinforced cement, which is cheaper but heavier. Asbestos cement linings are also available in some locations.

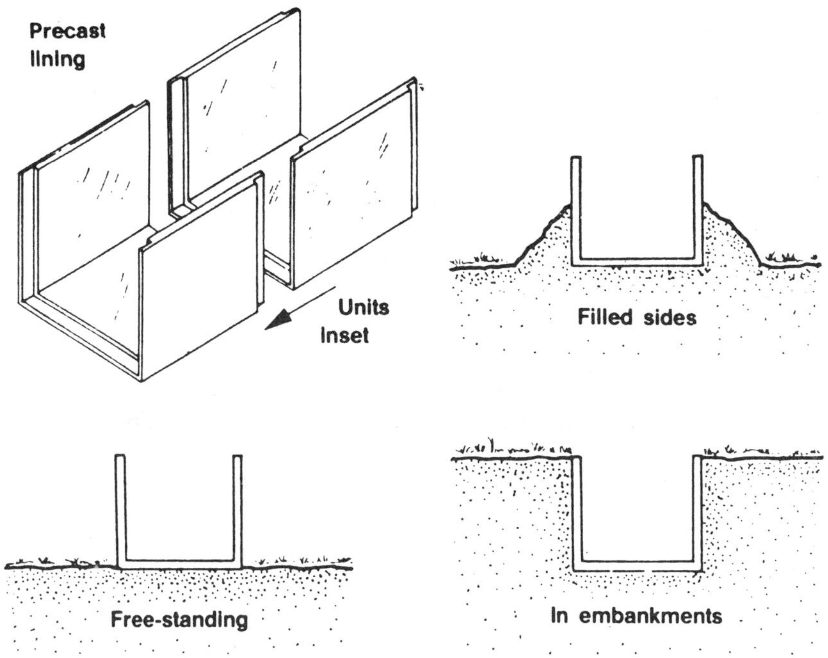

Using flexible linings

25. As mentioned earlier, several **flexible materials**, such as rubber or polythene, may be used. They have the advantage of good adaptation to ground shape and initial settlement or swelling, and can normally be laid relatively quickly in long continuous lengths. Adjacent lengths can be either heat sealed, cemented or simply overlapped. The sides are carried over the upper edges of the canal and dug into trenches, typically 30 to 40 cm deep and backfilled with earth or aggregate.

26. The main **disadvantage** is the ease with which they are damaged, through puncturing by sharp objects or coarse underlay material; once damaged the base material is easily washed out. Exposure to strong sunlight and high surface temperatures can make many of these materials more brittle and hence more prone to damage. Reinforced linings, although more expensive, are more durable. For this reason, these linings are often covered with soil, clay, brick or slab. Polythene sheet may also be useful as an anti-seepage lining for more rigid materials such as brick or slab.

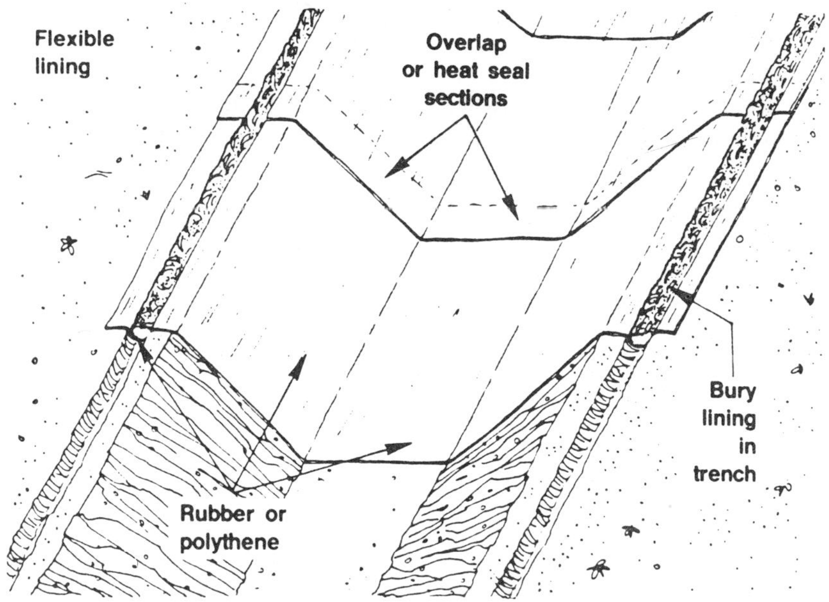

85 Drainage canals

1. Drainage canals are built to carry the water draining from the fish ponds away from the farm site, usually into a lower natural channel.

2 **The design** of a particular drainage canal should be based on the pond or series of ponds it will have to drain:

- **for small ponds**, (nursery ponds, for example), the drainage canal may be designed to drain more than one pond simultaneously, within a period of perhaps a couple of hours;
- **for medium-size ponds**, the drainage canal is usually designed to drain the ponds one by one within a reasonable period of time, from half a day to a full day;
- the design of the drainage canal also depends on the **type of pond outlet** and its water carrying capacity (see Chapter 10);
- **for large fish farms**, the total drainage time of all the ponds should not exceed 1 day per hectare (or 5 days for a 5-ha water surface and 25 days for 25 ha); the pond outlet structures should be designed accordingly.

3. Drainage canals are usually **unlined and trapezoidal** in cross-section. They are designed and built as described in Section 83.

4. Remember that to ensure good and complete drainage, **the lowest level of a drainage canal should be at least 20 cm deeper than the lowest point in the pond**.

86 Diversion canals

1. A diversion canal should be built to divert excess stream water **around a barrage pond** if the pond is built where there is a likelihood of floods (see Section 14, **Construction, 20/1**). Such a canal should therefore be deep and wide enough to evacuate the largest flood waters. The diversion canal starts from **a diversion structure** (see Sections 73 to 75).

2. Design and construction methods of diversion canals are similar to those given for earthen feeder canals (see Section 83). The following points are of particular importance:

- **the canal's initial bottom level** should be equal or slightly lower than the stream bottom level;
- **the canal dimensions** should be at least equal to those of the stream channel when in full flood;
- the canal should be at least **5 m away from the pond banks**;
- when staking out the canal axis, set **all the tops of the stakes at the same level**;
- it is best **to design the canal bottom with no slope** and to build **a series of drop structures** at regular intervals (see Section 87). The total height of these drops should be equal to the difference in elevation between the initial canal bottom and the original stream bottom at the junction point **J**;
- when digging the canal, **start from its downstream end**.

Note: consider constructing a diversion canal only if its dimensions are reasonable; if not, it might be best to select another site for a barrage pond or to study the possibility of having diversion ponds.

About drainage canals

A Drain small ponds two or more at a time
B Drain medium-size ponds one by one
C Drain large ponds one by one

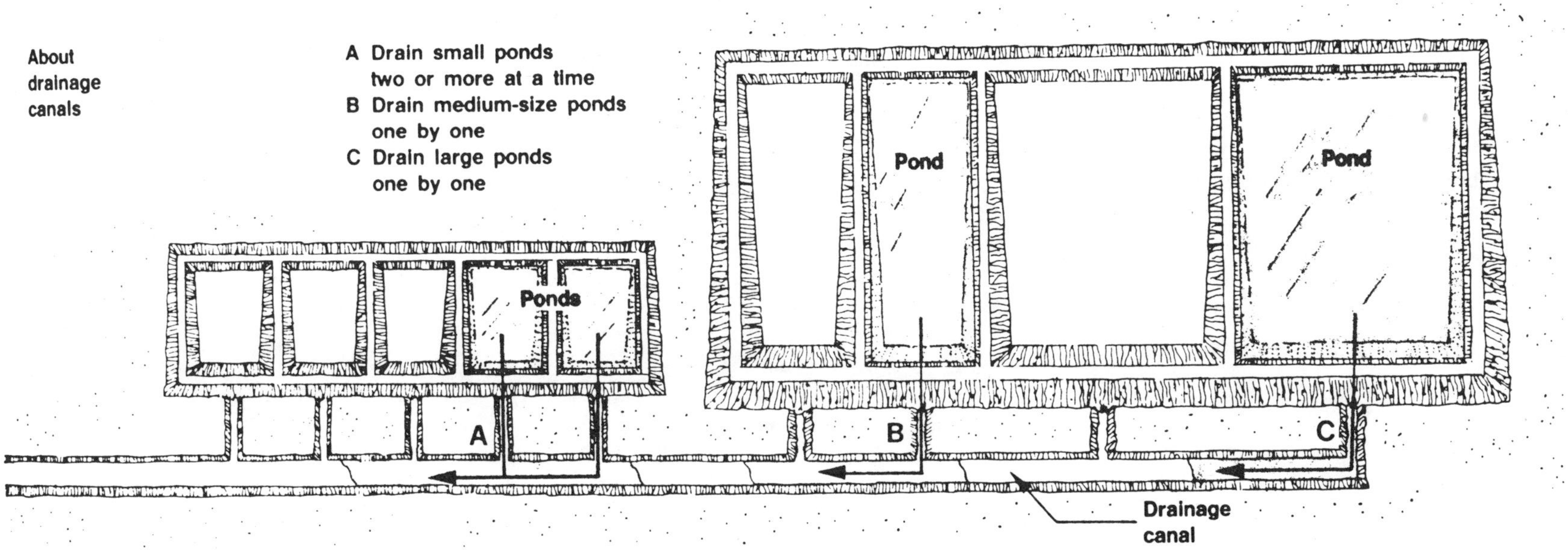

About diversion canals

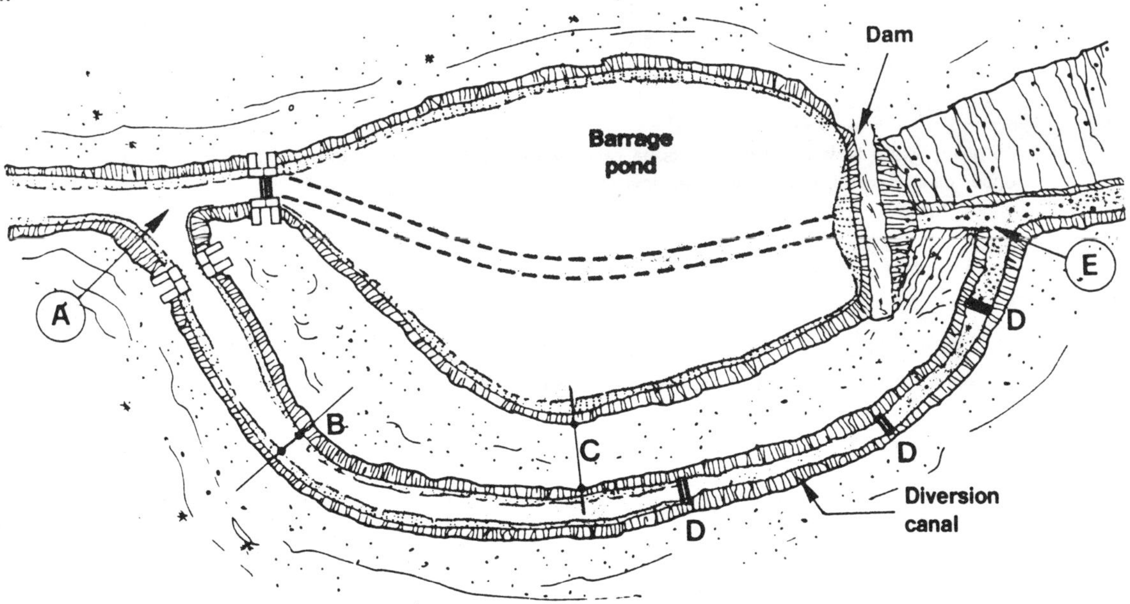

A Bottom level of inlet equal to or lower than bottom level of stream
B Canal width equal to stream width at times of flood
C Canal at least 5 m away from pond banks
D Drop structures
E Start digging canal at the downstream end

87 Water control structures for canals

1. Several kinds of water control structures are used in feeder canals for various purposes, see **Table 44**.

2. These structures can be built with a range of materials such as wood, fired bricks, concrete blocks or concrete, according to local availability and size of the fish farm. Refer to Chapter 3 of **Construction, 20/1** to select the right kind of material and to use it properly.

TABLE 44

Water control structures for canals

Structures	*Purpose*	*Type of canal (section: paragraphs)*
OVERFLOW GATE lateral type end type	To evacuate any water excess and to dry canal sections for maintenance/repair	Feeder canal (87: 3 - 5) (87: 21 - 24)
DIVISION BOX T-type X-type	To divide water flow into 2 or 3 parts, to raise water level in upper canal section, to control water flow at ponds intake	Feeder canal (87: 6 - 15)
DROP	To reduce canal slope and water velocity	Feeder canal Diversion canal (87: 16 - 20)
SETTLING BASIN	To remove the suspended soil particles from turbid water	Feeder canal (116)
STILLING BASIN	To slow water down	(117)

Water control structures for a small fish farm

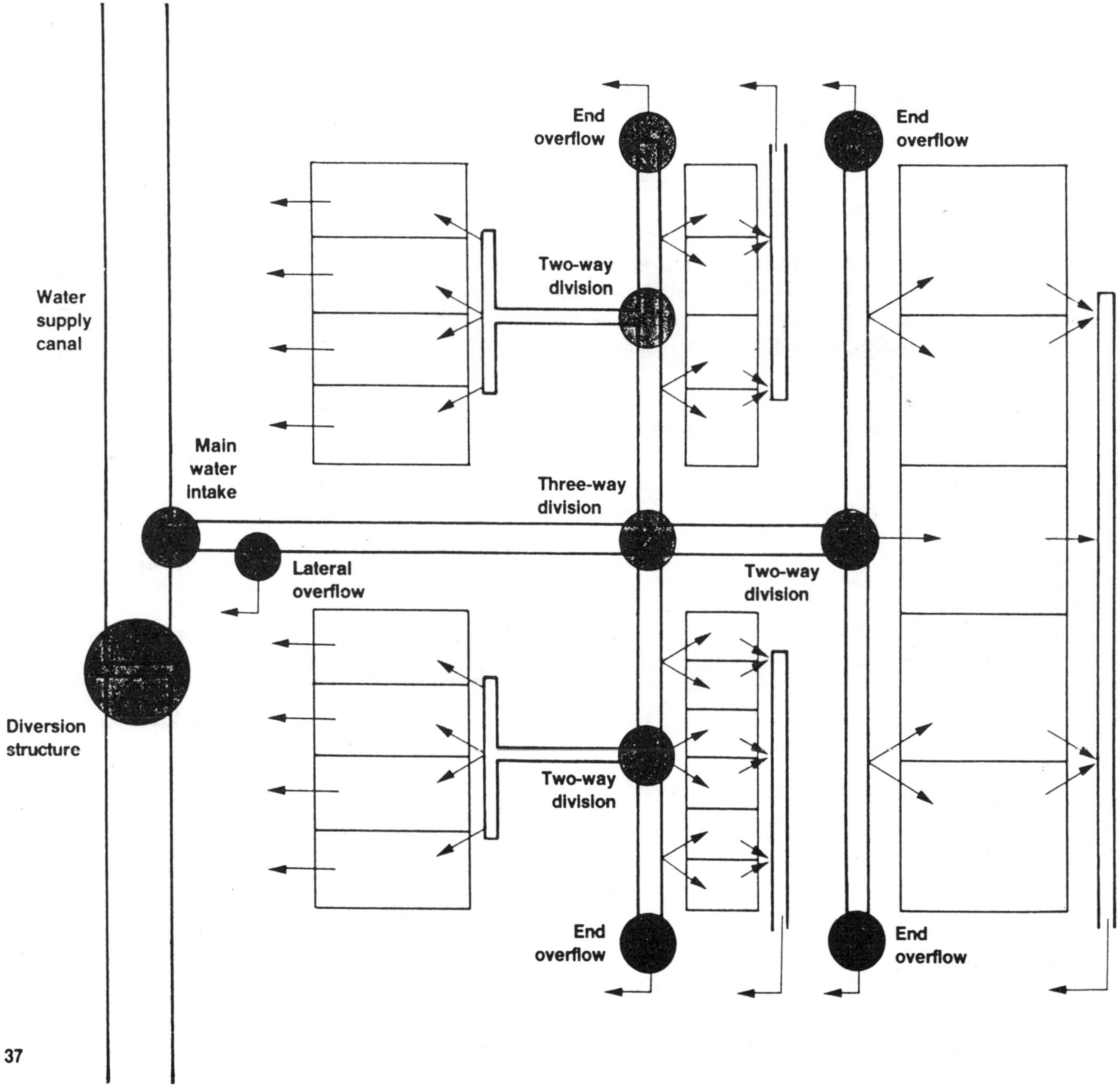

Note: see also page 37

Lateral overflow gates

3. To guard against damage from sudden rises of the water level in feeder canals, **a lateral overflow gate** should preferably be built immediately downstream from the main water intake. Other similar gates should also be built further down, on the various feeder canals.

4. Lateral overflow gates are built in one of the side walls of the feeder canal. They are usually **box-like structures** with a set of two grooves in their lateral walls. Wooden planks are placed into these grooves up to **a level slightly higher than the normal water level in the canal**. Clayey soil is well packed between the two rows of planks to control any water seepage. When the canal has to be completely drained for maintenance or repair, the earth and the planks are removed.

5. **Lateral overflow gates** can be built from wood, bricks, concrete block or reinforced concrete.

Plan of concrete block lateral overflow gate

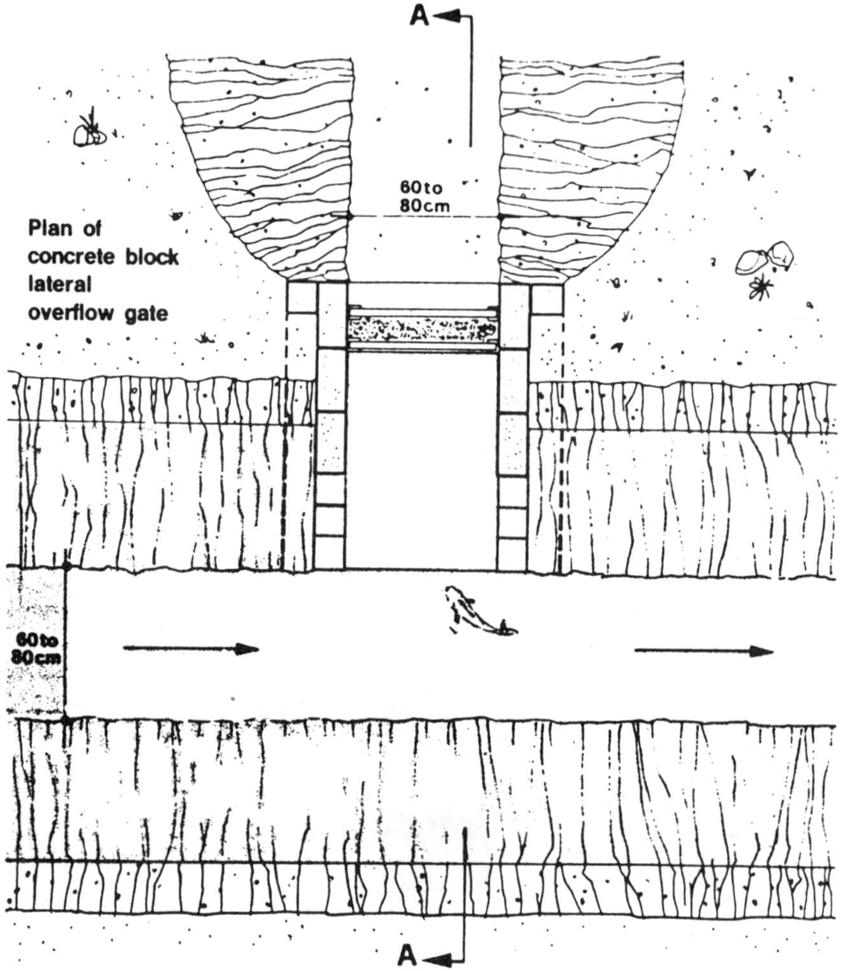

Section AA

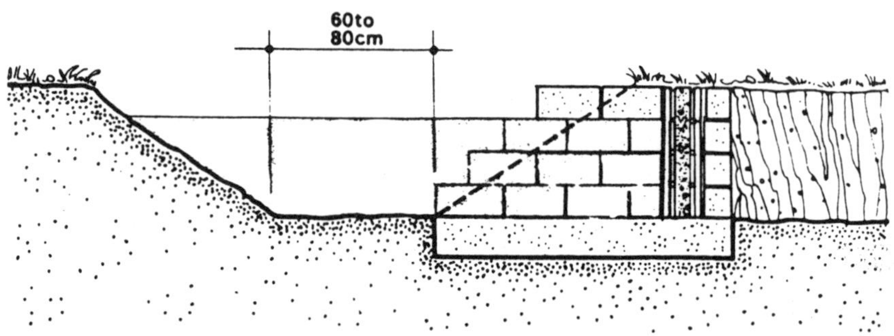

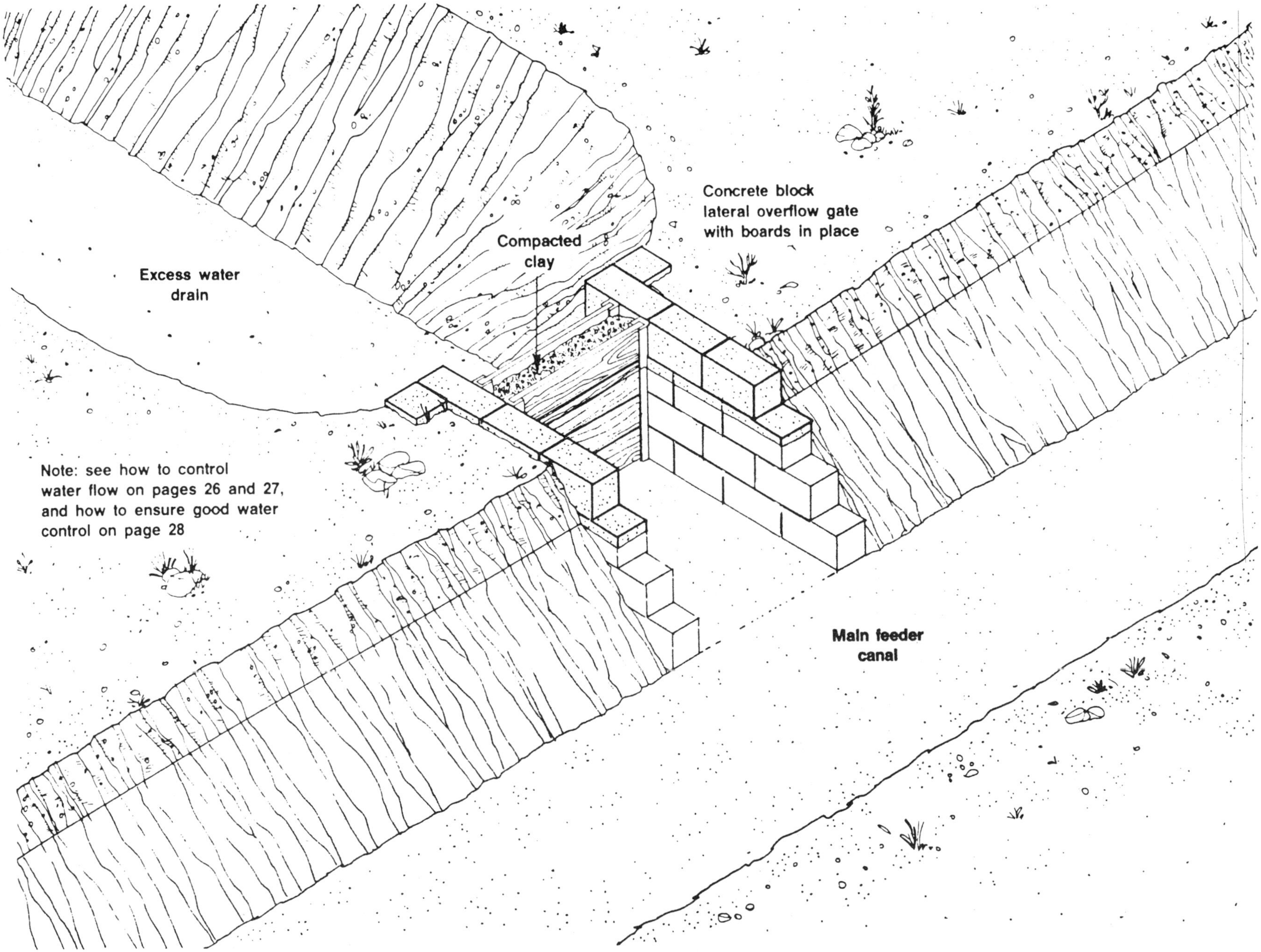
Concrete block
lateral overflow gate
with boards in place
Compacted
clay
Excess water
drain
Note: see how to control
water flow on pages 26 and 27,
and how to ensure good water
control on page 28
Main feeder
canal

Building a lateral overflow gate

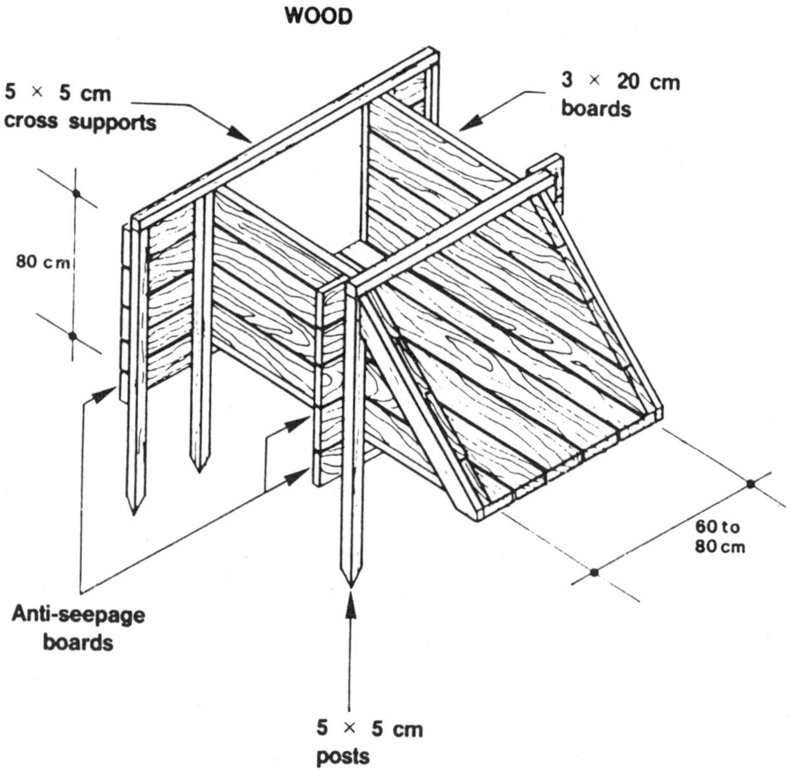

Note: the dimensions on this page and on page 87 are suitable for a medium-size pond system; dimensions may vary, however, depending on the size of your pond system

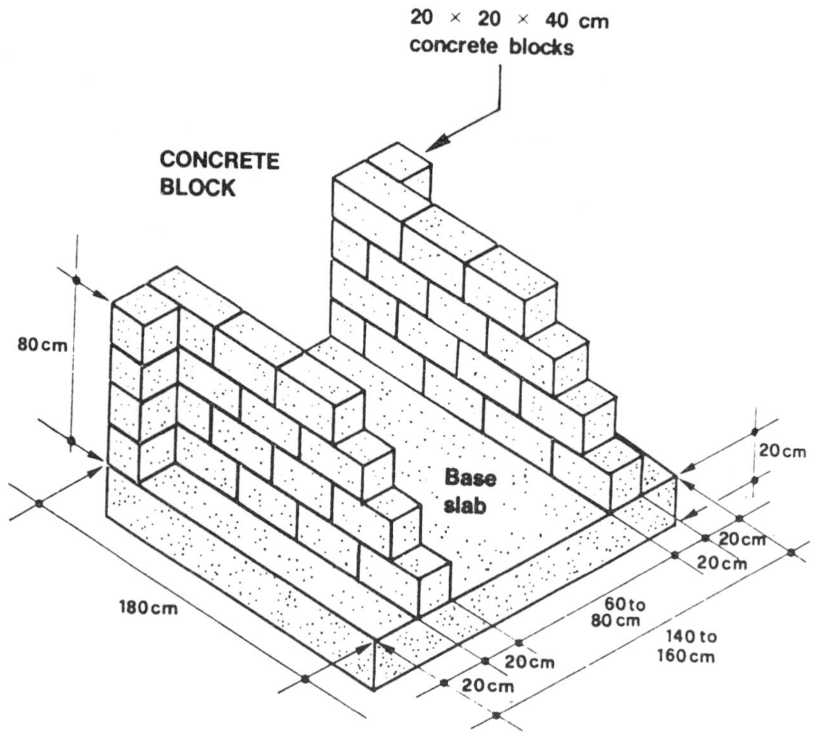

CONCRETE OR
REINFORCED CONCRETE

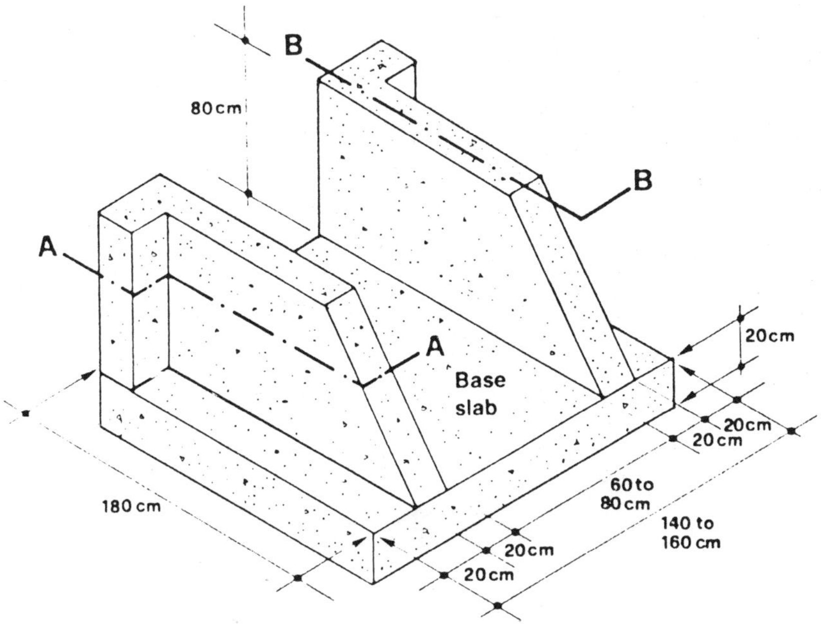

Placement of steel bars
for reinforced concrete

Plan section AA

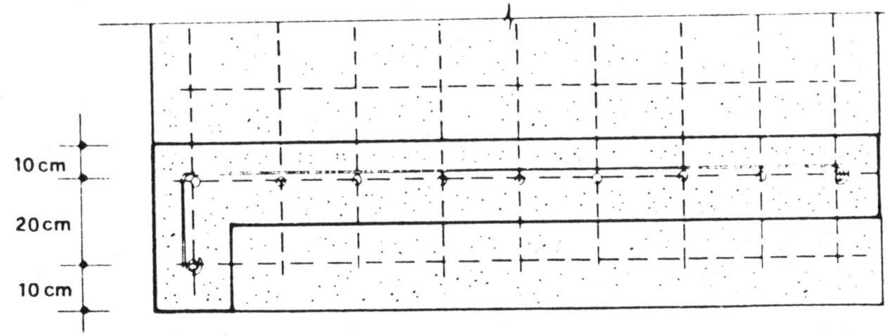

Vertical section BB

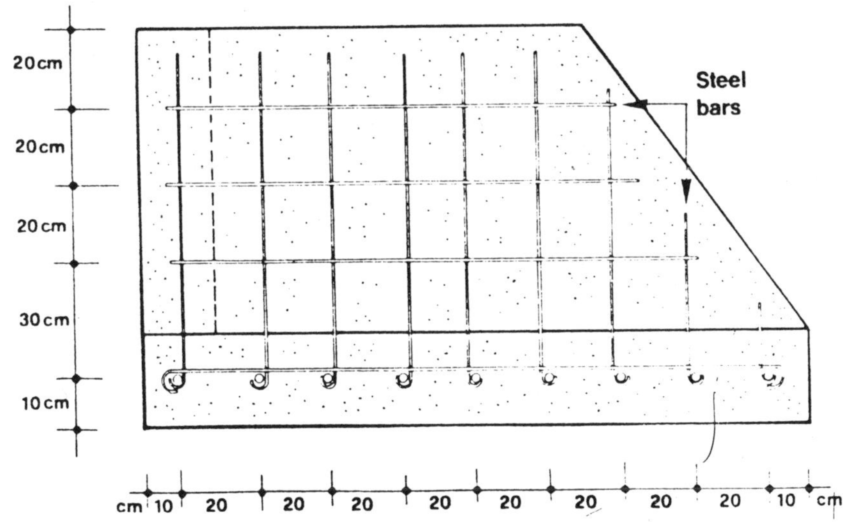

Three-way division boxes

6. A three-way division box **(X-box)** is used in a feeder canal **to divert** part of its water flow or its total flow into either:

- **one or two pond intakes**; or
- **one or two additional feeder canals**.

7. Such diversions are usually done at right angles. They offer the same possibilities as T-boxes (see paragraphs 11 to 15 in this section), but with an additional gate.

8. Three-way division boxes are built across the feeder canal and two diversions. They are **x-like structures** with three gates and sets of grooves, either single or double. **Wooden planks** make it possible to close or regulate the water flow through each gate independently.

9. **The width of each side gate** should be proportionate to the water flow they have to accommodate. **The width of the front gate** also varies according to the flow passing through it.

10. Three-way division boxes can be built of wood, bricks, concrete block or reinforced concrete (see Chapter 7).

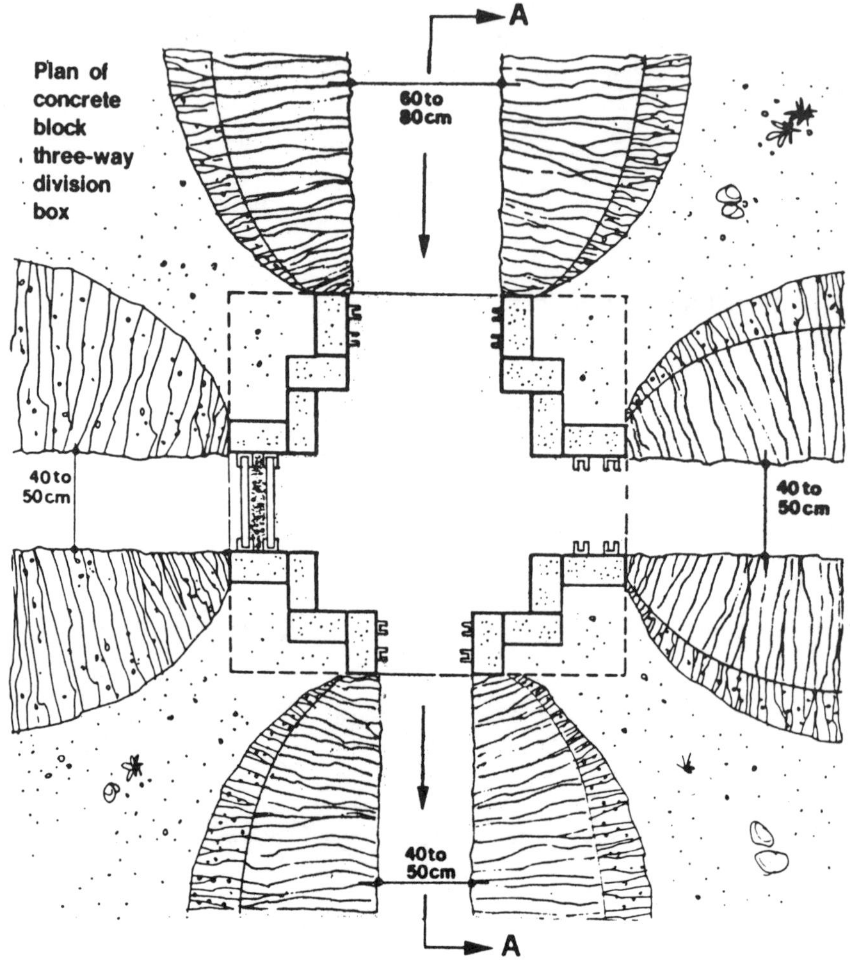

Section AA

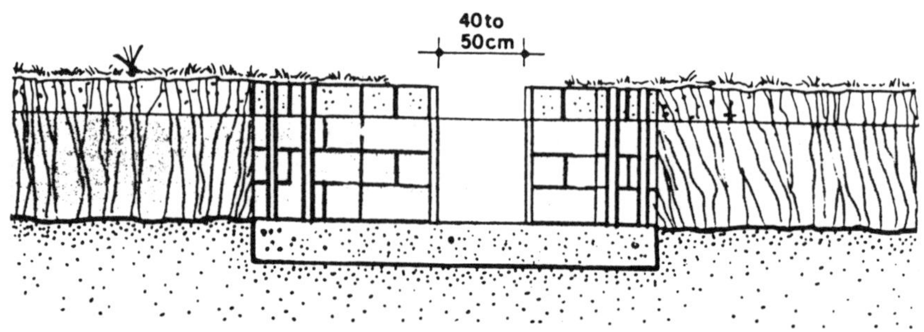

Building a three-way division box

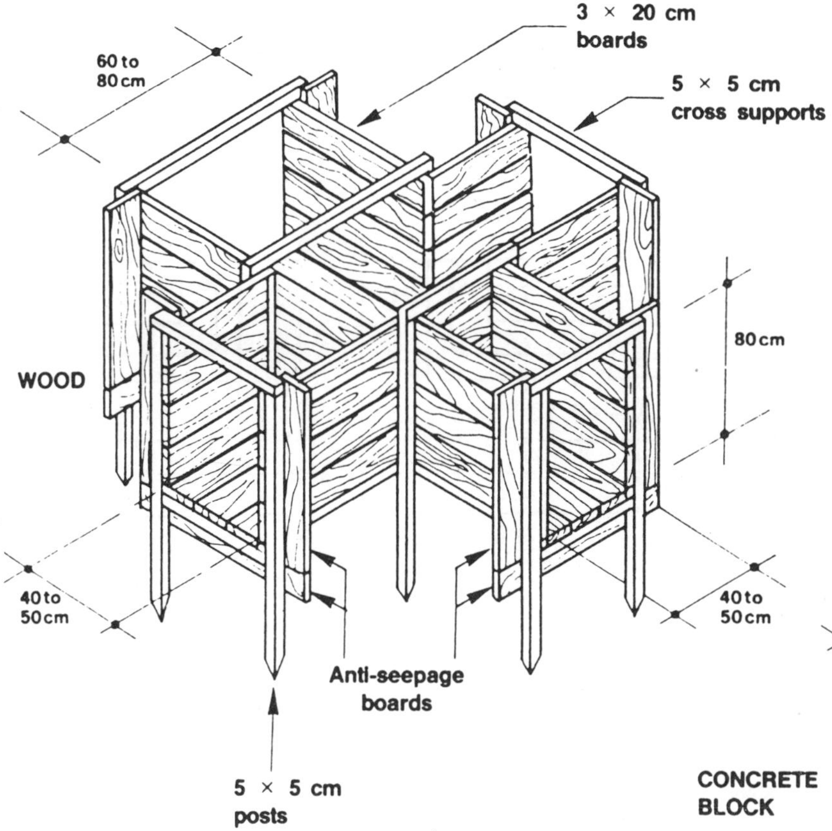

Note: the dimensions on this page and on page 90 are suitable for a medium-size pond system; dimensions may vary, however, depending on the size of your pond system

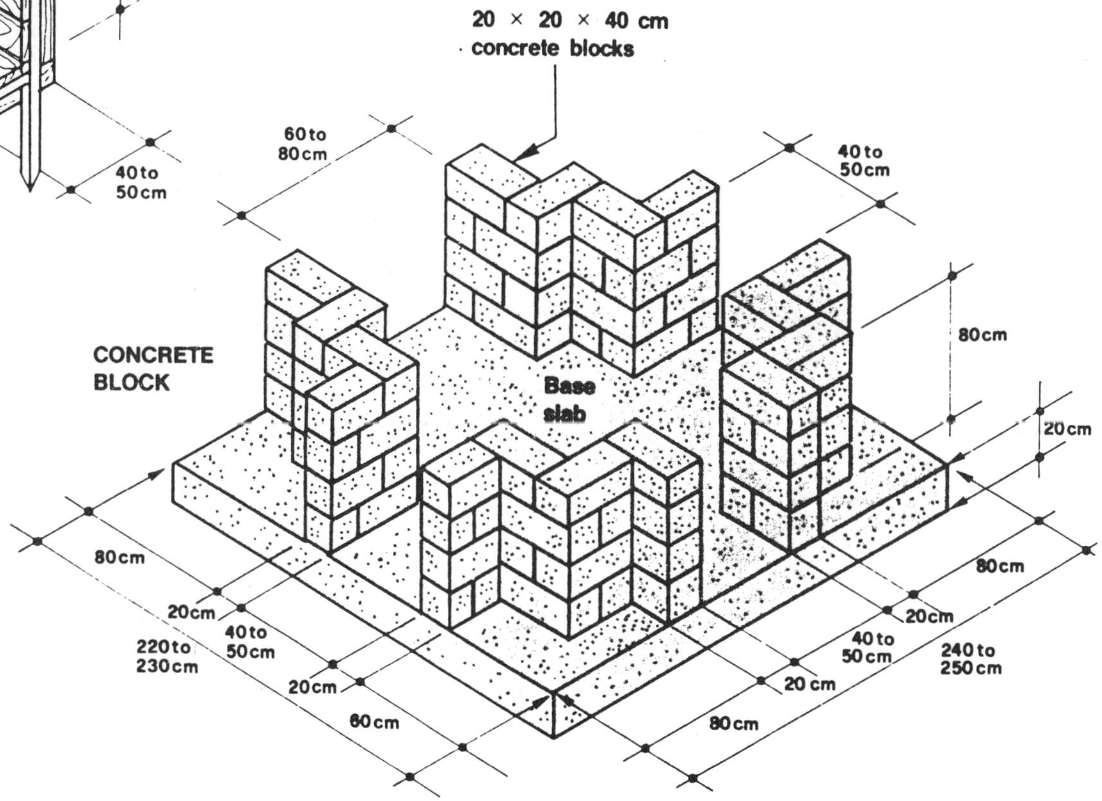

CONCRETE OR REINFORCED CONCRETE

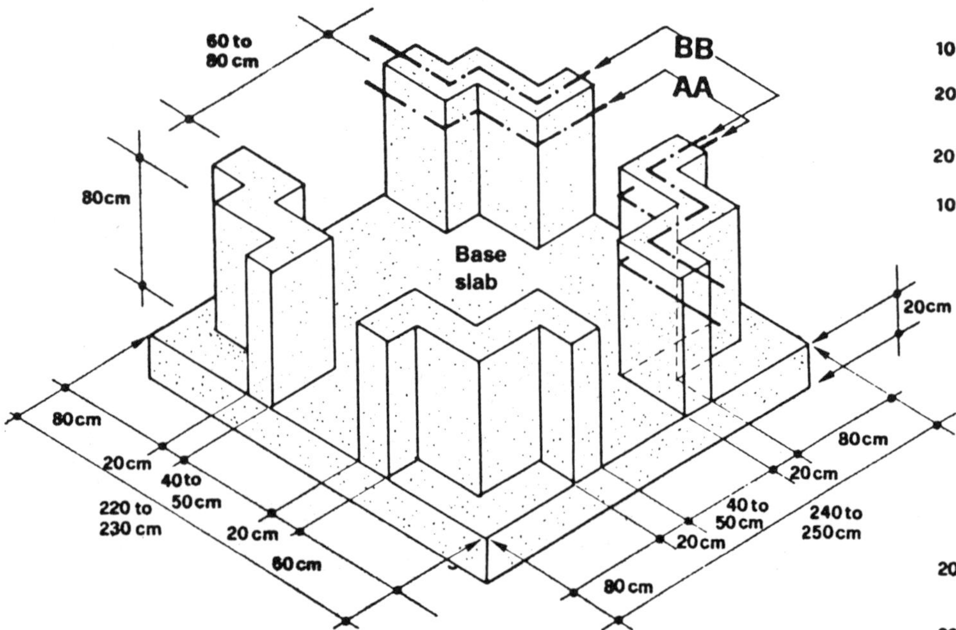

Placement of steel bars for reinforced concrete

Plan section AA

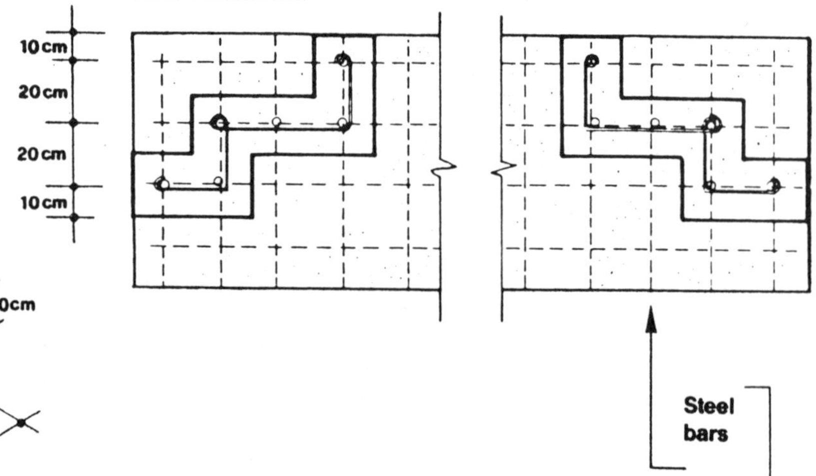

Vertical section BB

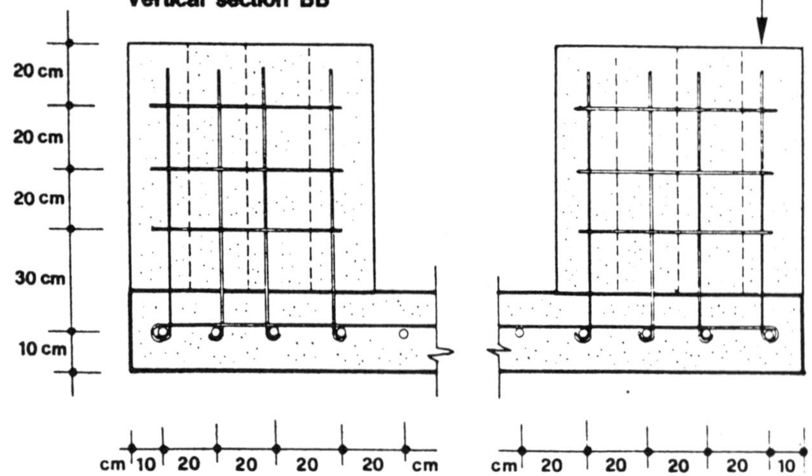

Two-way division boxes

11. A two-way division box **(T-box)** is used in a feeder canal **to divert** part of its water flow or even its total flow into either:

- **one water intake of a pond**; or
- **one additional feeder canal**.

12. Usually such a diversion is constructed at a right angle. The division box also makes it possible **to regulate the quantity of water flowing into a particular pond or canal** at any time. Maximum flow into the pond is obtained by raising the water level within the canal and fully opening the pond intake connection. No water is admitted into the pond whenever the pond intake connection is blocked.

13. Two-way division boxes are built across both the feeder canal and the pond water intake. They are **T-like structures** with two gates and sets of either single or double grooves. **Wooden planks** placed in these grooves serve to close or regulate the flow through each separate gate independently.

14. **The width of the side gate** should be proportionate to the water flow it has to accommodate. **The width of the front gate** usually does not differ from that of the box entrance.

15. **Two-way division boxes** can be built of wood, bricks, concrete block or reinforced concrete (see Chapter 7).

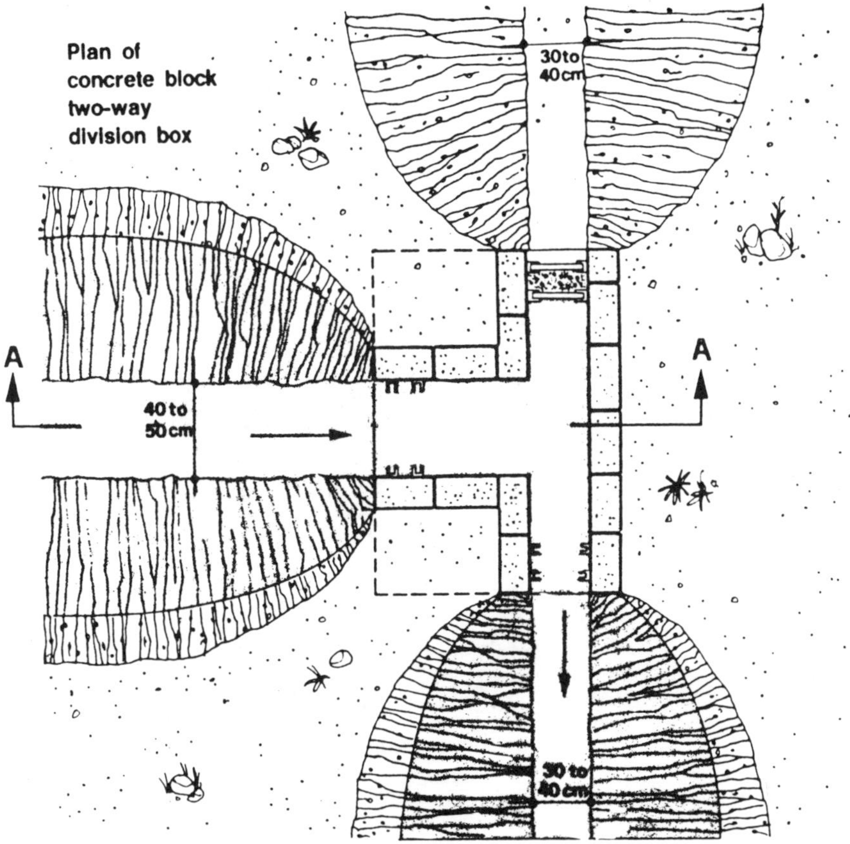

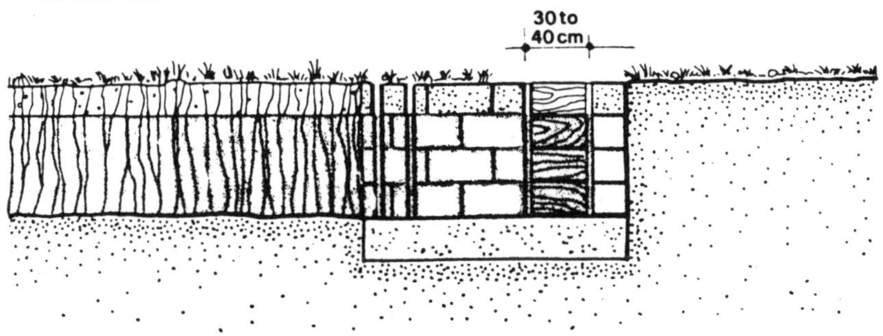

Building a
two-way division box

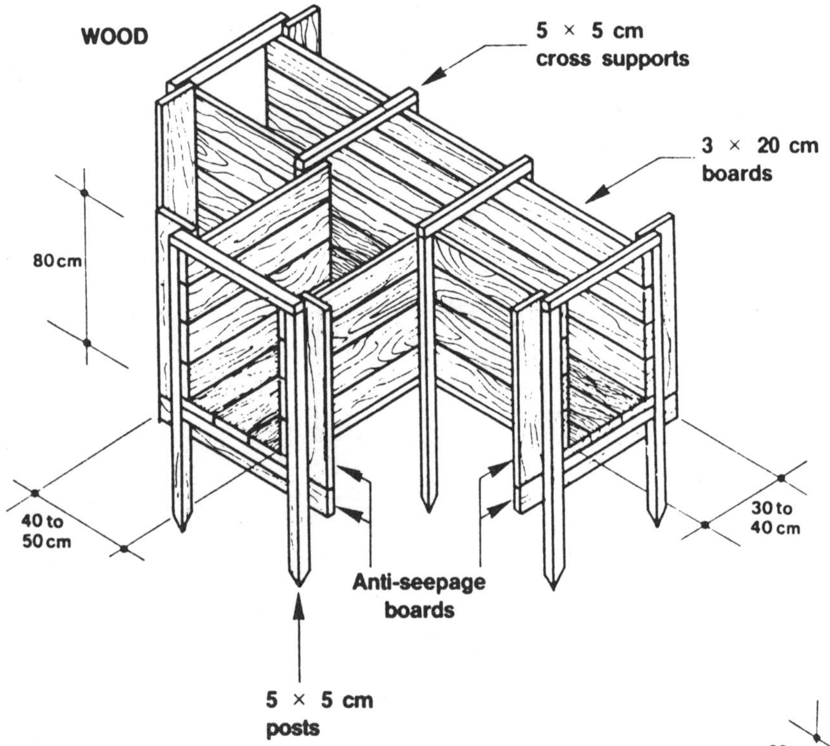

Note: the dimensions on this page and on page 93 are suitable for a medium-size pond system; dimensions may vary, however, depending on the size of your pond system

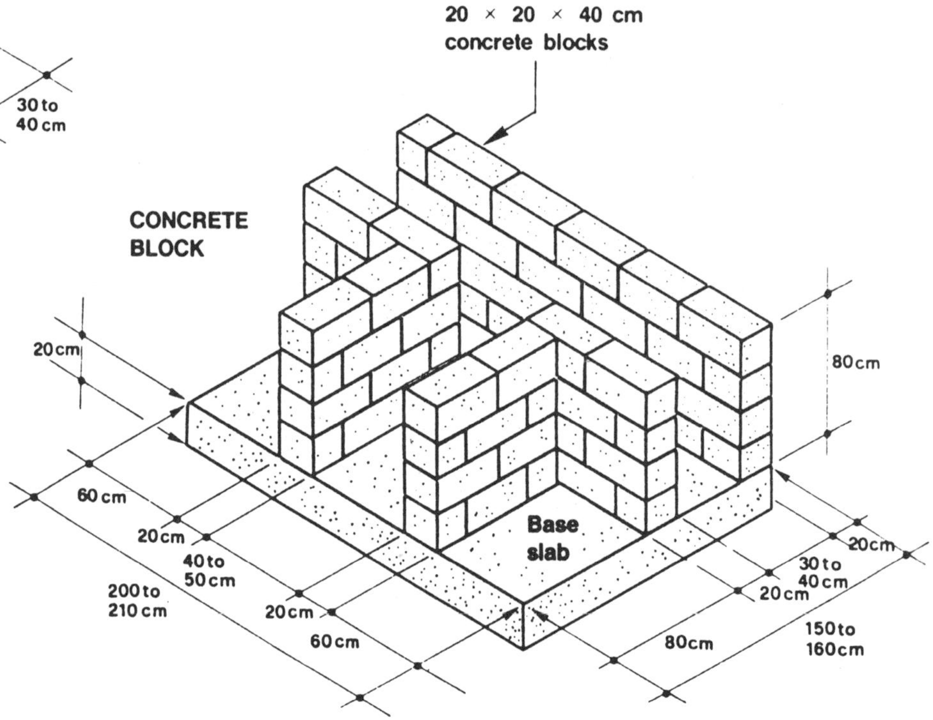

CONCRETE OR REINFORCED CONCRETE

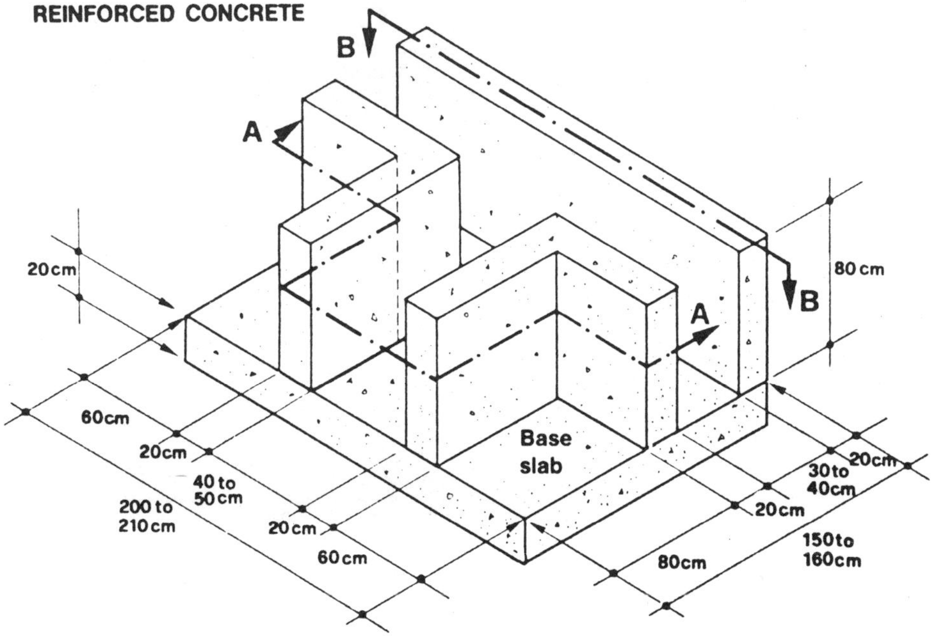

Placement of steel bars for reinforced concrete

Plan section AA

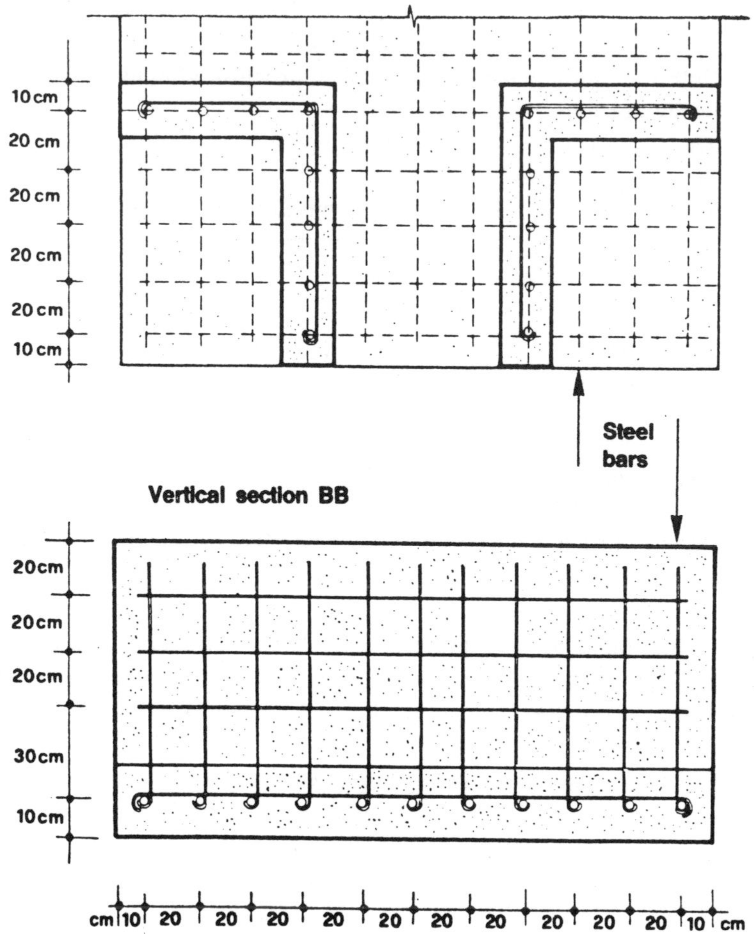

16. Drop structures should be used in feeder canals and diversion canals whenever **the bottom slope must be reduced** to slow down the water below its permissible velocity (see Section 82). When the water flow is relatively large, it is always better to have **a near-horizontal canal bottom** and to build drop structures whenever it becomes necessary to lower its relative elevation.

17. Drops can be built in various ways from wood or concrete. For their design and construction, remember the following points:

(a) Both the sides and the base of the structure **should be deeply and firmly set** into the soil.

(b) **The top level** should be slightly higher than the upstream level of the canal bottom.

(c) Widen the canal cross-section immediately downstream from the drop, deepen the canal bottom downstream and protect this part either with stones or concrete.

18. For a **wooden drop** use water-resistant wood (see Section 31, **Construction, 20/1**). You can use:

- lengths of tree limbs, 10 to 15 cm in diameter; or
- planks.

19. Make sure the structure is well bedded in and that the wood is securely fastened.

20. When building a **concrete or brick drop**, remember that because it is heavy it must be well built. Make sure **the foundation is well set up** and that there is no risk of erosion along or below its edge.

Note: you can also use **smaller drop structures,** usually made from wood or brick, for falls of less than 20 cm. They are simpler and can be more lightly constructed, but you will need more of them for the same overall drop.

Plan of drop structure

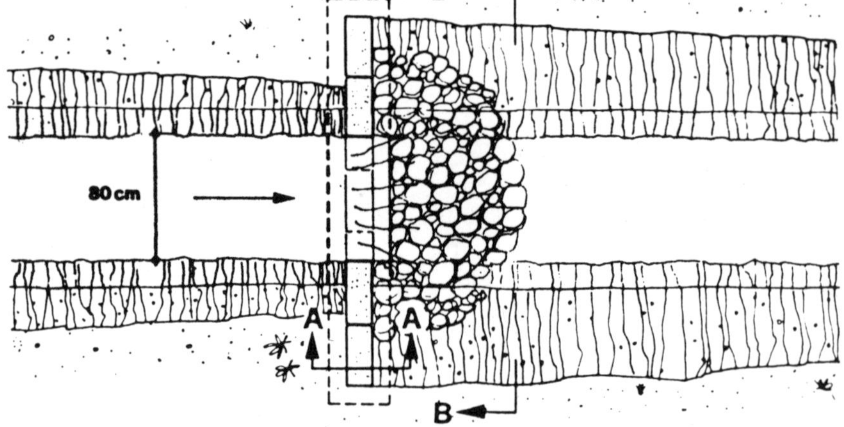

Section AA

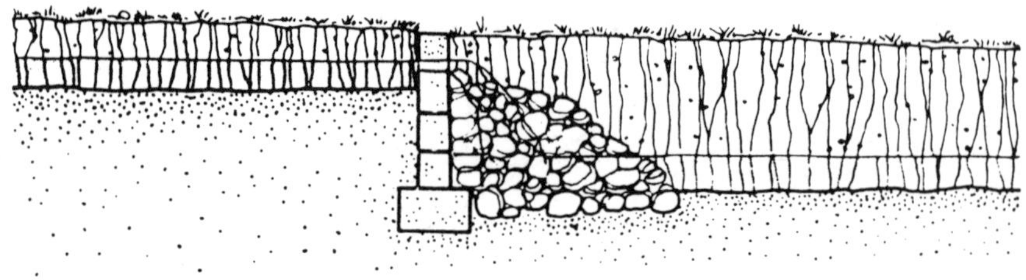

Section BB

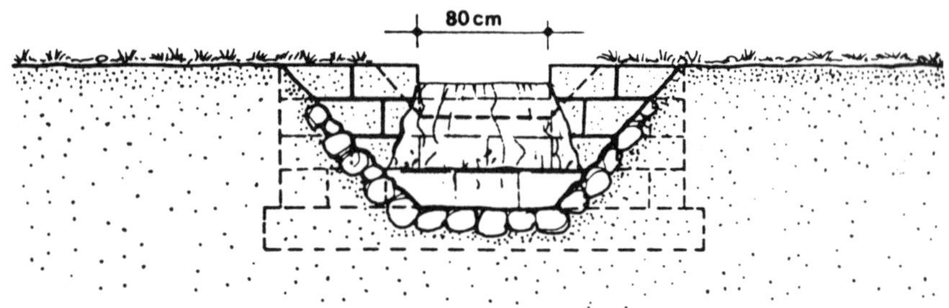

Building a drop structure

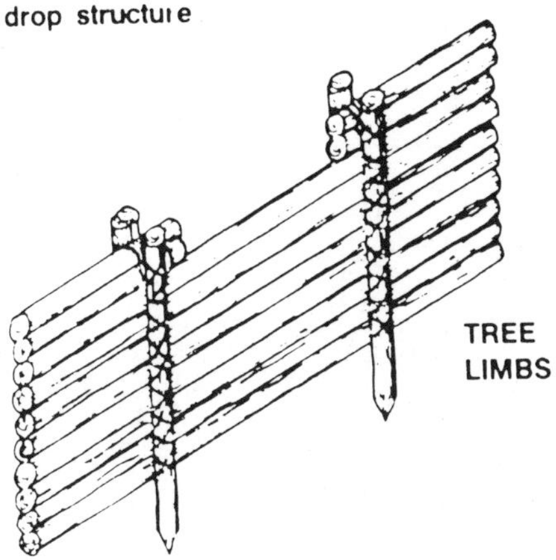

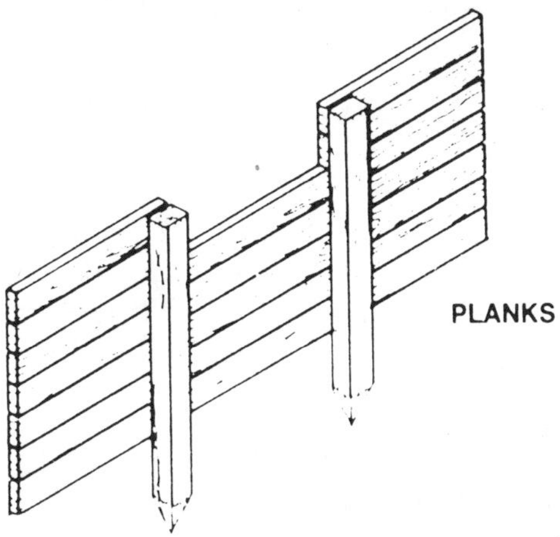

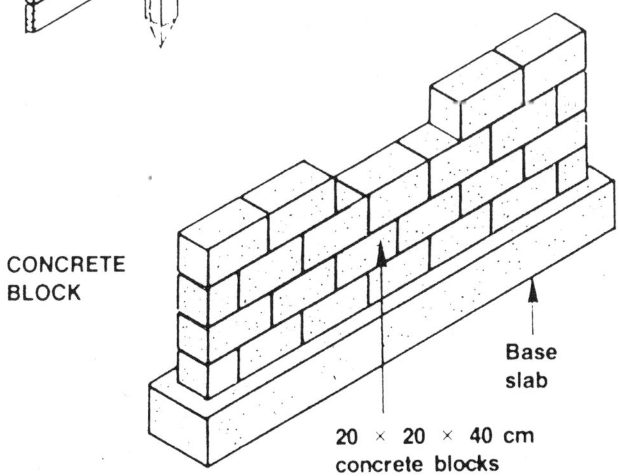

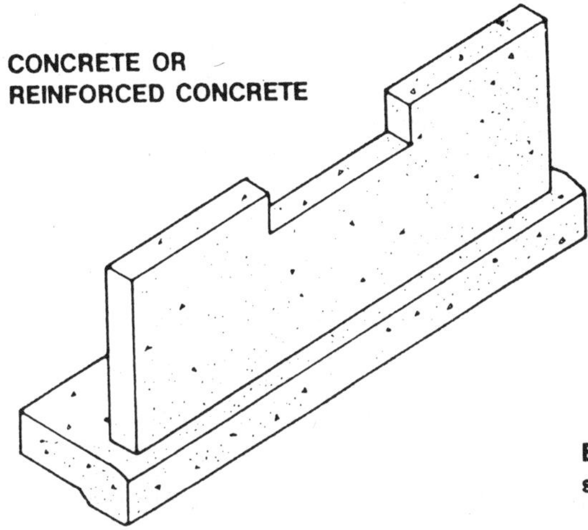

Placement of steel bars for reinforced concrete

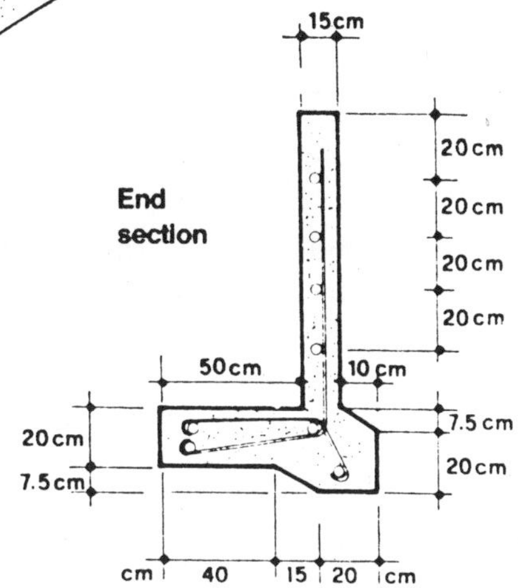

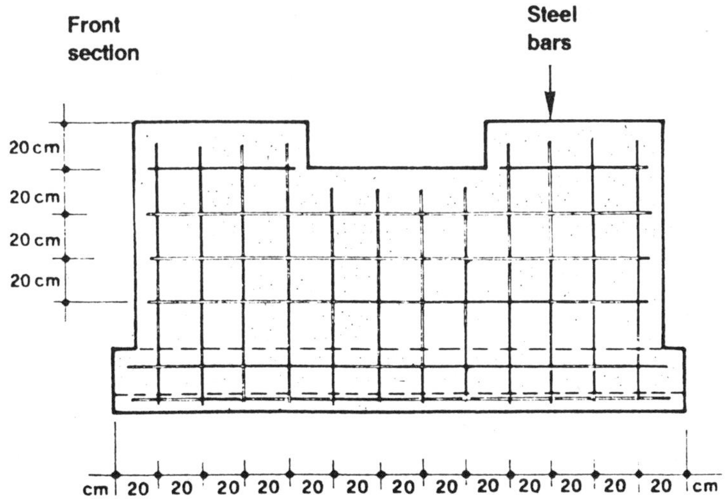

End overflow gates

21. End overflow gates should be built **at the end of every feeder canal** to drain any excess water away from the ponds, for example, in a drainage canal or a natural depression or channel.

22. A simple overflow gate can be made from a pipe placed at a level higher than the level of the pond inlets.

23. It can also be built within the feeder canal section as a **rectangular box-like structure** with **two sets of grooves**. It is used in the same way as lateral overflow gates (see Section 87, paragraphs 3 to 5).

24. **End overflow gates** can be built of wood, bricks, concrete block or reinforced concrete using the same methods.

Plan of concrete block end overflow gate

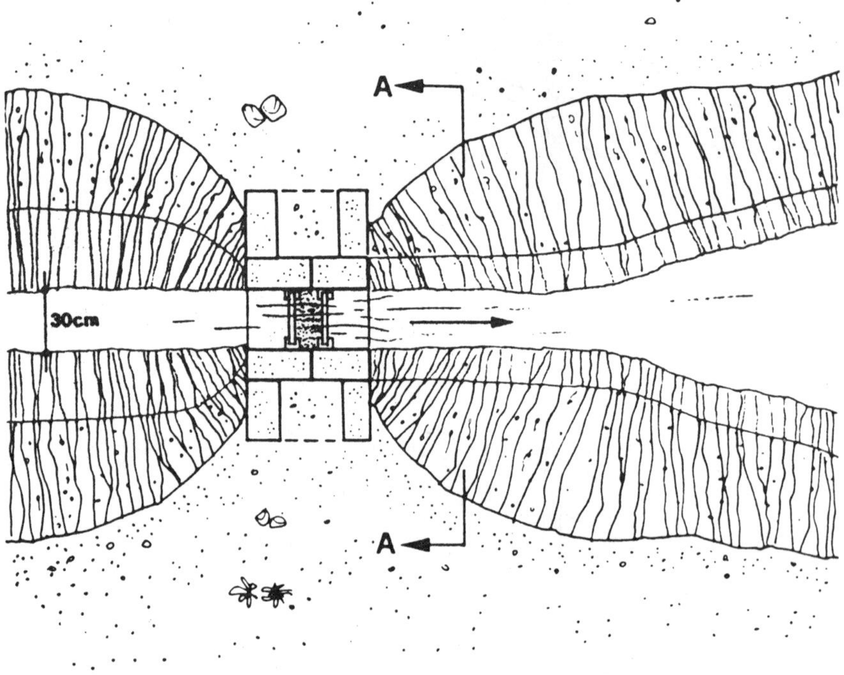

Section AA

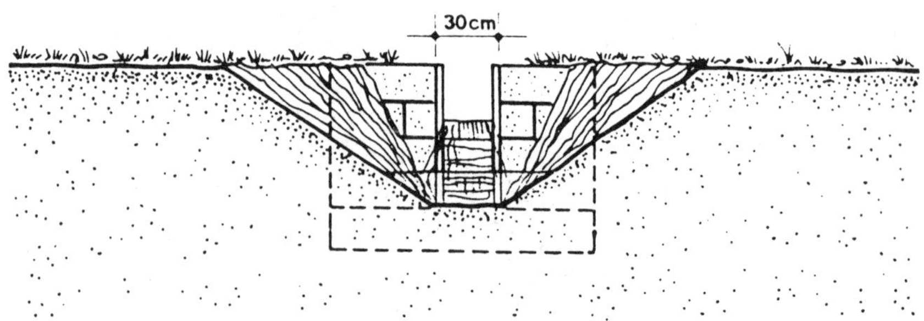

Building an end overflow gate

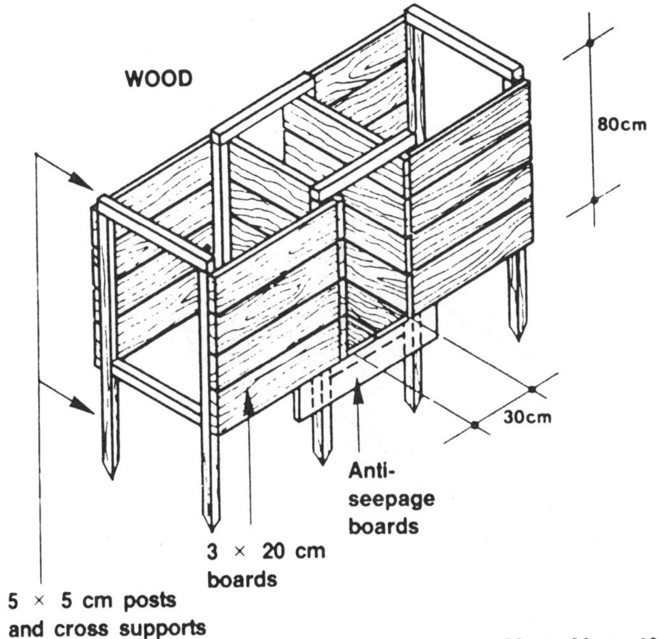

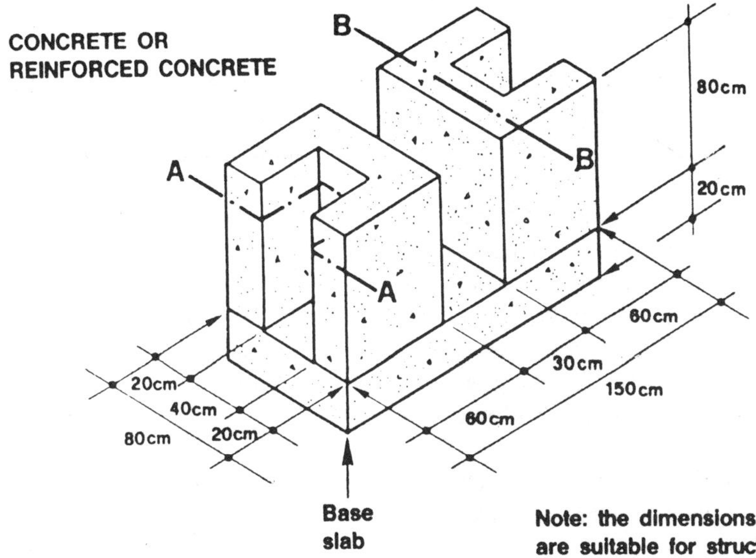

Note: the dimensions shown are suitable for structures for a medium-size pond system

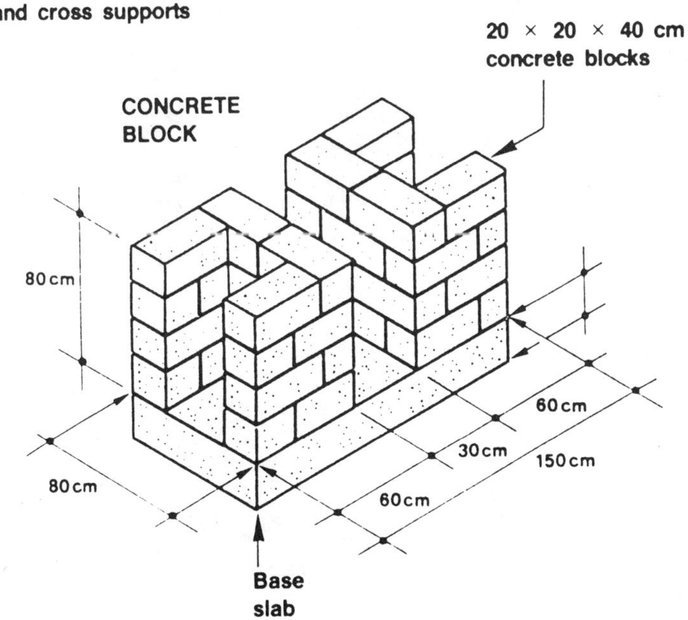

Placement of steel bars for reinforced concrete

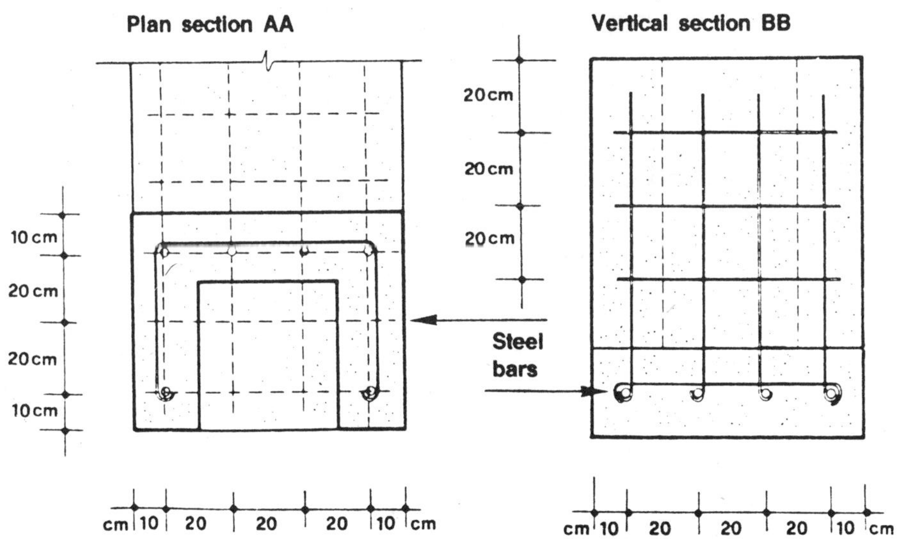

88 Simple aqueducts

1. Aqueducts are used on fish farms to transport water above ground level, for example a main feeder canal crossing over a small drainage canal.

2. A simple aqueduct can be built of **wooden boards**, assembled with wooden cross-bracing and supported by a wooden framework at regular intervals. Its cross-section is usually rectangular. **Maximum permissible average velocity** should be limited to 1.5 m/s **(Table 35)**. **The water carrying capacity** of such an aqueduct can be estimated as explained earlier (see Section 82), using a coefficient of roughness **n** = 0.014. Therefore (1 ÷ n) = 71.43 **(Table 37)**.

3. Another possibility is to use **200-litre metal drums** cut into two halves, welded or bolted together and fixed over a platform built of wood or rock.

4. **Corrugated metal sheets** can be easily assembled lengthwise with flexible tarred joints and welding points to form a semi-circular aqueduct, similar to those built of precast concrete **flumes** or plastic flumes for irrigation purposes (see Section 82). Supports can be similar to those described for half metal drums.

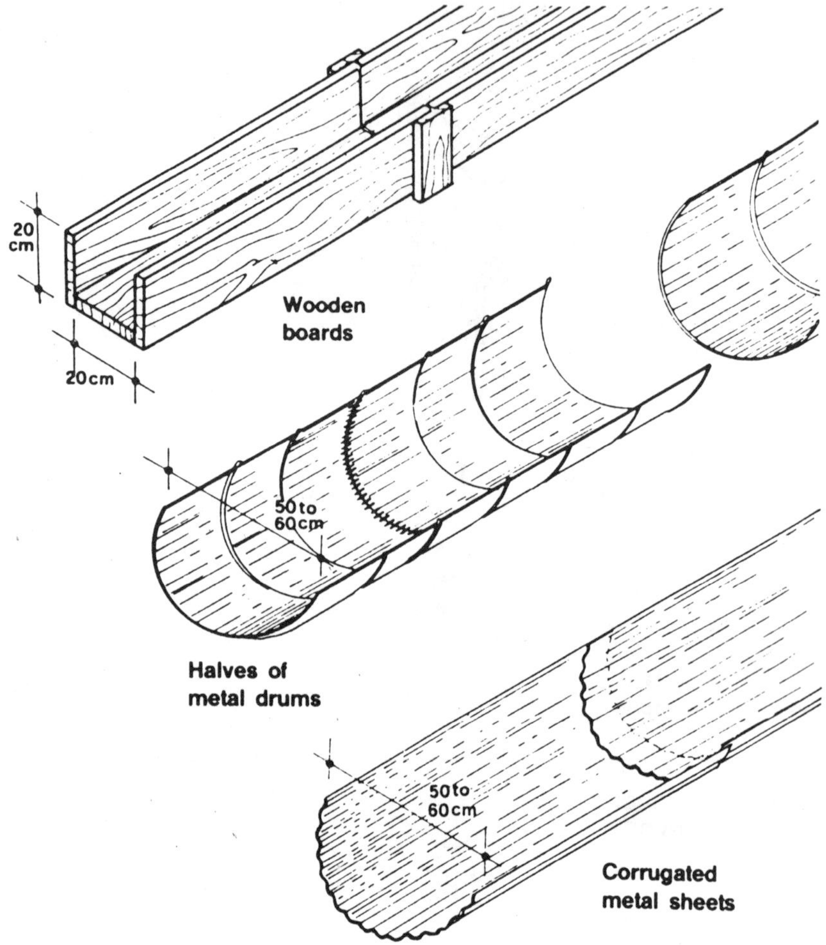

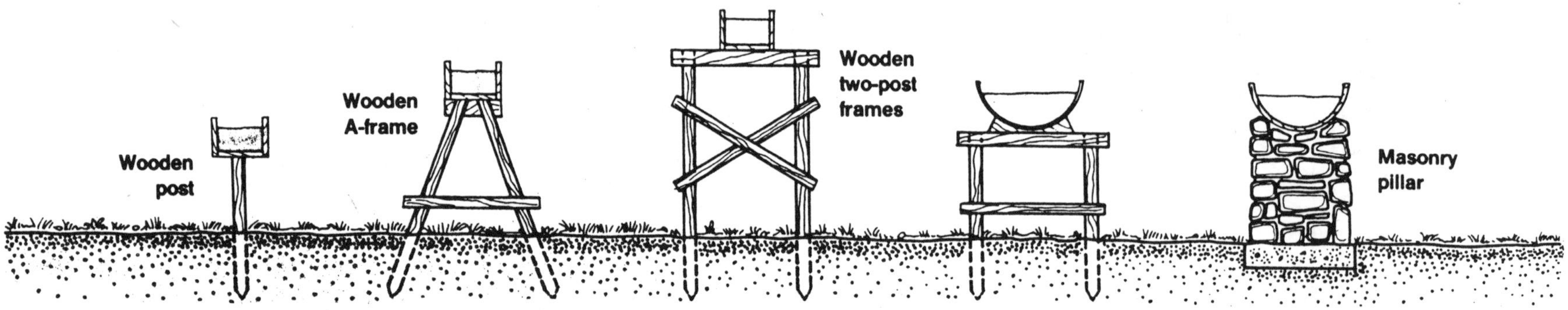

Various ways to support an aqueduct

Short pipelines

1. When water carried in open canals must **be conveyed under roads** or other obstructions, a short pipeline can be used. It should be strong enough to support the weight of vehicles passing over it. Precast concrete pipes buried at least 60 cm under the road surface are often used (see Section 38, **Construction, 20/1**). Particular attention should be paid to:

- the quality of the joints between the pipe sections; and
- the quality of the end connections between the pipeline and the open canal section on either side of the road.

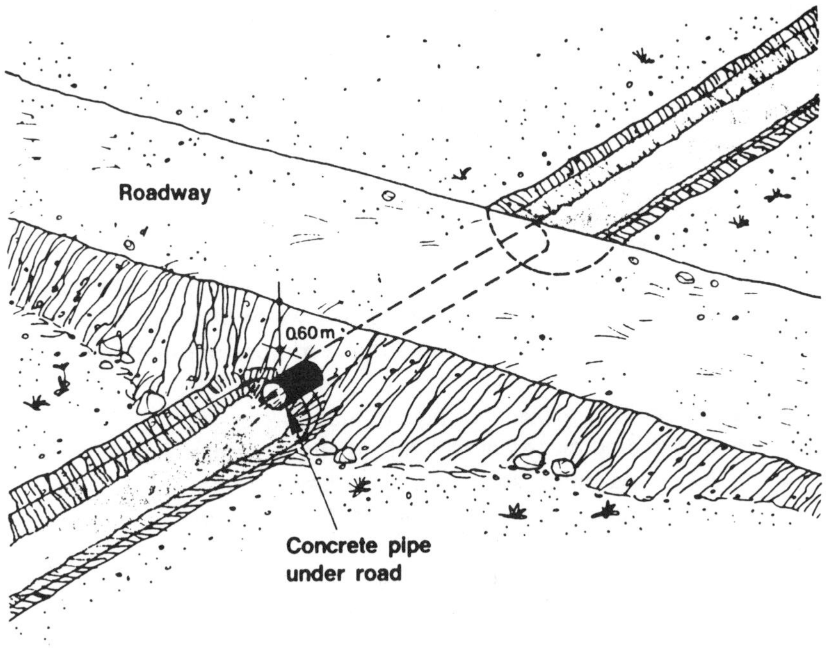

Example

Characteristics of short pipelines for road crossings

Water flow (l/s)	*Pipe inside diameter* (cm)	*Water velocity* (m/s)
120-140	40	1.00
230	60	0.80
480	80	0.95

Concrete pipes

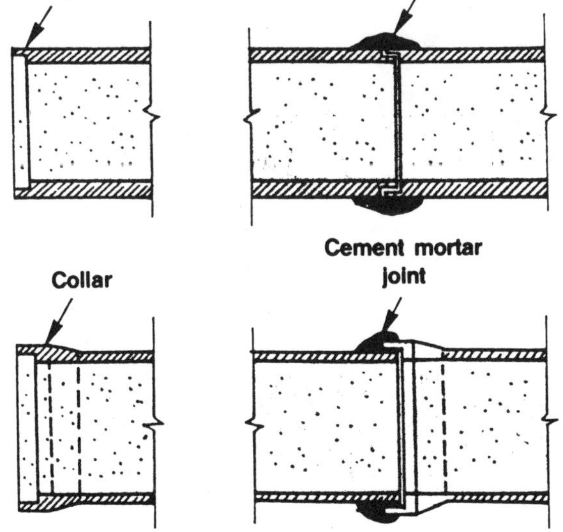

2. More details on laying pipes and preparing pipe foundations are given in Chapter 10.

3. When no provision is made for inlet or outlet structures in ponds, the water can be siphoned over the dike. This is also discussed in Section 38, **Water, 4**.

4. A siphon is **an inexpensive device** that can be carried to different points of the fish farm according to need. It can easily be made from a piece of flexible rubber or plastic hose. If the total length of **the siphon is relatively short**, for example for transferring water from a shallow feeder canal to a small pond, you can also make a more durable **rigid siphon**:

- by bending a piece of plastic tube;
- by bending pieces of galvanized sheet metal to shape them into tubes, carefully soldering all seams so they are airtight and assembling these segments to form a truncated V-shape by further soldering;
- by cutting and connecting (with cement or solvent cement, by welding, etc.) a metal or PVC pipe.

5. **To start a siphon**, remember that you have to fill it first with water. Place the discharge end of the siphon **at least 25 cm lower** than the intake end. Keep the intake end well under water and let the water flow out of the filled siphon. To be successful you may need some practice, especially if the diameter of your siphon is large (see also Section 102).

Draining by siphon

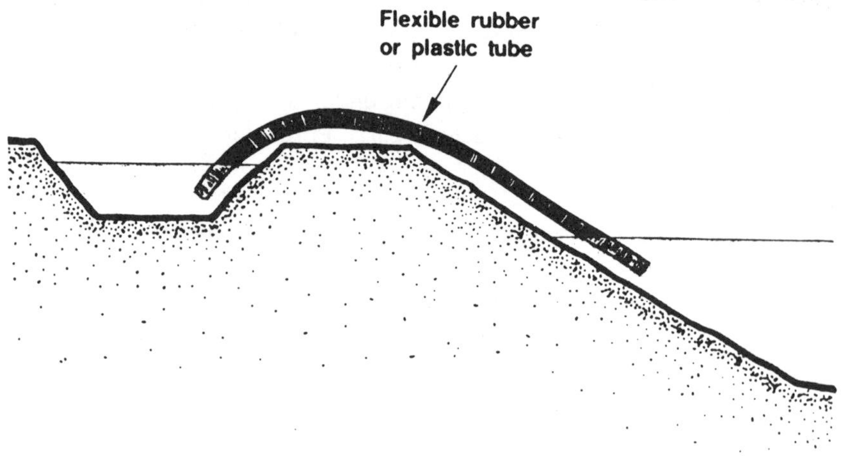

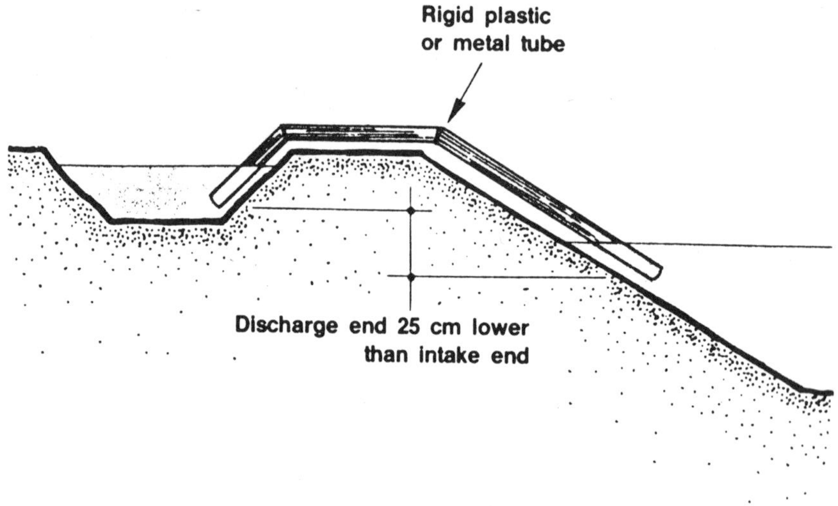

6. **The water discharge capacity** of a siphon depends on:

- **the inside diameter** of the tube;
- **the head** or difference in elevation between the water surface on the upper level and either the water surface on the lower level (if the siphon outlet is submerged), or the centre of the siphon outlet (if it is free flowing).

7. You can estimate **the water discharge capacity** of your siphon (in l/s) from **Graphs 11 and 12** on the basis of these two measurements. You can also use **Table 45** for small siphons and low pressure heads.

Example

The inside diameter of your siphon is 5 cm, and the head is 21 cm. From Graph 11 find the water discharge capacity equal to about 2.5 l/s.

The inside diameter of your siphon is 18 cm and the head is 27.5 cm. From Graph 12 estimate the water discharge capacity equal to 35 l/s.

Water discharge capacity

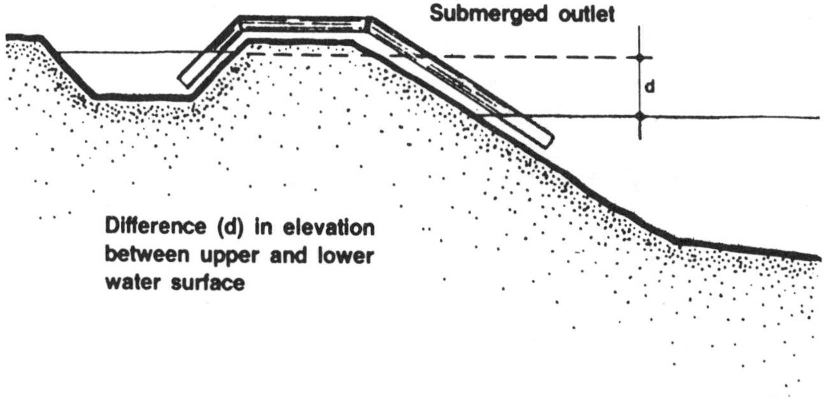

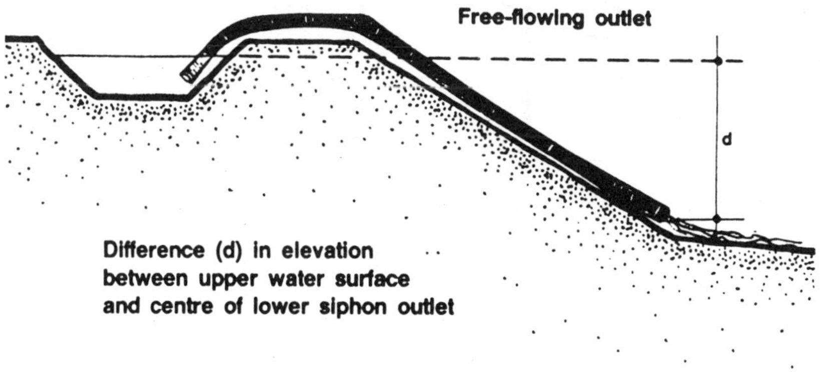

GRAPH 11

Siphons with an inside diameter smaller than 9 cm

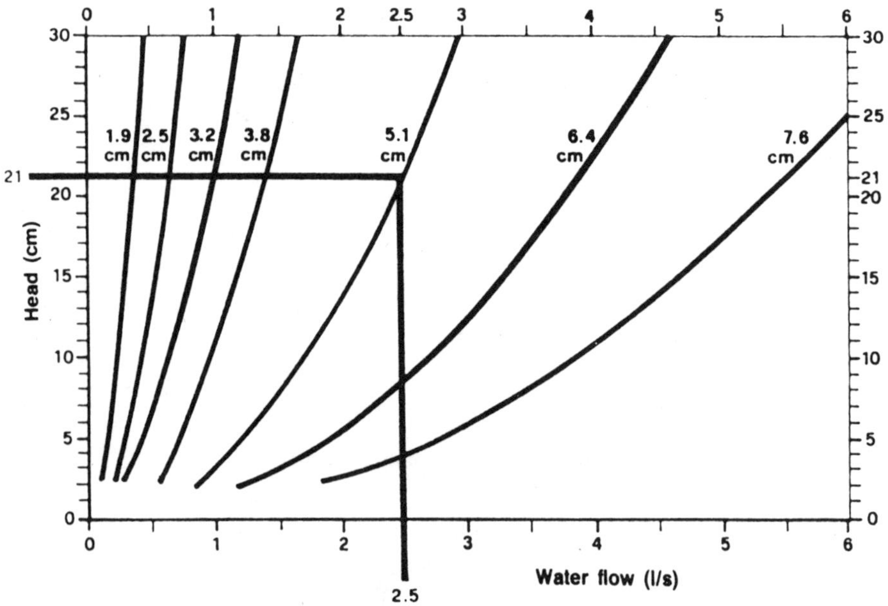

GRAPH 12

Siphons with an inside diameter larger than 9 cm

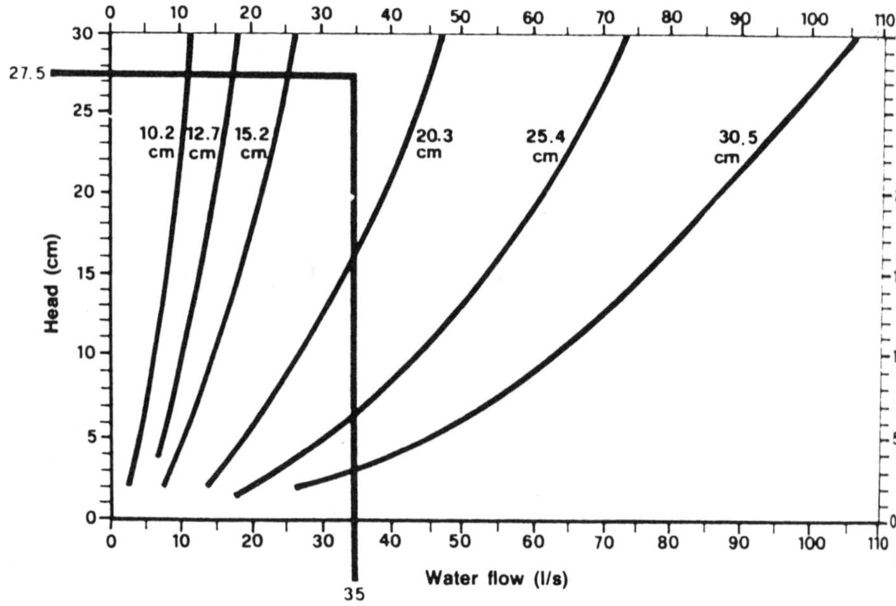

TABLE 45

Flow through small siphons under low pressure heads (in l/s)

Inside diameter of siphon (cm)	*Pressure head* (cm)						
	5	7.5	10	12.5	15	17.5	20
4	0.75	0.91	1.06	1.18	1.29	1.40	1.49
5	1.17	1.43	1.65	1.85	2.02	2.18	2.33
6	1.68	2.06	2.38	2.66	2.91	3.14	3.36
7	2.29	2.80	3.24	3.62	3.96	4.28	4.58
8	2.99	3.66	4.23	4.72	5.18	5.59	5.98
9	3.78	4.63	5.35	5.98	6.55	7.07	7.56
10	4.67	5.72	6.60	7.38	8.09	8.73	9.34

9 POND INLET STRUCTURES

When do you need an inlet structure?

1. Inlet structures are built to **control the amount of water** flowing into the pond at all times. The need for an inlet structure varies with the type of water supply being used to feed the pond.

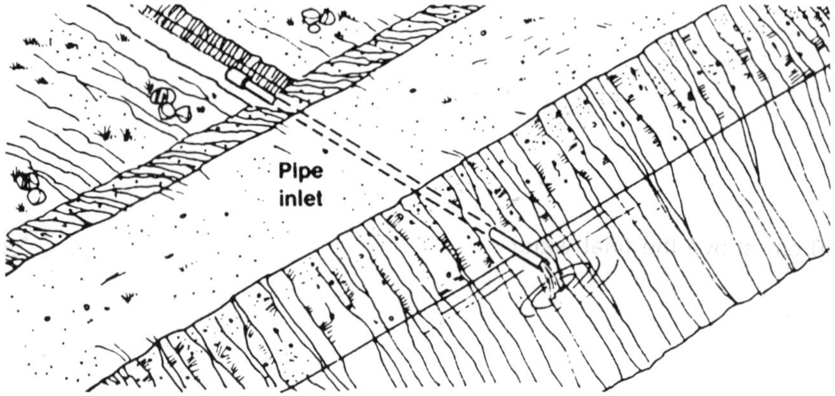

2. **There is no need for an inlet structure** for a pond supplied entirely by rain, surface runoff, groundwater or a spring which emerges within the pond, nor for a barrage pond built directly on the stream.

3. **An inlet structure may be built** for a pond supplied through a feeder canal, for example by diverted stream water, a spring outside the pond, a well or a pumped water supply.

Different types of inlet structures

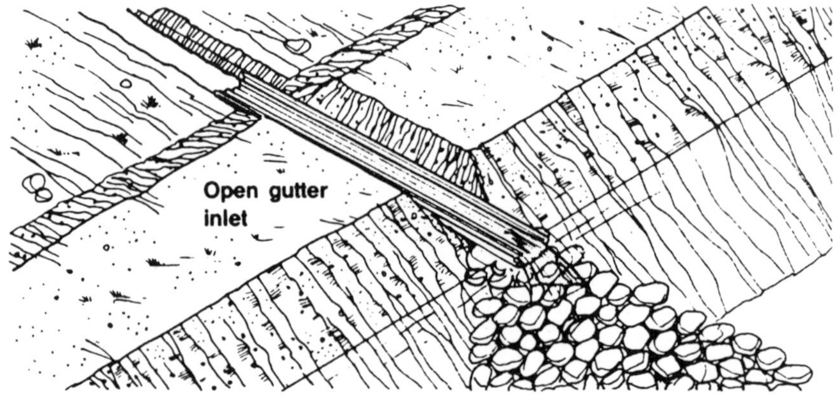

4. There are **three main types** of inlet structures:

- pipe inlets;
- open gutter inlets;
- canal inlets.

5. If **a division box** is included in the water feeder canal and it is close enough to the pond dike, the inlet structure can be built as part of the division box.

Note: if you have **to feed water into two adjacent ponds**, you can use a division box to regulate the water flow toward the two pond inlets. In this case, it is best to add a separate gate at each pond inlet.

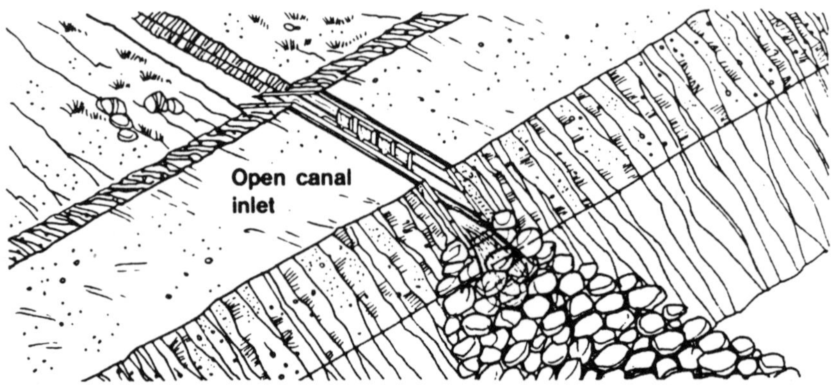

6. When designing and constructing an inlet structure, you should **pay particular attention** to the following points:

(a) Place the inlet **at the shallow end** of the pond.
(b) **Design its bottom level** to be at the same level as the bottom of the water feeder canal and ideally **at least 10 cm** above the maximum level of the water in the pond.
(c) Design the inlet structure **to be horizontal**, with no slope.
(d) Try to arrange the structure so that **water splashes** and mixes as much as possible when entering the pond.
(e) Remember that you have **to keep** unwanted fish **out** of your pond. Design your pond inlet accordingly.

Example

Your pond has an area of 200 m^2 and an average water depth of 0.75 m. You wish to fill it within 6 hours.

- Total water volume required: 200 m^2 × 0.75 m = 150 m^3 = 150 000 l
- Total time available: 6 hours = 360 minutes = 21 600 seconds
- Inlet capacity required: 150 000 l ÷ 21 600 s = 6.94 l/s ≈ **7 l/s**
- If you have this water flow available when you fill the pond, you can go ahead with this estimate and plan on filling your pond within 6 hours.

Note: design inlet for a water capacity large enough for the pond to be filled within a reasonable amount of time, from a few hours for a small pond to a few days for a large pond. The design also depends on the water flow available (see **Water, 4**).

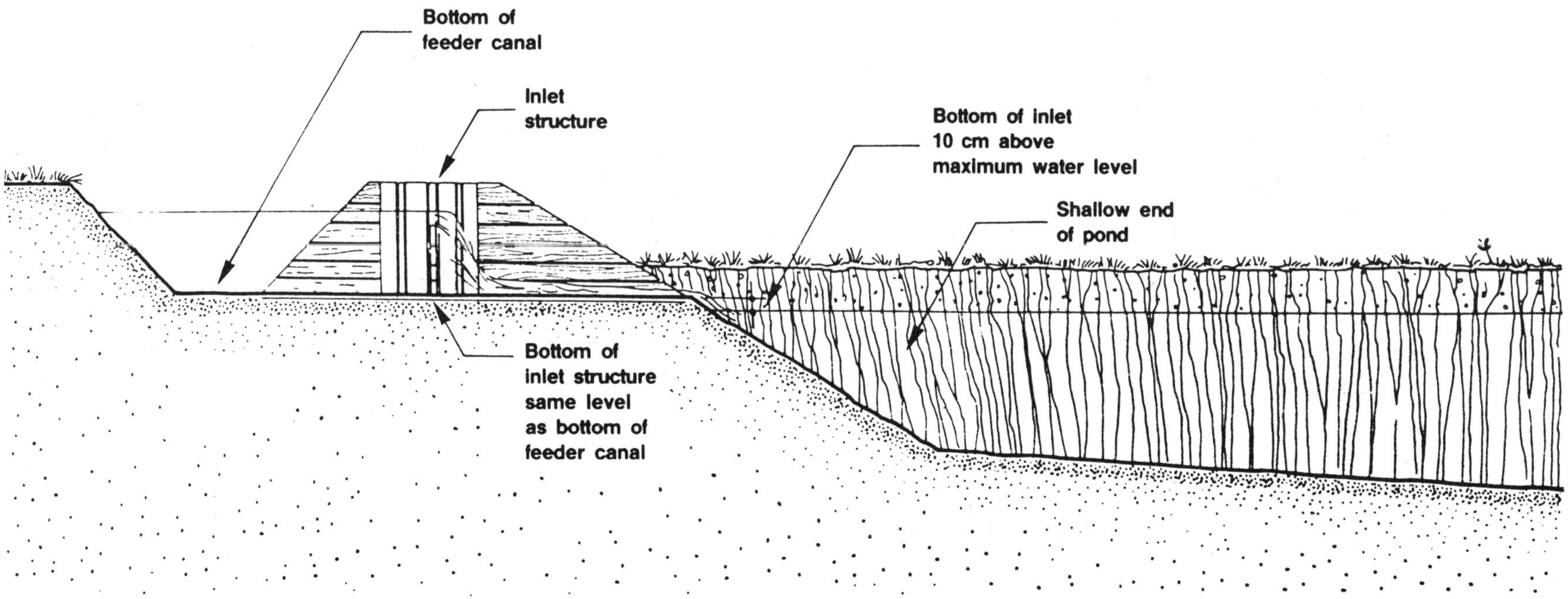

91 Pipe inlets

1. Pipe inlets can be made from **various materials**, depending on the water supply required and the inside diameter of the pipe. If the pipes you have are too small to provide the required water flow, you may have to use more than one pipe at each pond inlet. Usually, pipe inlets extend for about 0.60 to 1 m beyond the edge of the water surface of the pond when it is full, and they should be at least 10 cm above the final water level. Base the selection of the material of the pipe inlet on the following table:

Material	*Inside diameter of pipe*		
	Less than 10 cm	10-15 cm	Over 15 cm
Bamboo	yes	—	—
Galvanized iron	yes	—	—
Plastic	yes	yes	yes
Asbestos cement	—	—	yes
Concrete	—	—	yes

2. Remember that to estimate **the water carrying capacity** of a pipe you can use either **Graph 1, Table 13 or a mathematical formula** as explained in Section 38, **Construction, 20/1**.

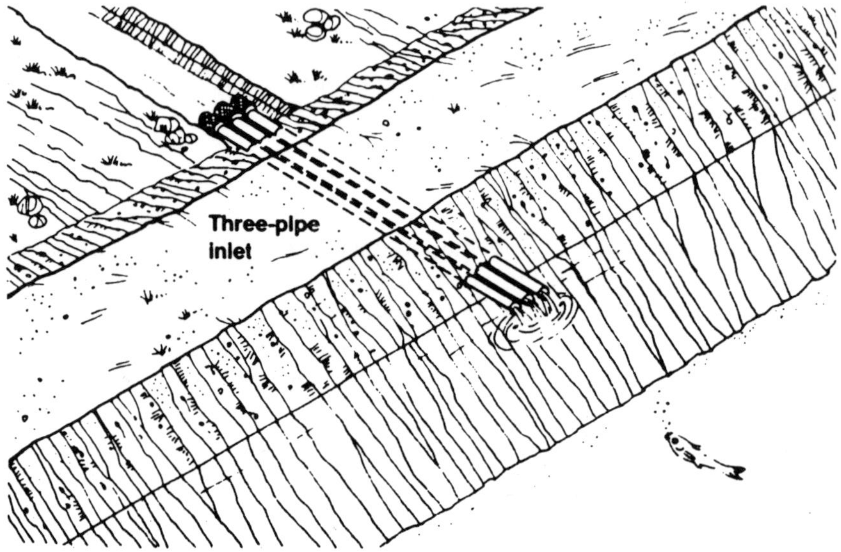

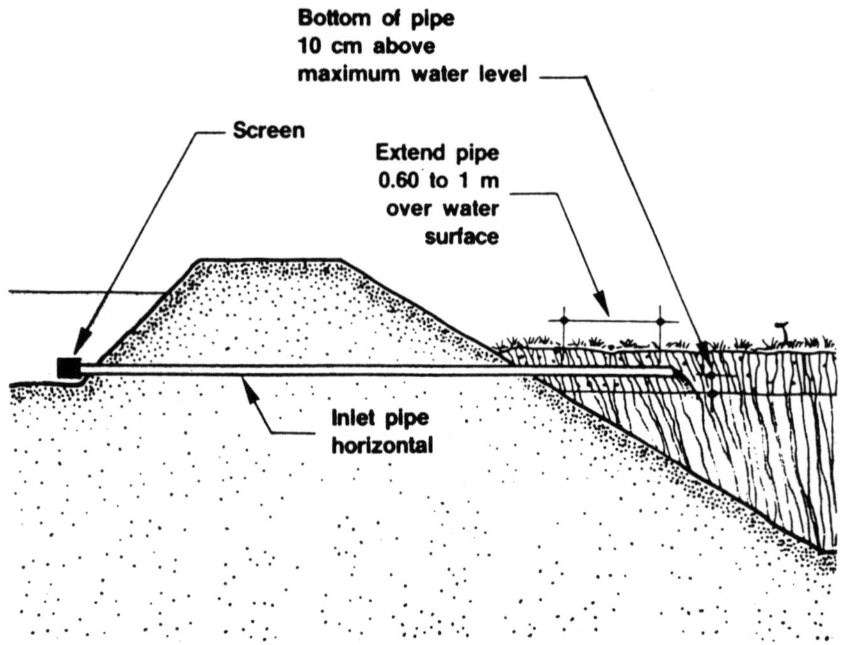

3. **Siphons**, as flexible or stiff pipes, can also be used for filling ponds (see Section 89). In this case, the pond requires no proper inlet. Pumps can also be used (see Section 39, **Construction, 20/1**).

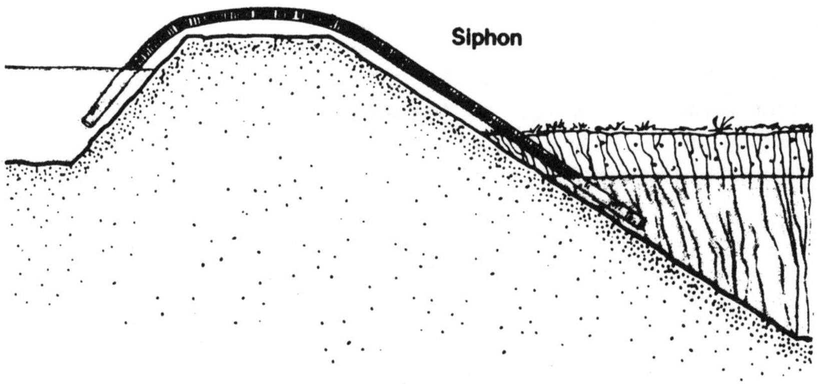

4. **Bamboo pipes** (see Section 31) make cheap and good inlets whenever locally available. They can be used in several ways for filling small ponds, for example:

- without modification, the water flow being regulated upstream;
- with the inclusion of a mobile plate for flow regulation;
- with modification for improving water quality.

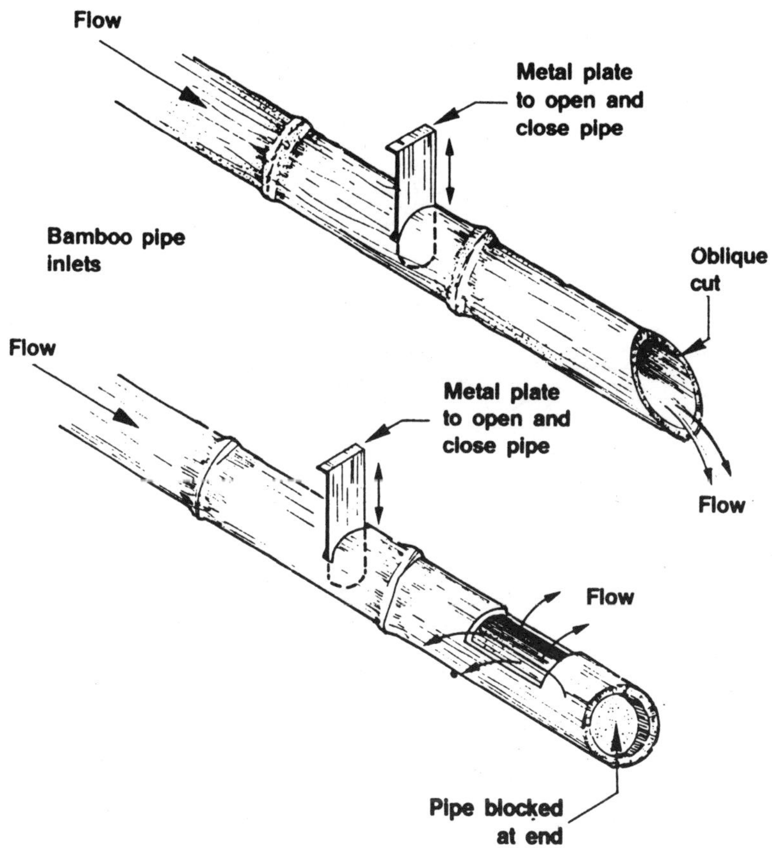

5. Small **galvanized iron pipes** and **plastic pipes** are more expensive but also much more durable than bamboo pipes. They can be fitted with **a mechanical valve** to regulate the water flow or can be set up with a **swinging arm** to let water into the pond. A simple pipe sleeve can be constructed to control the water flow.

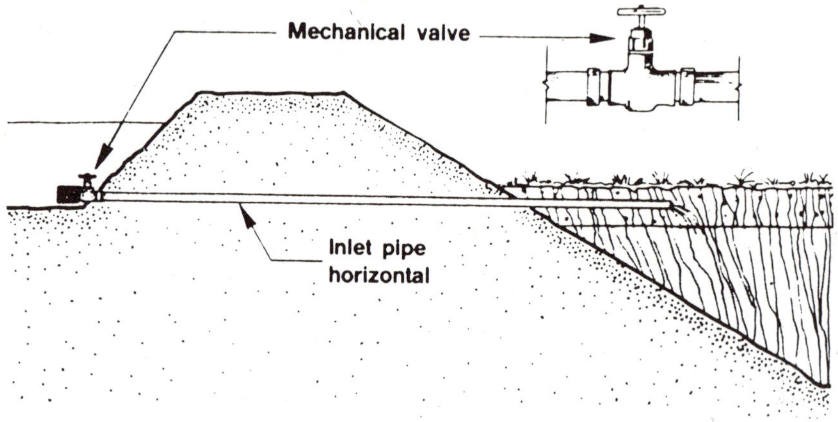

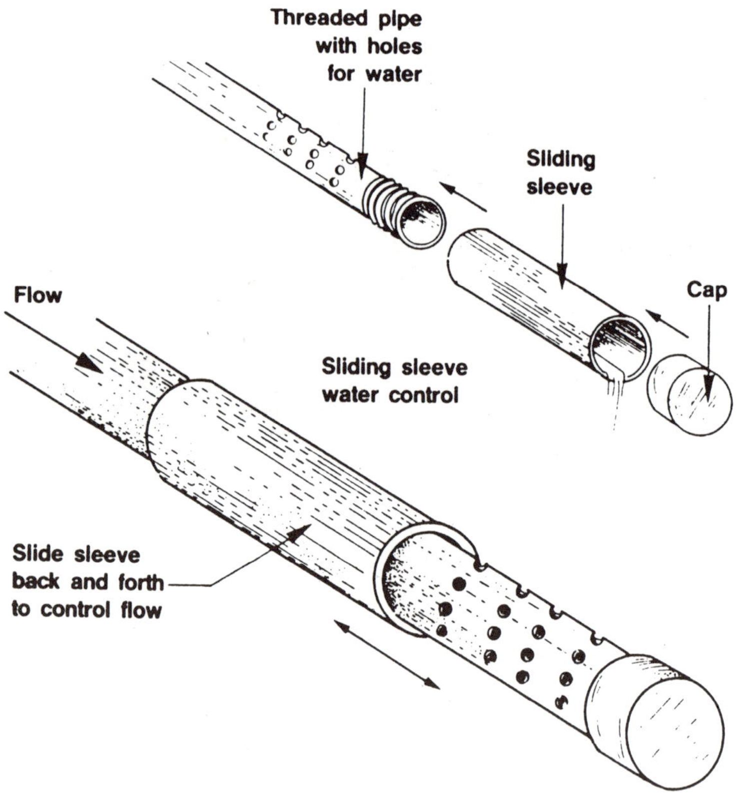

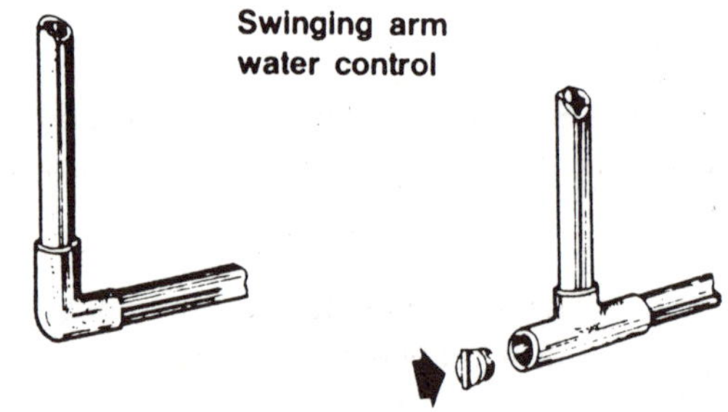

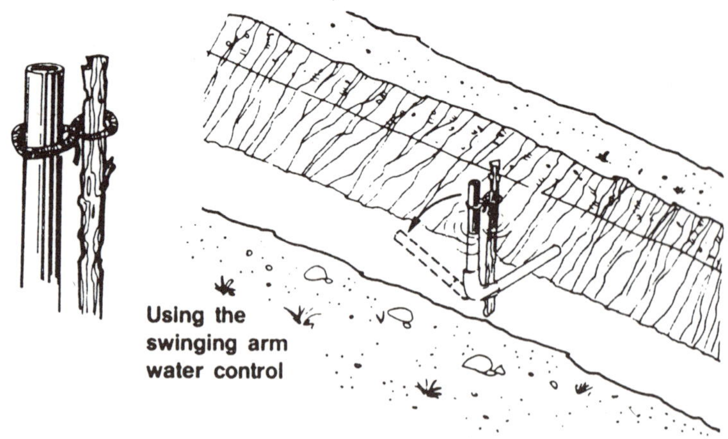

6. Larger **plastic, asbestos cement, and concrete pipes** are necessary to fill large ponds. They are generally used together with two-way division boxes, through which their water flow can be regulated.

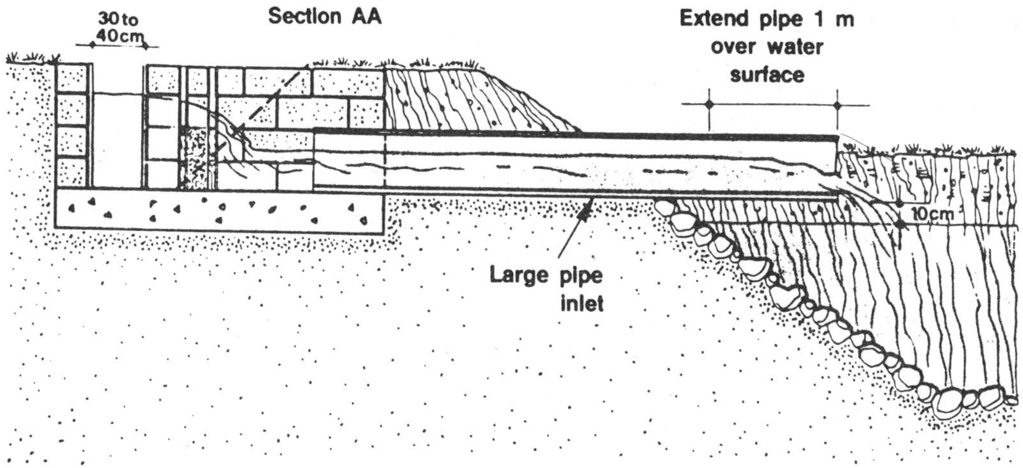

Concrete block
two-way
division box
with large pipe inlet

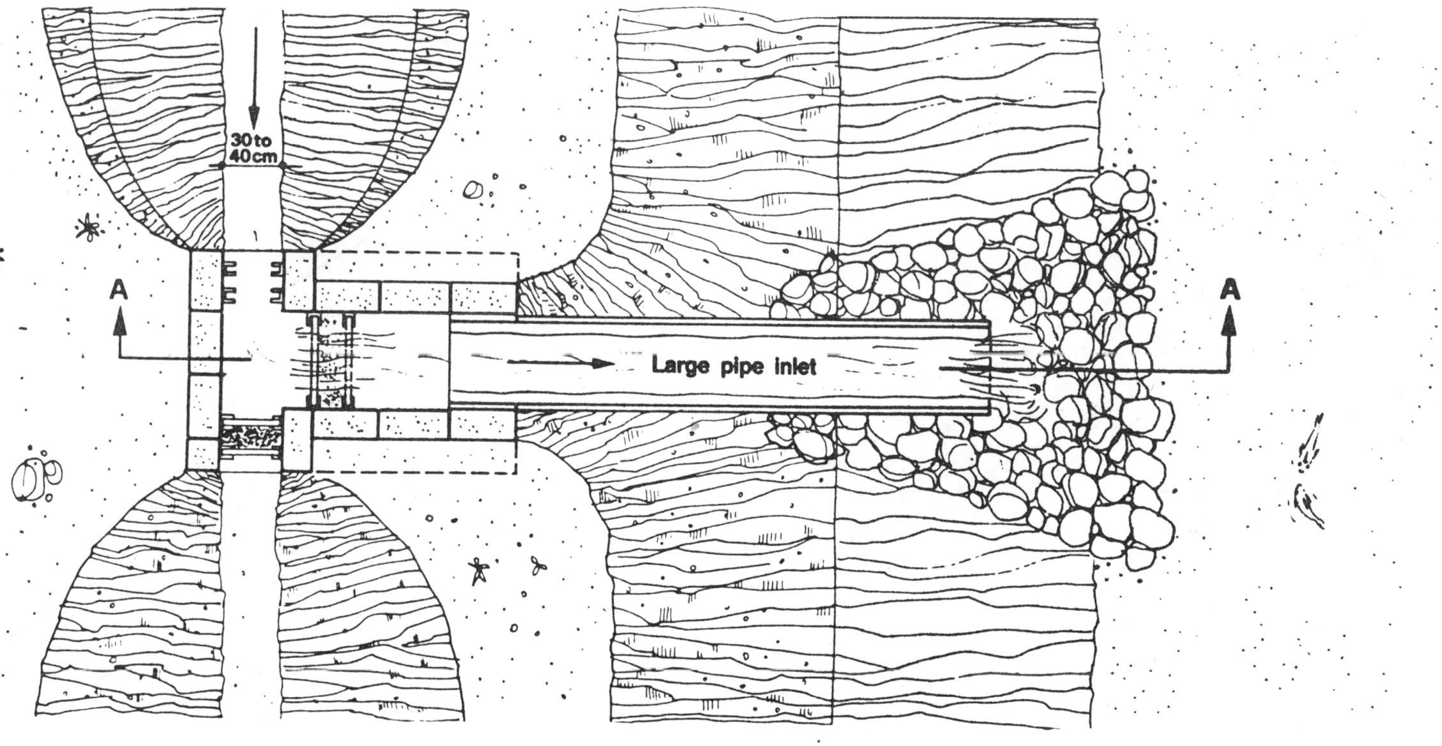

92 Gutter inlets

1 Gutter inlets usually extend for about 1 m over the water surface when the pond is full. They can be made simply from various materials such as:

- **bamboo**: by cutting a bamboo culm lengthwise in half and cleaning out the partition walls. The diameter is usually limited to 10 cm or less;
- **wood**: by assembling three boards to form a rectangular gutter. A flow-regulating gate can easily be added;
- **metal**: by bending lengthwise a galvanized iron sheet into a semi-circular gutter. The flow should be regulated upstream.

2. To estimate **the water carrying capacity**, use **a minimum head loss** of 0.2 mm along a 2 m length or slope **S** = 0.2 ÷ 2 000 mm = 0.0001 or 0.01 percent (see also Section 82).

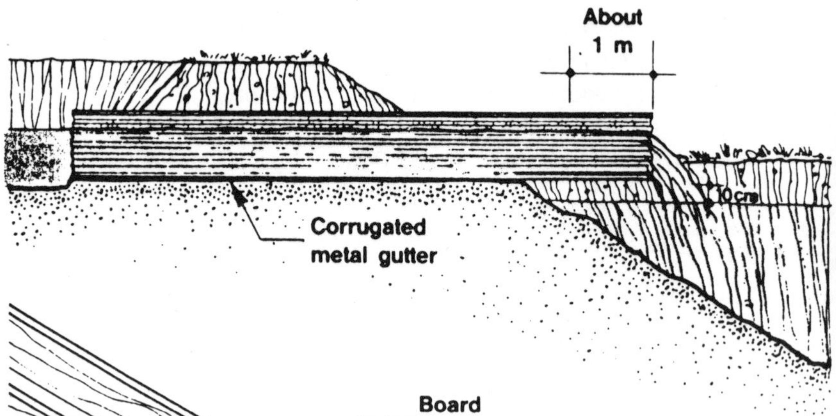

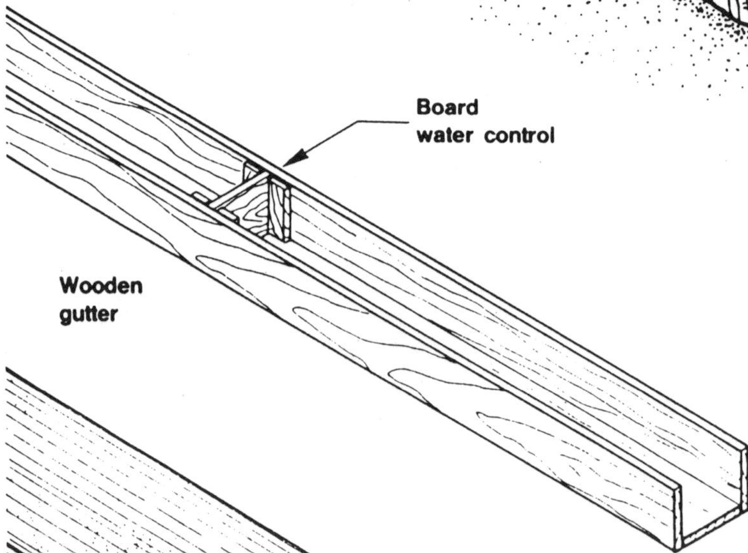

Wooden gutter

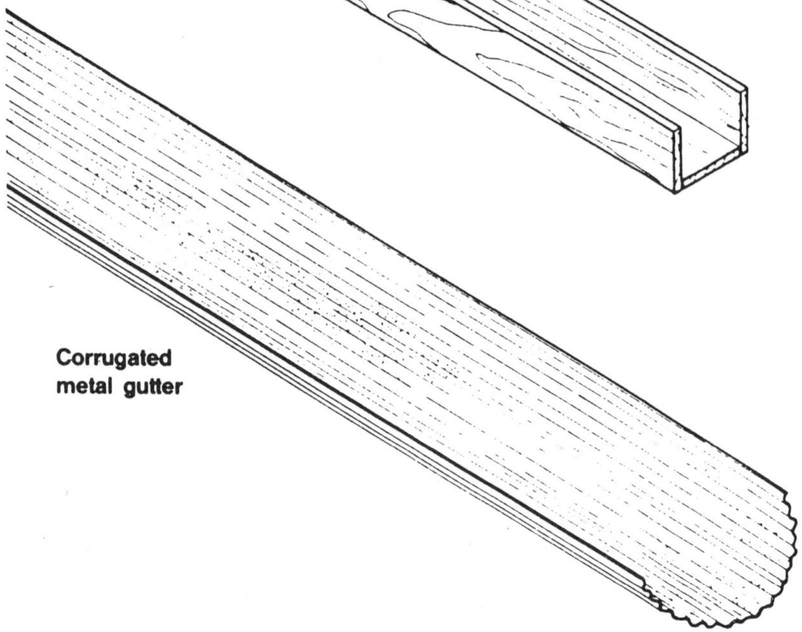
Corrugated metal gutter

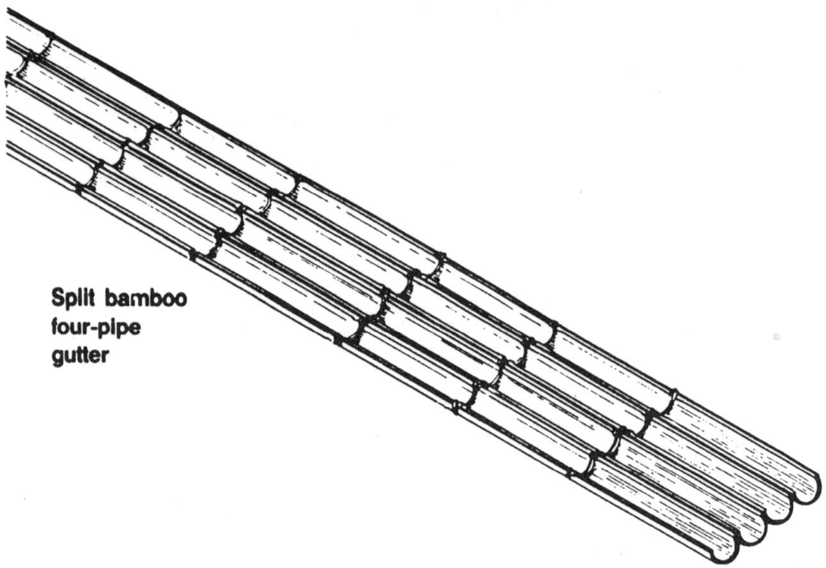
Split bamboo four-pipe gutter

93 Canal inlets

1. A small open canal can be built to connect the water feeder canal to the pond, usually from a division box. There are several possibilities such as:

- digging **a small earthen canal**, with a trapezoidal section;
- building **a small lined canal**, with a rectangular section and using either **wood, bricks** or **concrete**. Small parallel walls are built on a light foundation along the sides of the canal. If necessary, two pairs of grooves are added to regulate the water flow with thin boards and to keep unwanted fish out with a sliding screen.

2. To estimate **the water carrying capacity** use **a minimum slope value** of **S** = 0.0001 as above (see also Section 82).

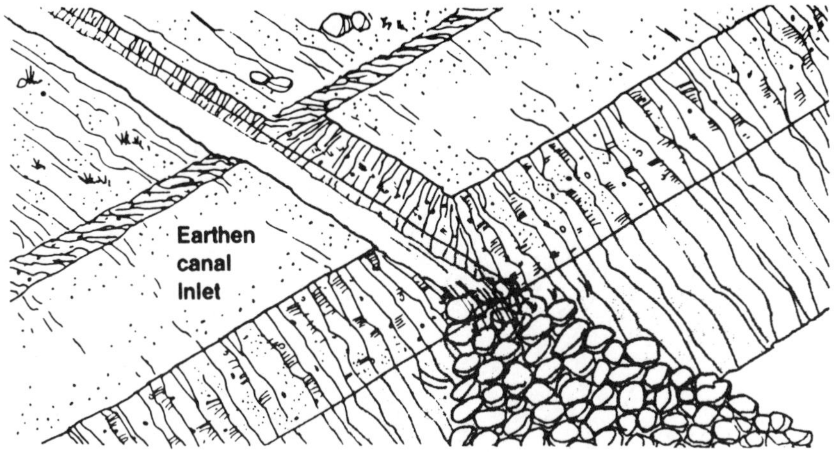

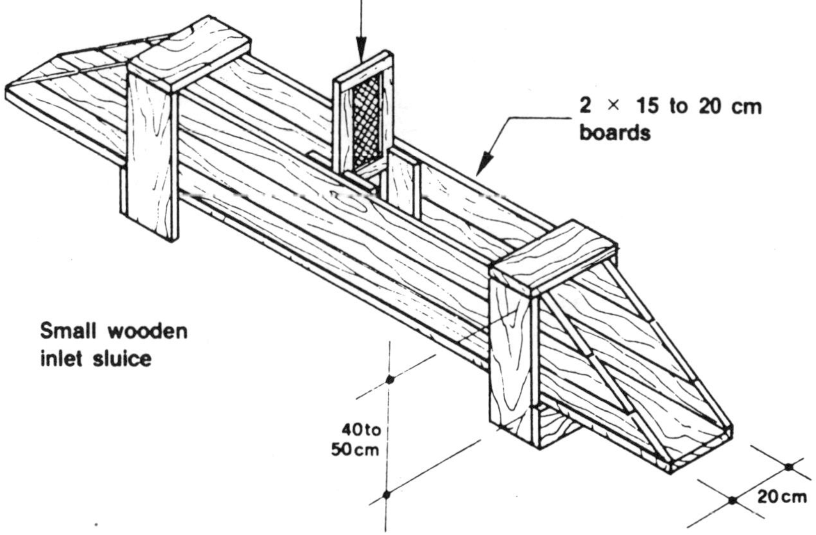

Small wooden inlet sluice

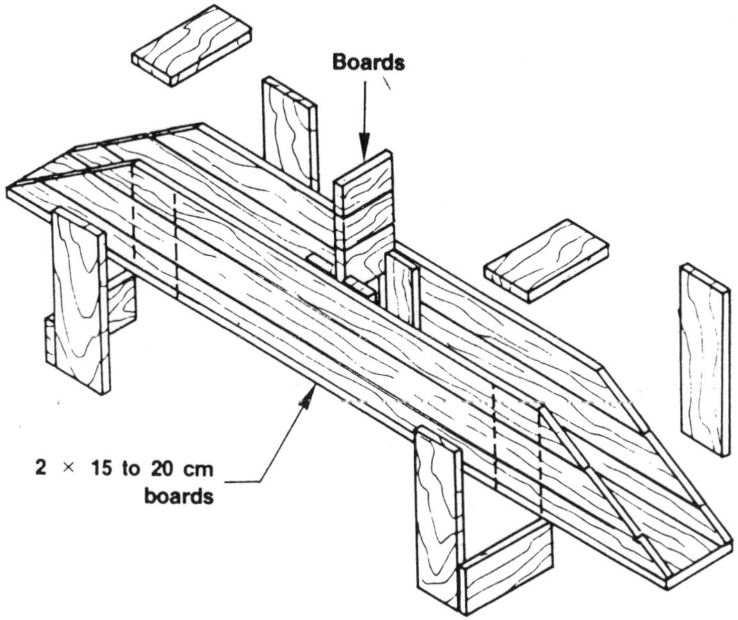

Parts of small wooden inlet sluice

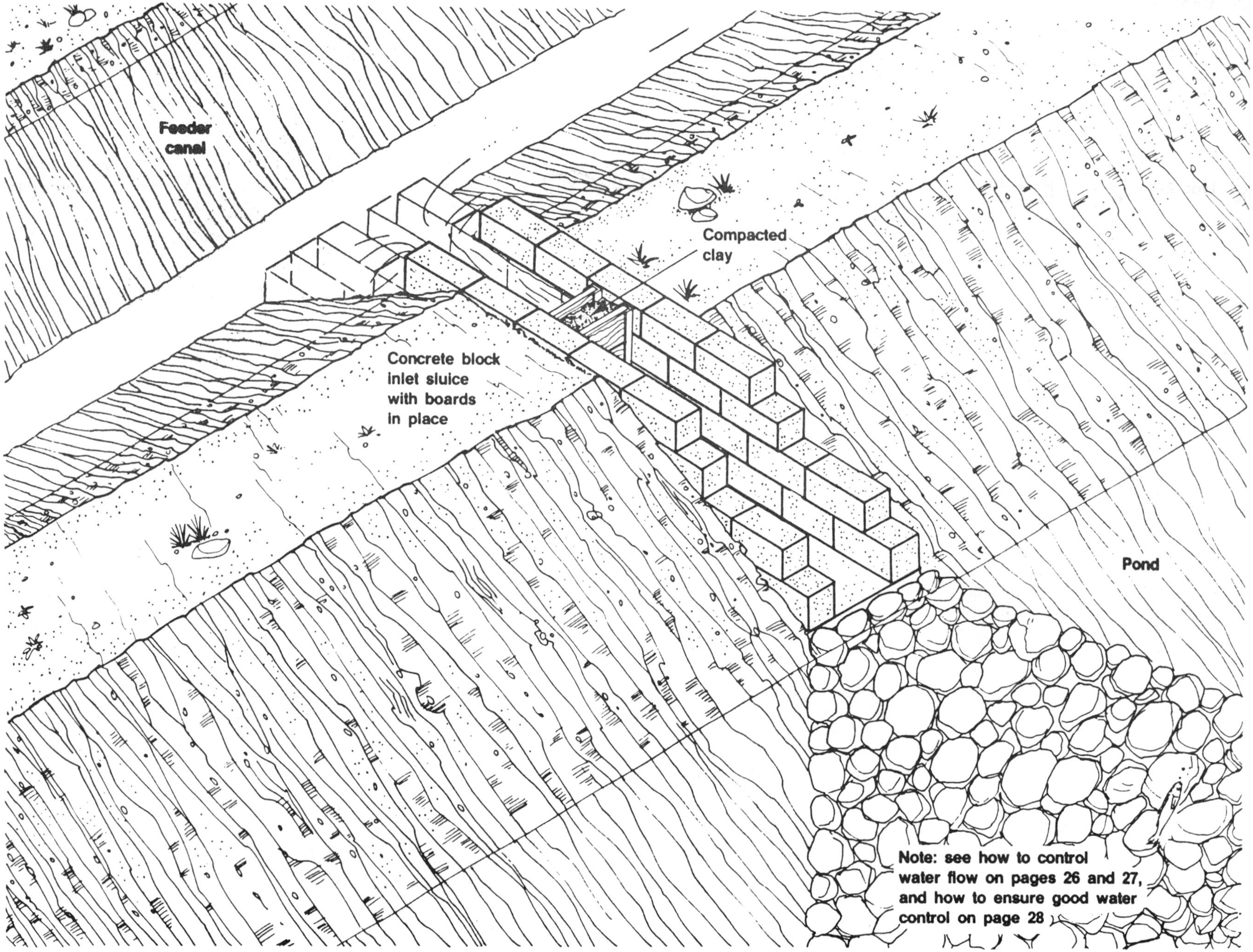
Feeder canal
Compacted clay
Concrete block inlet sluice with boards in place
Pond
Note: see how to control water flow on pages 26 and 27, and how to ensure good water control on page 28

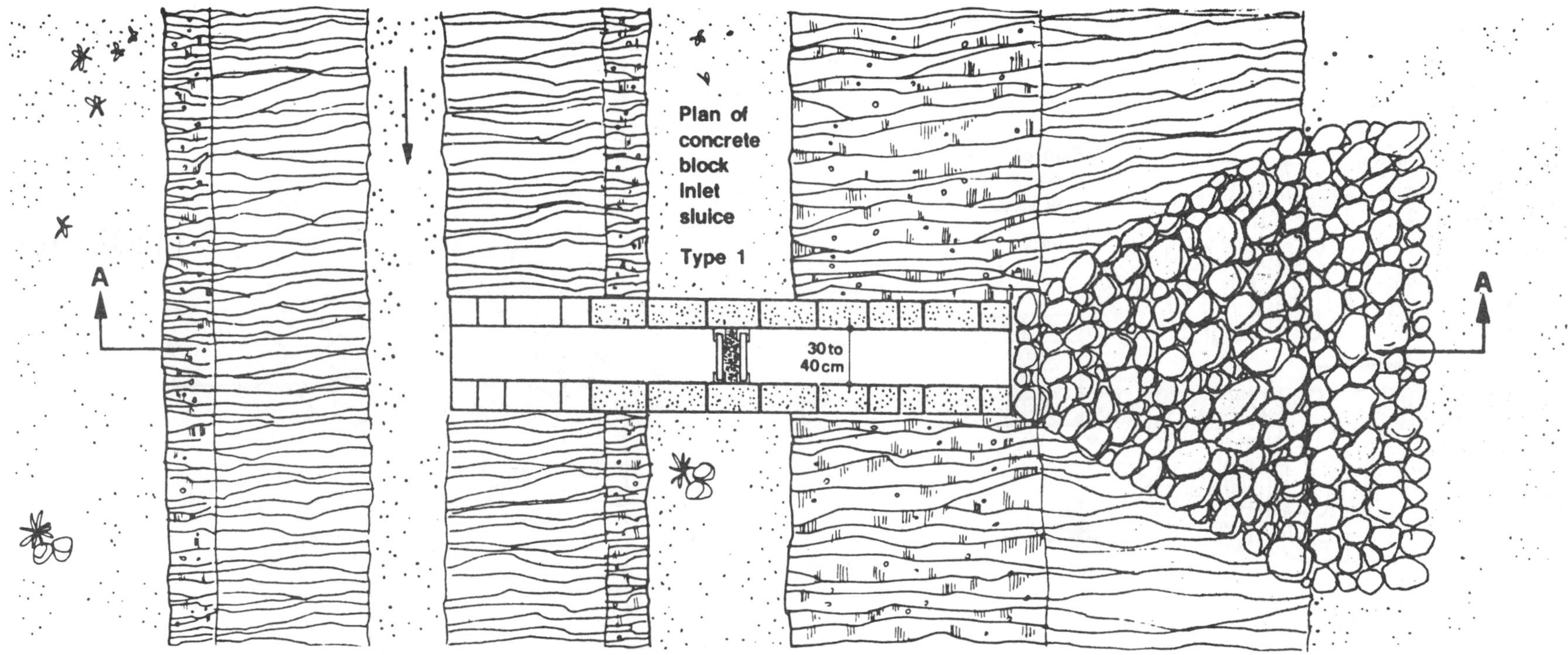

Plan of
concrete
block
inlet
sluice
Type 1
A
A
30 to
40 cm

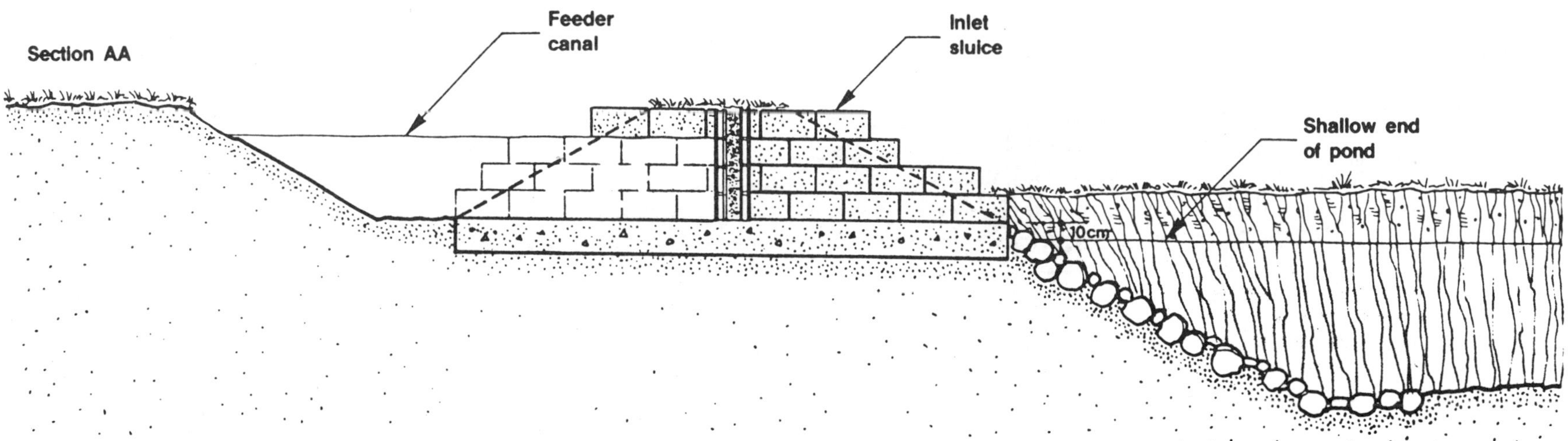

Section AA
Feeder
canal
Inlet
sluice
Shallow end
of pond
10cm

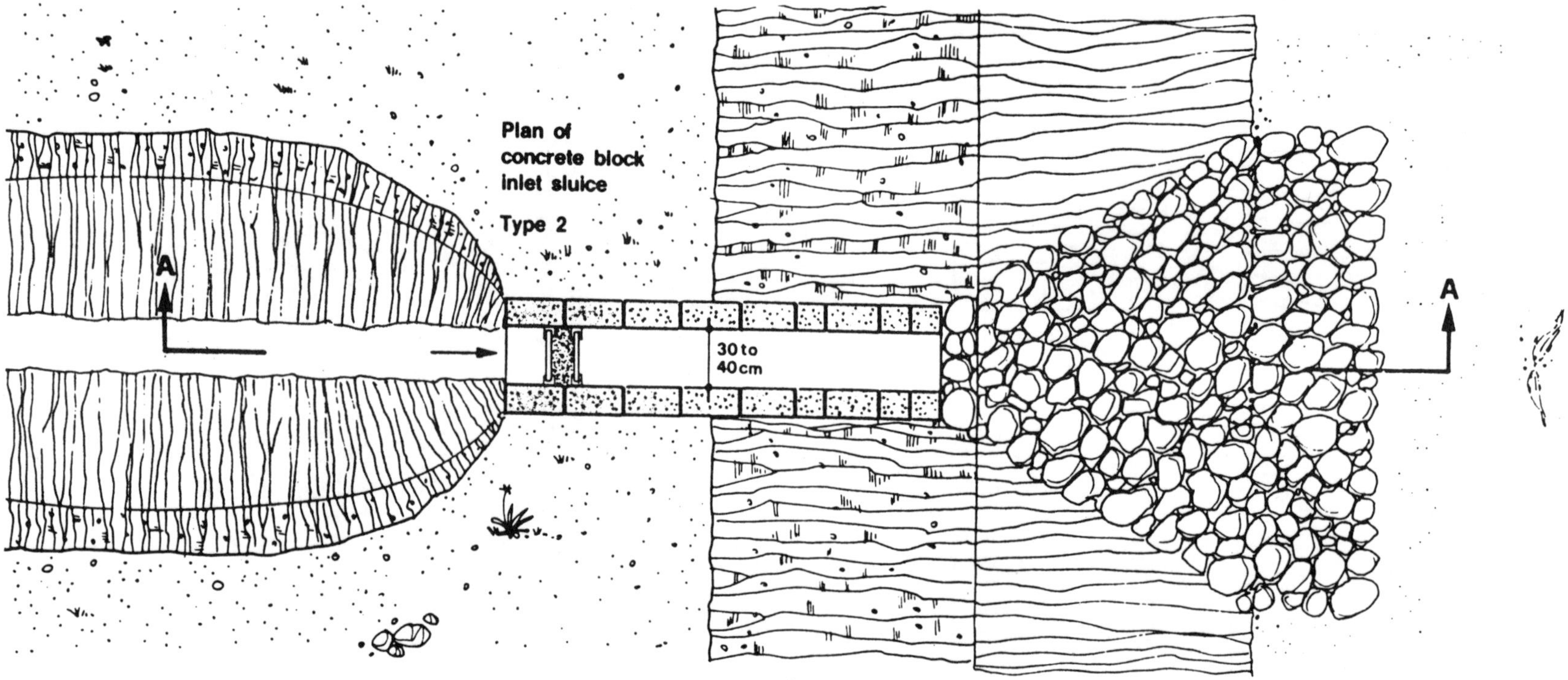
Plan of
concrete block
inlet sluice
Type 2
A
A
30 to
40cm

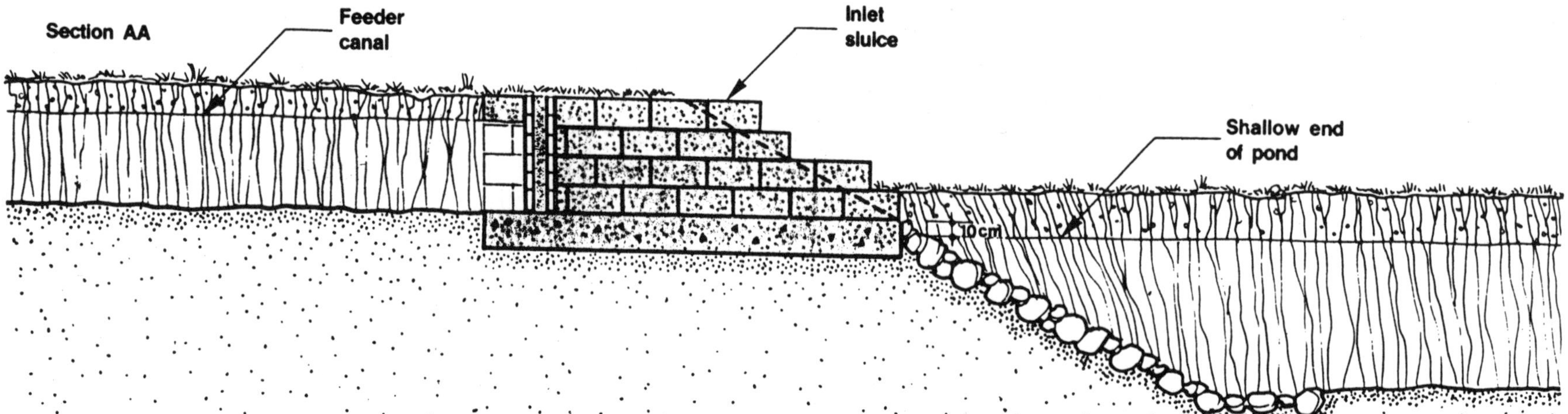
Section AA
Feeder
canal
Inlet
sluice
Shallow end
of pond
10cm

Building an inlet sluice

CONCRETE BLOCK
Type 2

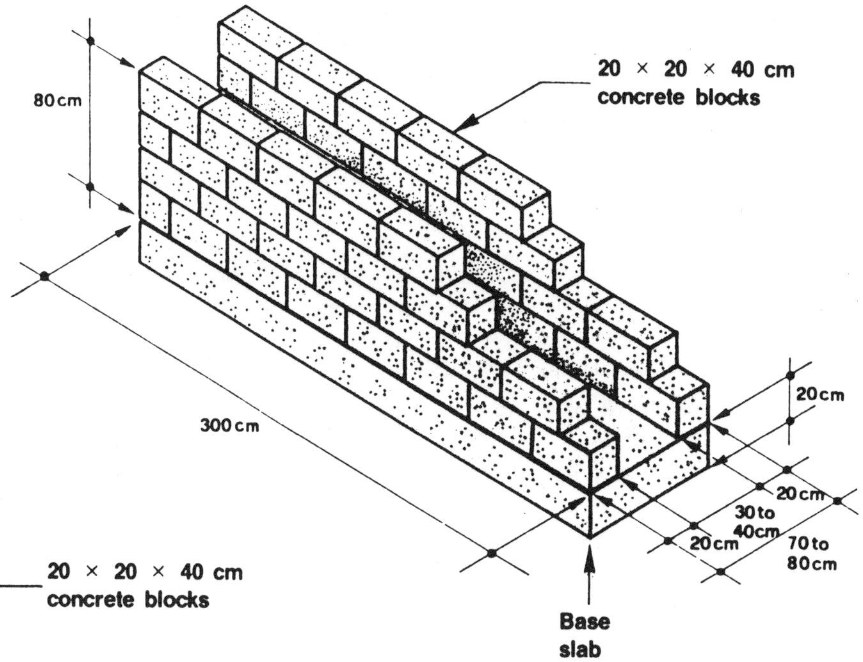

CONCRETE BLOCK
Type 1

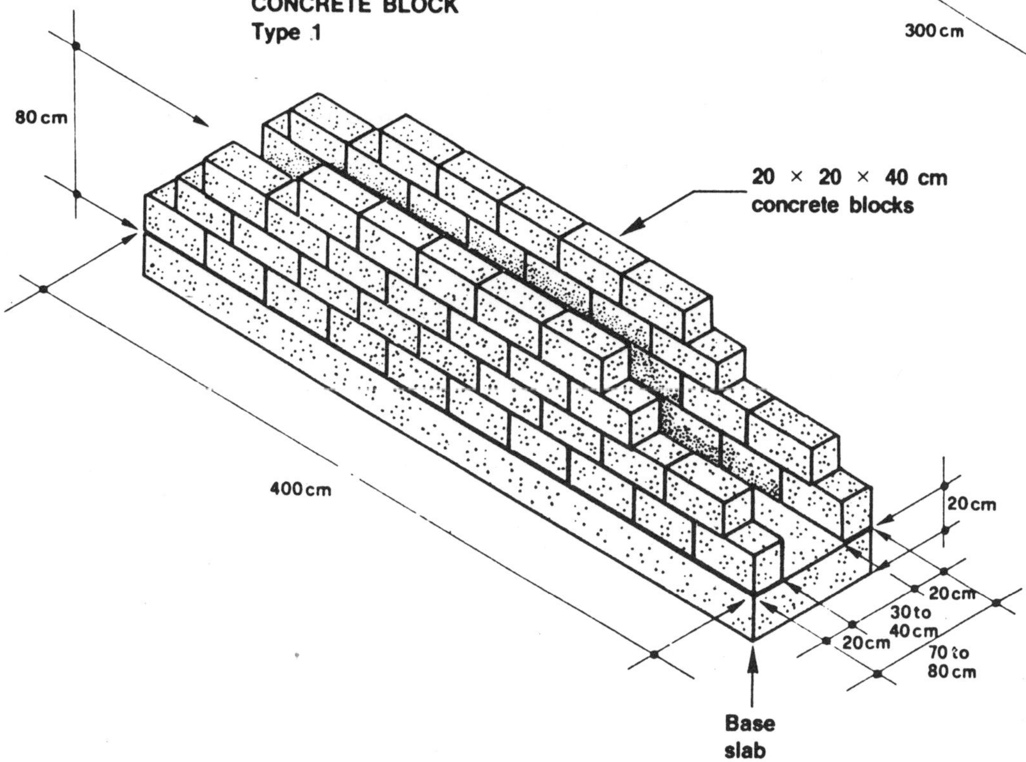

Note: the dimensions shown are suitable for structures for a medium-size pond system; dimensions may vary, however, depending on the size of your pond system

94 Aerating and mixing incoming water

1. Unless the pond is operating with large flows of water, the effect of aeration of inlet water on the overall pond conditions is not great; to fill the pond, or to top it up, it helps to ensure that **incoming water is well aerated**. Similarly, it can help to mix incoming water well with existing water. Several useful devices help to do this such as:

- **inlet weir**: this structure spreads out the incoming water and is particularly useful for smaller ponds where the splashing, together with the spreading, help to mix and aerate at the same time;
- **inlet box**: this structure can be built of wood, brick or concrete and acts simply to drive the incoming water below a sluice board, so that it enters the pond below the normal water level, thus helping to mix lower pond water levels;
- **combined system**: a weir can be arranged to supply aerated water, which is then forced downwards into the lower pond levels, at the same time pulling in extra air under pressure.

2. You will learn more about water aeration devices in the next manual in this series, **Management, 21/1**, Sections 26 to 28.

95 Dike protection at a pond inlet

1. Earthen pond dikes are very susceptible to erosion by running water, particularly when they are not protected by a grass cover and when they are in direct contact with falling water. It is important that you protect the part of the dike situated under the pond inlet.

2. You can do this in various ways, for example:

- by fixing **a layer of brushwood** along part of the dike's side and pond bottom;
- by making a simple **reinforcement** from wood, planks, bamboo or matting around the inlet;
- by **piling rocks** under the inlet pipe or gutter to break the fall of the water;
- by **paving** part of the wet side of the dike and of the pond bottom with a layer of rocks, gravel or wooden boards.

Note: when you start filling your pond you should be especially careful to avoid eroding your dike with running or dropping water or both.

10 POND OUTLET STRUCTURES

100 Outlet structures

1. Outlet structures are built for two main reasons:

- to keep the **water surface** in the pond **at its optimum level**, which usually coincides with the maximum water level designed for the pond;
- to allow for **the complete draining of the pond** and harvesting of the fish whenever necessary.

2. In addition to these major functions, **a good outlet** should also ensure as far as possible that:

- **the amount of time** necessary to drain the pond completely is reasonable;
- **the flow of the draining water** is as uniform as possible to avoid disturbing the fish excessively;
- there is no **loss of fish**, especially during the draining period;
- water can be drained from **the top, bottom or intermediate levels** of the pond;
- any reasonable **excess of water** is carried away;
- the outlet can be **easily cleaned** and serviced;
- the **construction cost and maintenance** are relatively low.

3. In most cases, **outlets have three main elements**:

- **a collecting area** on the inside of the pond, from which the water drains and into which the stock is collected for harvest;
- **the water control itself**, including any drain plugs, valves, control boards, screens and gates;
- **a means for getting the water to the outside of the pond** such as a pipe or a cut through the wall, and/or an overflow structure. In both cases, a protected area on the outside of the wall must prevent the drain water from scouring the walls or drainage channel.

4. Pond outlets can be built in various ways, using **different materials** such as bamboo, wood, bricks, cement blocks or concrete. There are **four main types**, which will be discussed in turn:

- simple cuts through the dike (see Section 102);
- simple pipelines and siphons (see Sections 102 and 103);
- sluices (see Section 104); and
- monks (see Sections 105 to 109).

5. The main principles of design, including the catchpits, pipes and overflows, are discussed first in the next section.

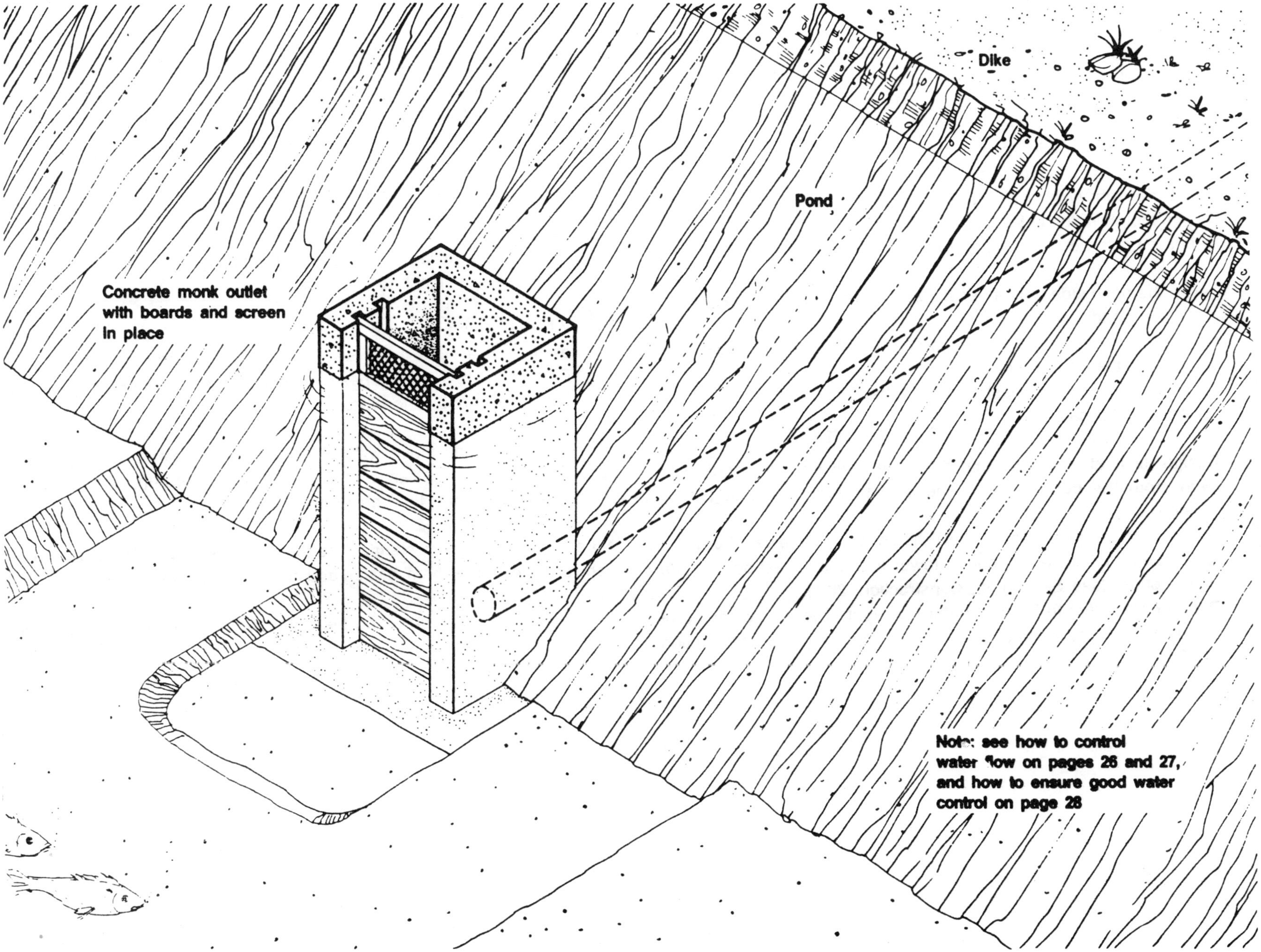
Dike
Pond
Concrete monk outlet with boards and screen in place
Note: see how to control water flow on pages 26 and 27, and how to ensure good water control on page 28

101 Important points when designing outlets

1. When designing and constructing an outlet structure, you should pay particular attention to the following points:

(a) **Design a water discharge capacity** large enough so that the pond can be drained within a reasonable amount of time, from a few hours for a small pond to one to two days for a large pond. (To select the right size of pipes see later in this section.)

(b) **Design the elevation of the bottom of the outlet structure** low enough to ensure complete drainage of the pond, so that:

- on the pond side, it is at least 10 cm lower than the lowest point in the pond;
- it slopes away from the pond, preferably with a slope equal to or greater than 1 percent;
- at its end, it is at least 20 cm higher than the bottom of the drainage canal.

(c) **Design its total length** so that the water will be discharged well **away from the outside toe** of the dike to avoid damaging it by erosion.

(d) **For a barrage pond**, the outlet should be built **away from the streambed** if possible and dug in lower than the lowest point of the pond bottom (see Section 66, **Construction, 20/1**).

(e) **Plan for the outlet to be built** before or right after the beginning of the dike construction, depending on the type of pond (see Section 26, **Construction, 20/1**).

(f) If the outlet structure is heavy, be sure to build it on **very well compacted soil only** (see Section 62, **Construction, 20/1**). You may also need to support it with simple piling.

(g) If there is **an outlet pipe through the dike**, it is always best to build **at least one anti-seep concrete collar** around it:

- place it so that it will be integrated with the dike;
- build it perpendicular to the pipe;
- extend it at least 15 cm from all sides of the pipe;
- make it at least 10 cm thick.

(h) When building **the dike above the pipes**, be sure to compact the earth well around them (see Section 62, **Construction, 20/1**).

(i) Remember that **small pipes can become easily blocked**, particularly inside the pond. Therefore:

- avoid using pipes with a very small diameter;
- protect the entrance of all pipes with a screen, and keep this screen clean through regular checks.

Note: you should try **to standardize the type of outlet** to be built on your fish farm, to make them easier and more economical to use.

2. To assist you in selecting the right type of outlet, consult **Table 46**. You should also take into account how much you can invest and which kind of materials are locally available.

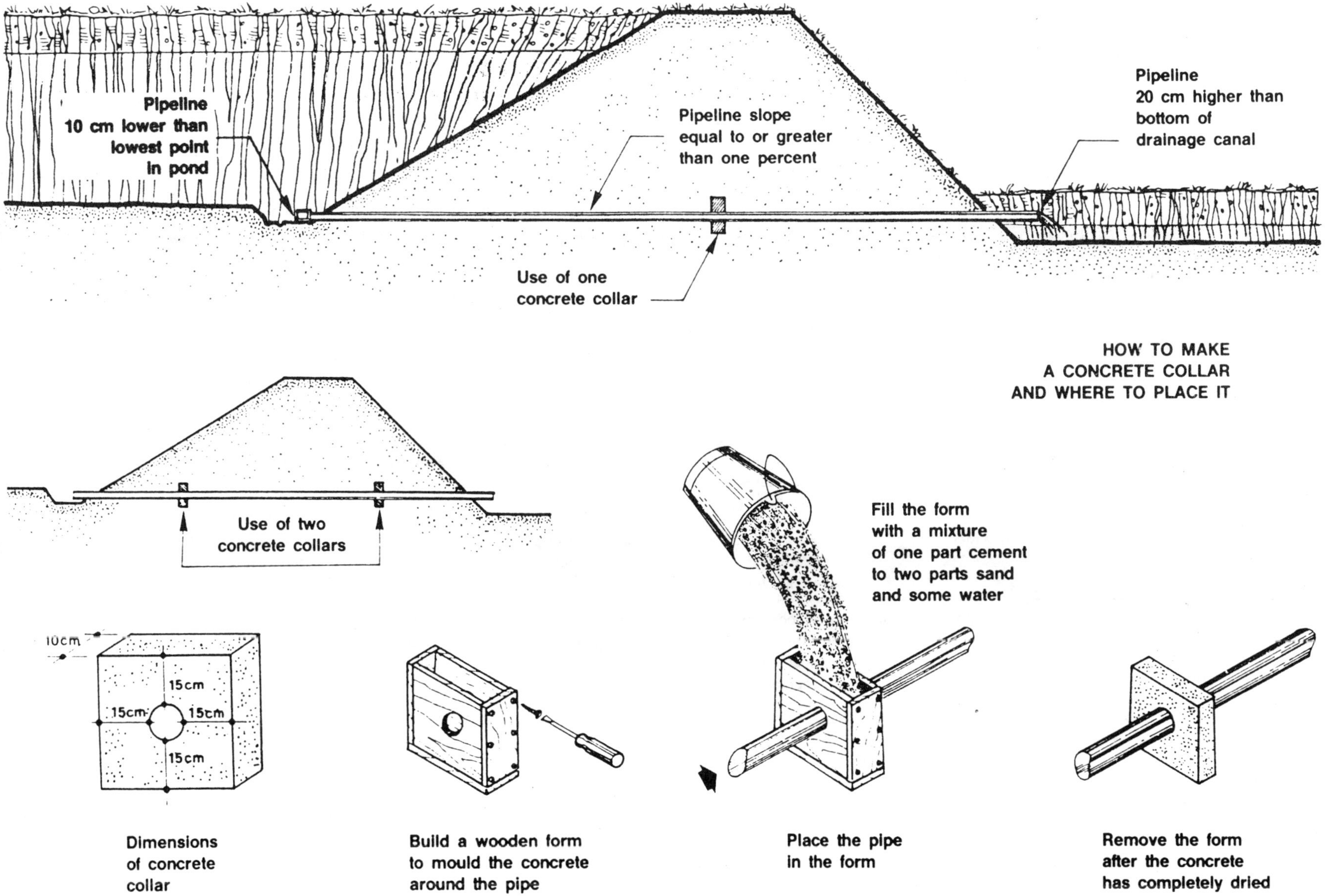

HOW TO MAKE
A CONCRETE COLLAR
AND WHERE TO PLACE IT

TABLE 46

Characteristics of various pond outlets

Type of pond outlet	*Pond size*	*Pipe quality and diameter*	*Water level control*	*Required materials*	*Section*
Siphon	Small	Rubber/plastic Diameter less than 10 cm	Nil	Flexible tubing	102
Closed pipe	Small	Bamboo, galvanized iron or plastic Diameter less than 10 cm	Nil	Hard pipes	103
Flexible stand-pipe	Small	Combined Diameter less than 10 cm	In pond Continuous	Flexible and hard pipes	103
Turn-down pipe	Medium	Hard plastic Diameter up to 20 cm	In or out of pond Continuous	Hard pipes, 90°-elbow	103
Monk	Medium to large	Wood, plastic, asbestos cement, concrete Diameter more than 10 cm	In pond Continuous	Wood, bricks, cement blocks, or concrete and pipes	105-109
Sluice gate	Small to medium	No pipe Max 80 cm wide	In pond Continuous	Wood, bricks, cement blocks, or concrete	104

Sizing outlet pipes

3. The inside diameter of outlet pipes will determine **the water discharge capacity** of the outlet structure.

4. Select the right **size and quality of the pipes** to be used, according to **the size of the pond** (see **Table 47**) and the **size of the pipes**:

- for very small diameters, you can use bamboo, galvanized or plastic pipes;
- for small diameters, plastic pipes are preferred;
- for medium diameters, asbestos cement pipes are preferred;
- for large diameters, reinforced concrete pipes are preferred.

5. **The water carrying capacity** of the selected pipes can be estimated from **Graph 1, Tables 13 and 14, or mathematical formulas** (see Section 38, **Construction, 20/1**).

TABLE 47

Sizes of outlet pipes for diversion ponds

Pond size (m^2)	*Pipe, inside diameter* (cm)
less than 100	5-7.5*
100-200	7.5*-10
200-400	10-15
400-1 000	15-20
1 000-2 000	20-25
2 000-5 000	25-30*
more than 5 000	40 or more

Not for monks where pipes should be at least 10 cm in diameter (see Section 117)

Note: for barrage ponds fed directly by a stream, you might need larger pipes than usual if the excess water inflow has to be discharged continuously through the pond outlet. In most cases, it will be safer and more economical to build **a lateral overflow structure** (see Sections 113 and 114).

Placing and fixing outlet pipes

6. In general, in stable soil with well-constructed dikes and smaller pipes, no special precautions are required, apart from ensuring that **pipe trenches are level** and that the **pipe will not be damaged**.

7. For longer and larger pipes such as those used in monks, some fixing or foundation may be required (see Section 38, **Construction, 20/1**). For additional information on monks see Section 105.

Additional overflow structures

8. For security reasons, you should always **ensure that the water level in the pond does not rise above the designed maximum level** and flow over the top of any dike (see Section 111 on the discharge of excess water). This circumstance might result not only in the loss of most of your fish but also in heavy repairs before you could use your pond again.

9. **In a diversion pond**, where most of the excess inflow waste is discharged at the inlet diversion structure, a pond outlet such as an open stand-pipe, a monk or a sluice should discharge any excess automatically. But you have to ensure that **all screens are kept clean**.

10. For **a barrage pond directly fed by a stream**, however, you need additional security such as a mechanical spillway for the continuous discharge of the excess water and possibly an additional security spillway for the occasional discharge of flood water (see Sections 113 and 114 for information on pipe overflows and mechanical spillways).

11. **If you do not use an outlet or if you use one which does not regulate the pond water level** (a closed pipe, for example), you will definitely need a protection structure to discharge any occasional excess water (see Chapter 11).

102 Simple outlets for small ponds

Using a cut in the dike

1. Very small rural ponds can be harvested by **cutting the dike open** at one of the deepest points of the pond. It is rebuilt when the pond has to be filled again. In such a case:

- the dike next to the cut can be damaged, especially if the water in the pond is relatively **deep and the water current too strong**;
- repairing the dike creates additional work;
- the quality of the dike can be impaired and the risk of it breaking away increased.

2. If for some reason, you cannot build an outlet for your pond, **you can limit the damage to your dike**:

(a) **Do not build your pond too deep**, and make the cut in stages, draining off the upper levels of water first.

(b) When draining the water, **use a piece of canvas**.

(c) When building the dike, **include two rows of strong wooden poles in it**. Between them, close the gap by carefully compacting clay soil up to the maximum water level. Protect the top of this gap with stones or gravel, so that any excess water can flow from the pond over it without damage. Each time you drain the pond, cut open the gap between the two rows of poles and rebuild it to fill the pond with water.

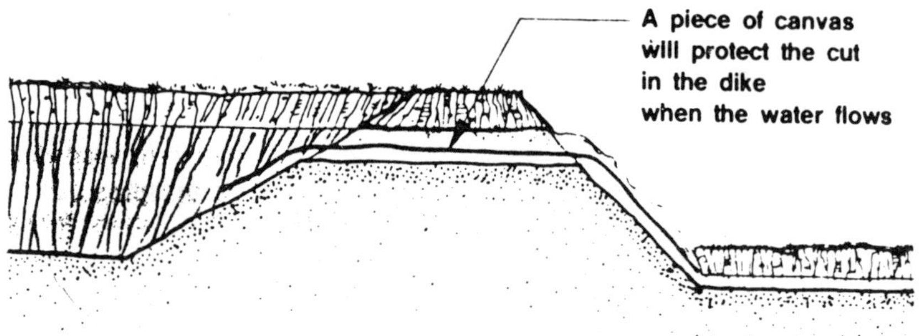

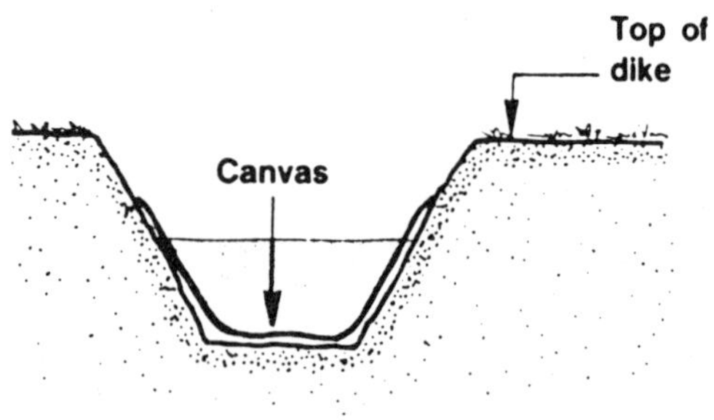

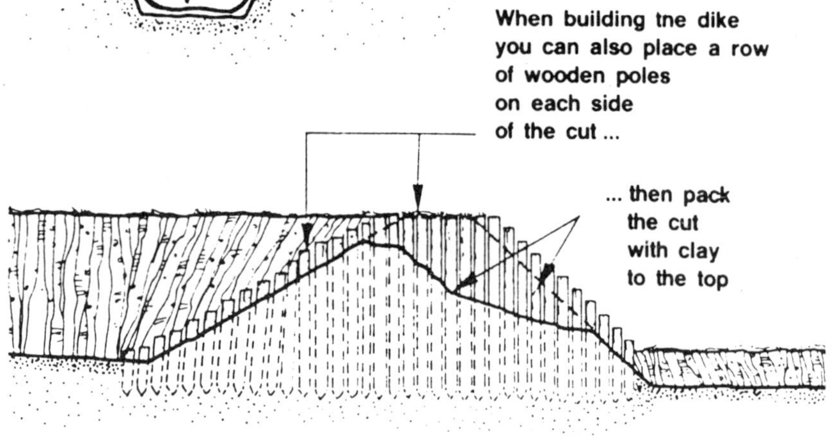

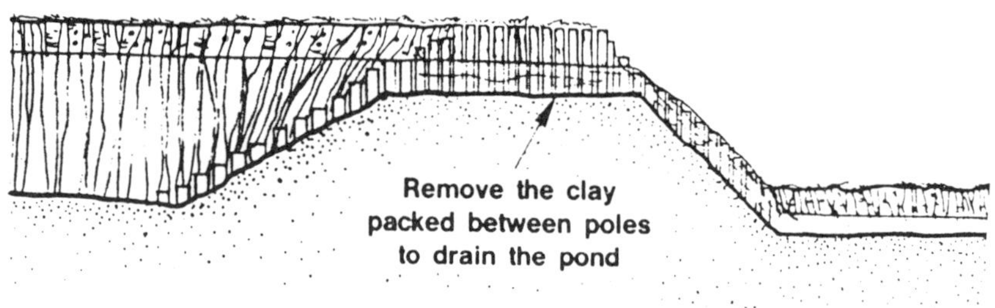

Using a siphon as a pond outlet

3. Small ponds can be drained either partly or fully using a siphon (see Section 89 for use of siphons). One of the limitations of this method is that the outlet of the siphon has to be at least 20 cm below the level to which you want to drain.

4. **The total length of the siphon** should be at least equal to the wet side of the dike plus the dike top width plus the dry side plus 30 cm. **The siphon diameter** should preferably not be larger than 2 to 3 cm, so that it will be relatively easy to start the siphon flowing.

5. **To start a siphon** flowing, proceed as follows, with the help of another person:

(a) Ask the other person to block the end of the siphon tube at its lowest end outside the pond, using either their hand or a plug.
(b) Position the siphon tube so it can be completely filled from the other end (which means taking the other end up to the top of the pond wall). Be careful not to have any kinks which may trap air.
(c) Fill the siphon with water from the other end; when it is full, block this end either with your hand, or better, with a second plug.

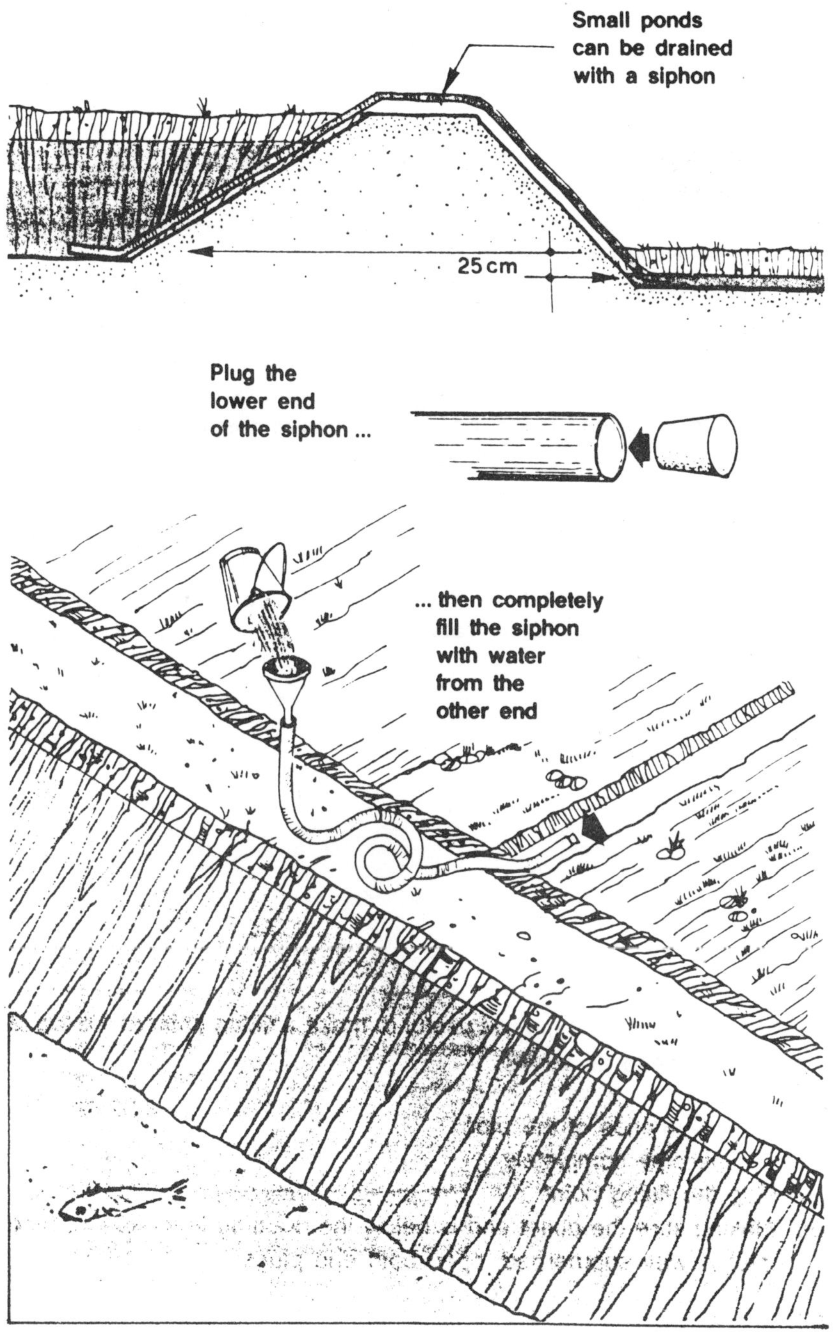

(d) Bring down the end of the pipe slowly and submerge it fully into the pond, keeping it blocked all the time.
(e) Ask your assistant to open the other end of the siphon, ensuring that it remains at its lowest level.
(f) At the same time, open your end of the siphon, ensuring that it remains well under water. The water should start flowing through the siphon.
(g) As the water keeps flowing continuously and its level drops, ensure that **the pond end of the siphon remains under water** and that the outer end is kept below the inner water level.

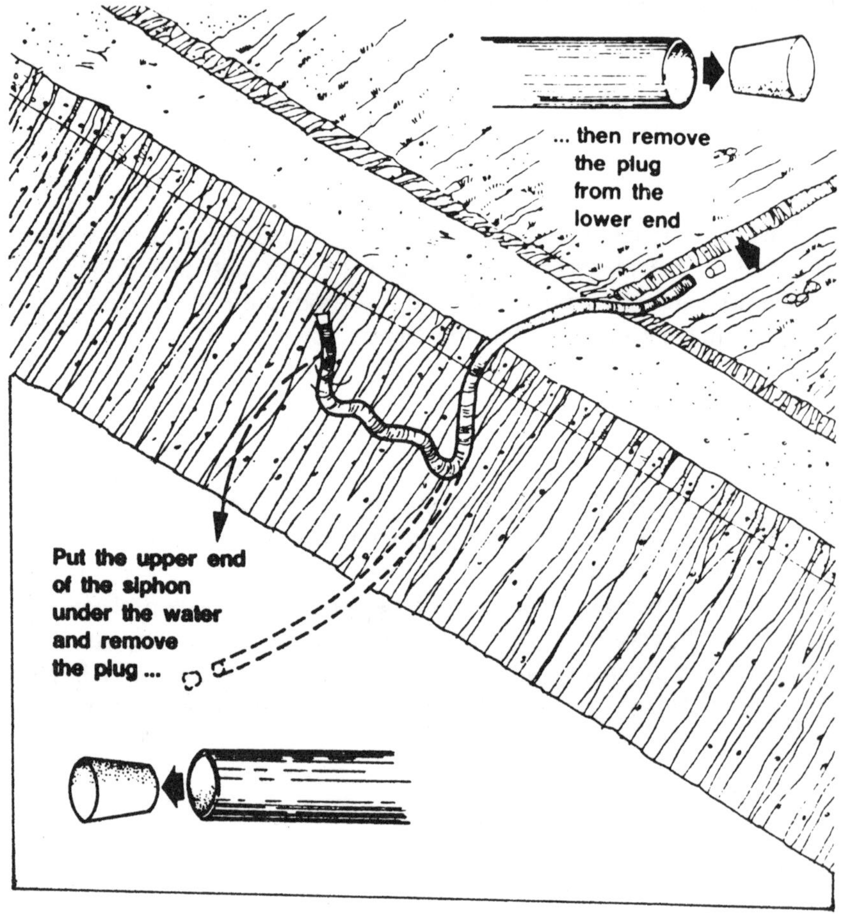

6. **To estimate how long it will take you to drain your pond,** you can use either **Graphs 11 and 12** or **Table 45**, together with the pond water volume (see **Water, 4**).

Note: in some cases, it may be useful to make **a filling point at the top of the pipe**. Proceed as follows:

(a) Plug both ends of the pipe.
(b) Fill the pipe completely.
(c) Plug the filling point.
(d) Making sure the outlet end is below the draining level and that the inlet is well submerged, open both end plugs.

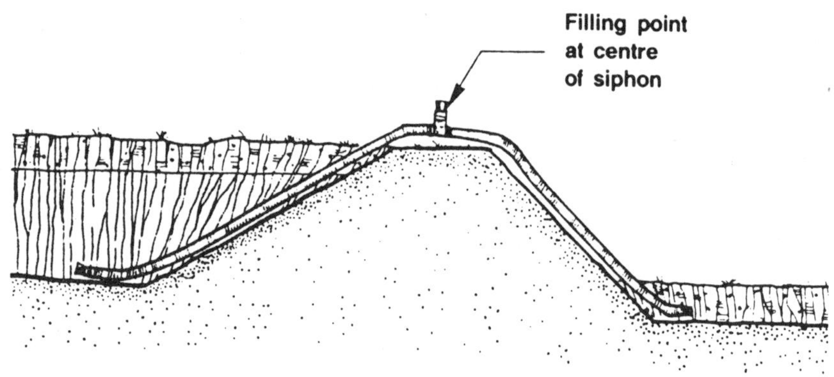

103 Simple pipe outlets

Using a simple pipe and stopper

1. **For small-size and shallow ponds**, a straight pipeline with a small diameter can be used as a water outlet. It is important that the pipes be laid down at the lowest point of the pond before the dike is built. You can select one of the following pipes according to availability and cost:

- **bamboo pipes** (see Section 31, **Construction, 20/1**);
- **galvanized iron pipes** (see Section 38, **Construction, 20/1**);
- **plastic pipes** (see Section 38, **Construction, 20/1**).

2. **The pipeline should be closed at one end** before you start filling the pond with water. You should preferably close the end of the pipe that is under water so that it does not become blocked by debris or even a fish. You can use, for example:

- a simple wooden plug;
- a screwing cap fitting the galvanized or plastic pipe; or
- a mechanical valve fitted to these pipes.

3. In some cases, you can also use a **stand-pipe**, which can be adjusted or renewed as required. This alternative is discussed in detail on the next pages in this section.

4. **When you wish to drain the pond**, proceed as follows:

(a) Remove the plug or cap, or open the valve; if you are using a plug, it may be useful to fix to it a metal or cord loop, which you can pull with a long-handled hook. You can even attach a pulling cord if you affix it securely.

(b) Immediately **put a screen** on top of the pipe end so that it does not get blocked.

(c) Keep cleaning the screen as necessary.

(d) When the pond is drained, either remove the screen and **close the pipe immediately** or remove the screen later, when you start filling the pond.

A simple pipe outlet

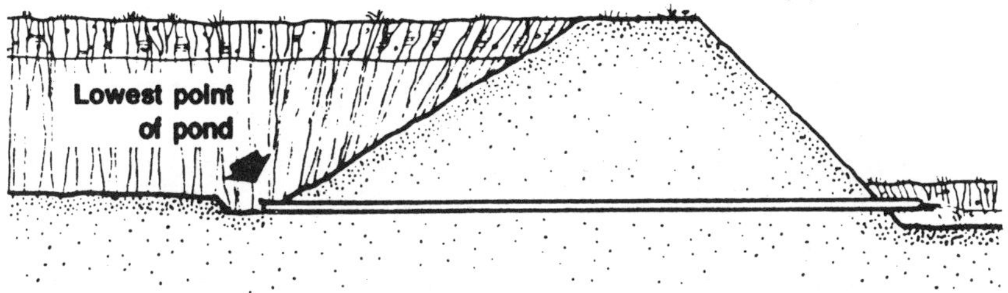

Wooden plug with a nail pull

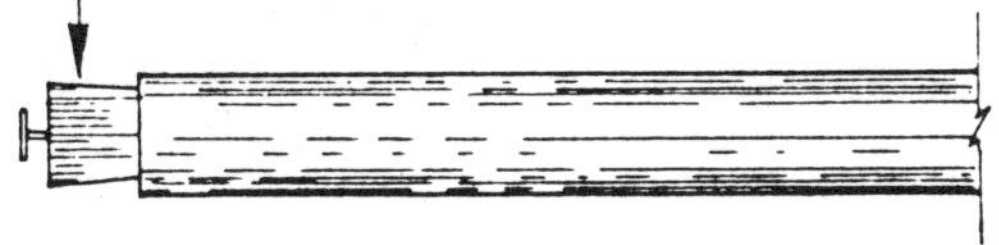

Screw cap

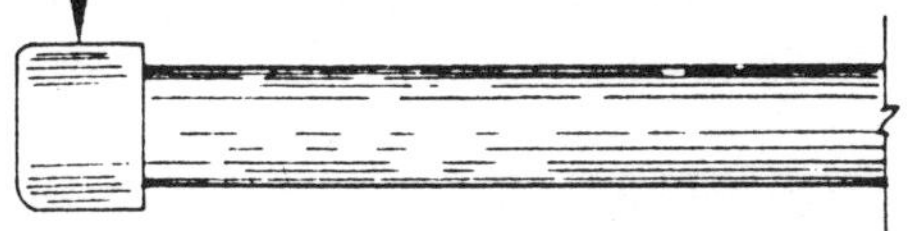

Mechanical valve

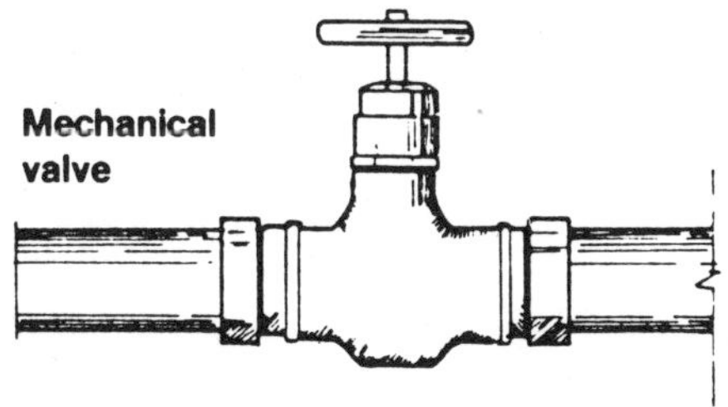

Screen tied in place around end of pipe when draining

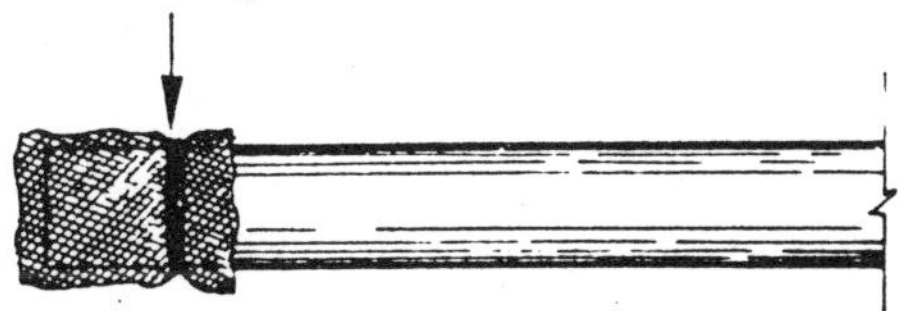

The flexible tube stand-pipe

5. This type of outlet is made of two parts:

- **a slightly sloping rigid pipe** such as a bamboo or a plastic pipe, running through the base of the dike; and
- **a vertical flexible tube** connected inside the pond to the rigid pipe and reaching up to maximum water level.

6. Remember that:

- to keep the tube vertical and avoid the accidental draining of your pond, you must **secure its top end to a strong wooden pole**;
- **to avoid any blockage of the tube, you should put a tightly fitted screen on the top end** of the vertical pipe;
- to be able to drain your pond completely, be sure **that the tube is connected to the pipe** at least 10 cm lower than the lowest point in the pond.

7. **To drain the pond**, detach the top end of the vertical tube from the pole. **Lower it progressively** as the water level drops. Remember to keep the screen over the end of the tube at all times.

Note: you can also use this type of outlet for regular overflow. Remember to keep a screen securely fitted.

8. You can also arrange this system with **the flexible tube on the outside**. In this case you need a screen at the inside end of the rigid pipe. Otherwise the principle of its use is the same.

Flexible stand-pipe inside pond

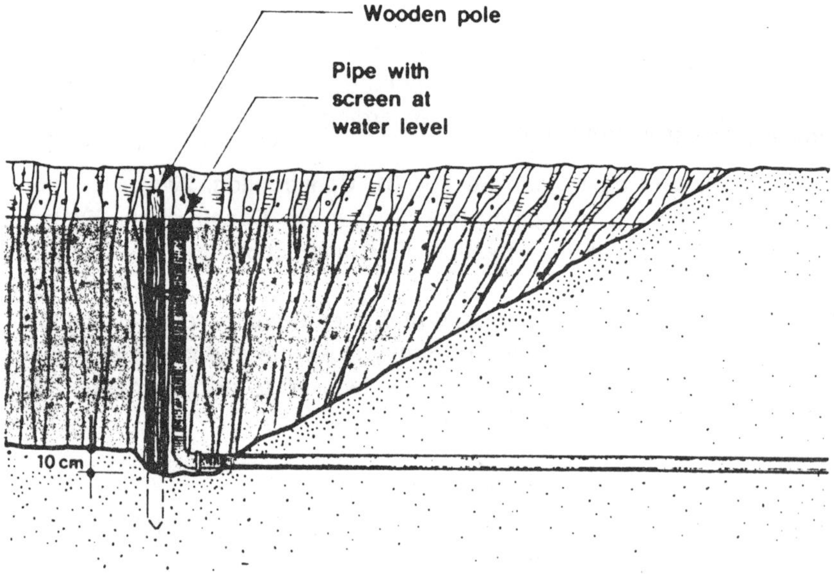

Flexible stand-pipe outside pond

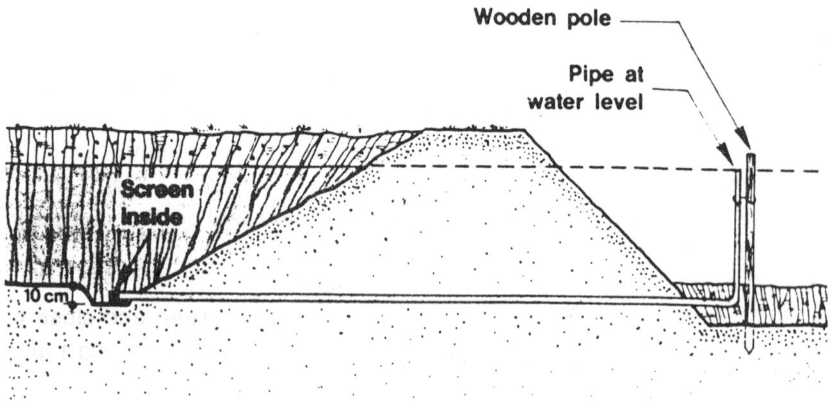

The turn-down stand-pipe

9. Similar to the flexible stand-pipe, this pond outlet is made of three rigid plastic parts:

- **a slightly sloping base pipeline**, made for example of one or more PVC pipes running through the dike;
- **a vertical pipe**, which reaches up to the maximum water level;
- **a 90°-elbow**, which connects these two pipes. It can be glued to the vertical pipe with plastic cement, but need not be unless the fit is very loose. **The connection to the base pipe is unglued**, but can be greased with a suitable material such as mineral grease, lard or palm soap.

10. To ensure that **the 90°-elbow does not become separate from the horizontal pipe**, either by accident or for poaching, protect it as follows:

(a) **Drive a treated wooden stake** about 3 to 5 cm thick and 50 to 60 cm long well **into the pond bottom, right in front of the vertical pipe**. Choose a water-resistant wood (see Section 41, **Construction, 20/1**). Drive the stake firmly into the pond base, making sure it is long enough to reach up above the top edge of the upright elbow pipe.

(b) Right at the back of this stake and at its centre, **drive a steel rod** or a small diameter steel pipe into the pond bottom to a depth of about 1 m and attach it to the stake. It should be long enough to reach up to the top of the vertical pipe.

(c) Weld a chain or tie a rope to the steel post, about 10 cm from its top end.

(d) Near the top of the vertical plastic pipe, strongly fix a small device for attaching this chain or rope. You can use, for example, a steel hook, a swivel eye or a small screw-shackle.

11. This type of outlet can be set up either:

- **inside the pond**, in front of the dike; or
- **outside the pond**, at the back of the dike, in which case you need a screen at the inner end of the base pipe.

12. It is usually best to have the vertical pipe inside the pond to reduce the risk of blocking the horizontal pipe and to control leakage.

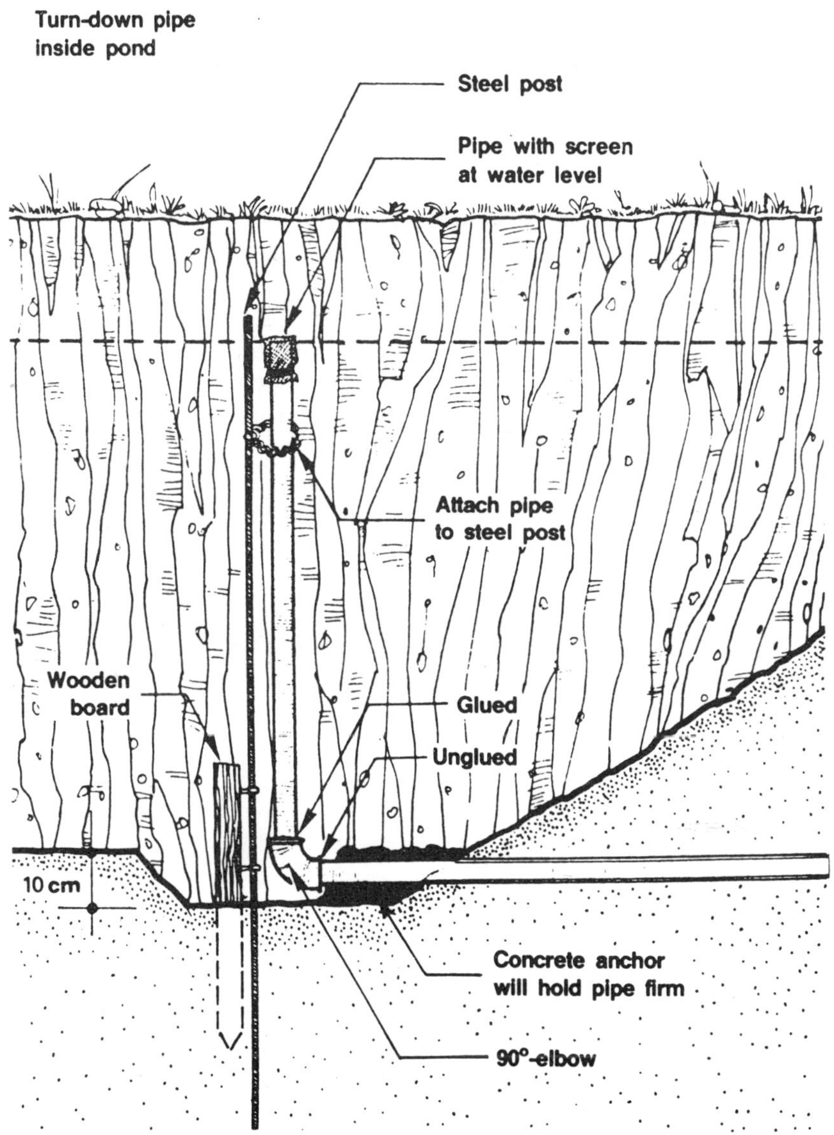

13. When using such a turn-down pipe, remember the following points:

(a) If possible, **design the opening of the horizontal pipe** to be at least 10 cm below the lowest point in the pond.

(b) For additional protection, you can **cover with concrete the section of the horizontal pipe that sticks out** in front of the dike inside the pond.

(c) Always **secure the vertical pipe well to the steel post** in front of it with the rope or chain.

(d) **Tightly fit a screen** on top of the vertical pipe.

14. **To regulate the water level in the pond**, set the pipe at the required angle by turning it up or down. Fix it in the set position with the chain or rope.

15. **To drain the pond**, turn the vertical pipe down progressively, following the water level as it drops. When it has reached the horizontal position, remove the elbow pipe from the end of the horizontal pipe to complete the draining and harvest the fish.

Note: keep the screen on the vertical pipe until you detach the elbow from the horizontal pipe. At that moment, immediately transfer the screen to the front end of this pipe.

16. As with the flexible pipe, you can use this system for handling **normal overflow water**, because any surplus in the pond above the selected pipe level will automatically drain.

Using the turn-down pipe

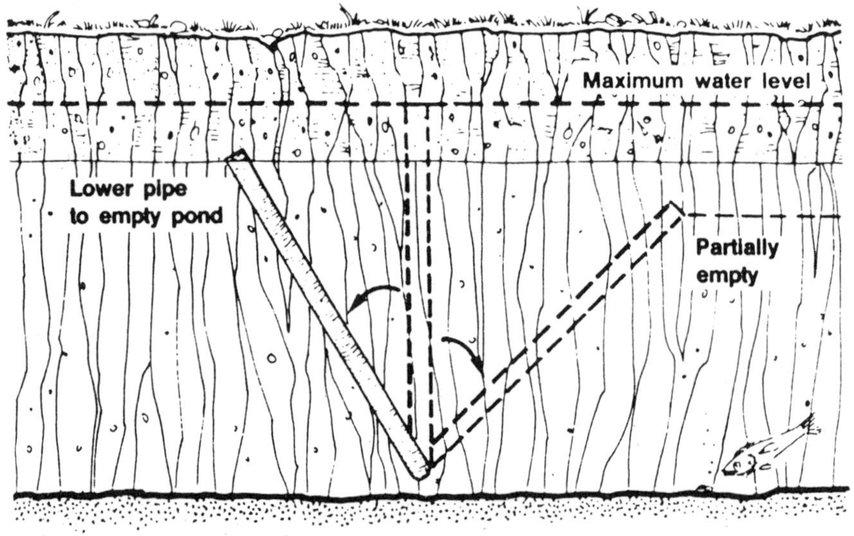

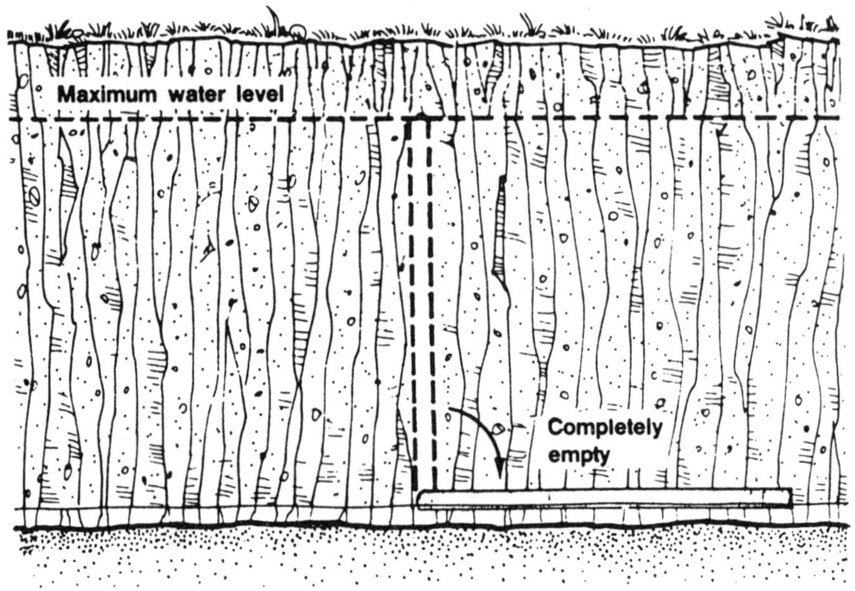

104 The sluice gate

1. A sluice gate consists of **a protected opening in the pond dike** that can be easily closed with wooden boards to regulate water level and can be screened to avoid fish losses. Whenever required, it can discharge excess water continuously.

2. As the sluice gate is **an open outlet structure**, it does not require a base pipe through the dike. It has the same functions as a monk (see Section 105), but has **some advantages**:

- a sluice gate is easier to build;
- less water leaks within the dike;
- it frightens the fish less and makes their harvesting easier;
- its water discharge capacity for a given size is usually greater.

3. However, its **major disadvantage** is that it is **more expensive to build**:

- it is built of bricks or cement blocks, concrete or reinforced concrete because, to last for a long time, it requires very good quality masonry;
- because more material is needed, this difference in cost rapidly increases as the size of the dike increases, as shown in the illustration.

4. Therefore, sluice gates are usually preferred to monks **for small- to medium-size ponds only**, when the dike to be crossed is relatively narrow.

Example

For a pond dike 2 m high with side slopes 1:1.5, you will need the following **quantities of concrete** (in cubic metres) to build sluice gates 0.30 to 0.50 m wide. For an adequate monk and its drainage, you would need less than two cubic metres of concrete and 7 to 8 m of 30 cm pipes.

Inside width of sluice gate (m)	*Width of dike top*	
	1 m	2 m
0.30	4.1	6.0
0.40	4.3	6.3
0.50	4.5	6.5

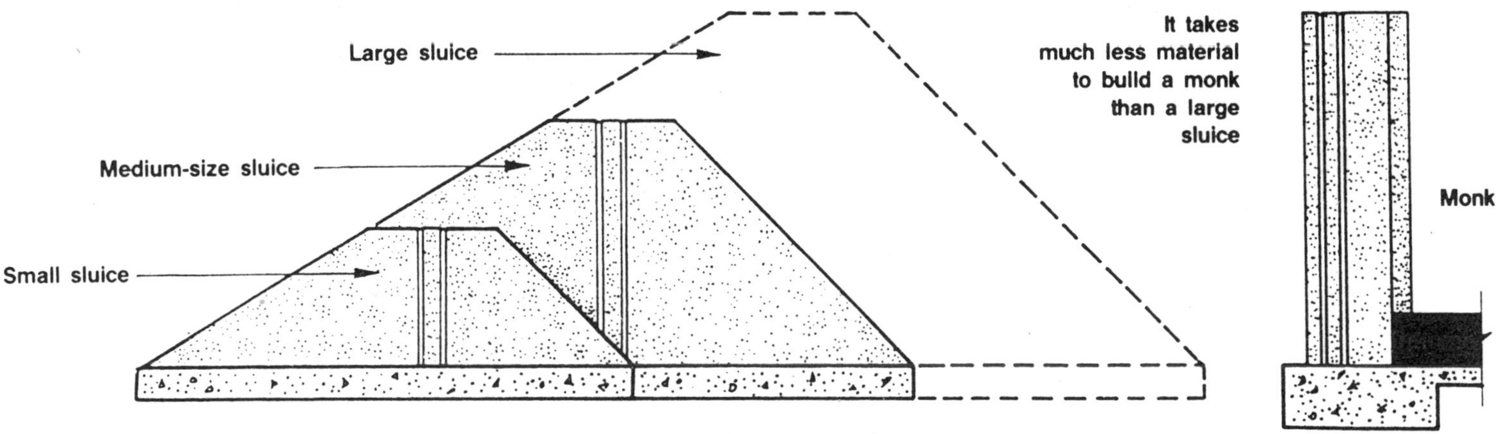

5. A sluice gate essentially consists of:

- **a horizontal floor**;
- **two vertical walls** parallel to each other;
- **grooves** to fix wooden boards and screen; and
- **side wings** that can be added to reinforce the construction and to reduce the risk of water leaks along the side walls.

6. It is best **to limit the inside width of the sluice gate to 0.80 m** at the most. More than that and it becomes difficult to handle the wooden boards, particularly when the pond is full.

7. **The number of grooves** can vary according to the size of the pond and the method used for harvesting the fish.

8. **If the pond is small**, or if the harvesting is done in a catch basin built behind the sluice gate, **one set of grooves** built into the centre of the sluice gate may be sufficient.

9. **If the pond is larger** and the harvesting is done in the sluice gate itself, **two sets of grooves** will make harvesting easier. Build them as follows:

- **a set of two pairs of grooves in the front of the central part** for one set of boards and one screen, for pond operation;
- **a set of two pairs of grooves in the rear part of the gate** for one set of boards and/or screens, for drainage of the pond and harvesting.

10. It is useful to build **a simple bridge over the top of the sluice gate** to allow the passage of people and light carts or barrows. It can easily be made by assembling wooden planks together, or by making a simple reinforced slab top, if you are using concrete.

11. To find out **how much water** you should be able to discharge through typical sluice gates, use **Graph 6 and Tables 32 and 33** in Chapter 7.

12. **Sluice gates can be built of various materials** such as wood, bricks, cement blocks or reinforced concrete. You will now learn how to use each of these materials to build a simple sluice gate.

Building a wooden sluice gate

13. Use strong, **resistant wood.** Pretreat it with anti-rot compounds before use (see Section 31, **Construction, 20/1**).

14. Determine the size of the sluice and the pieces of wood required according to the size of your dike. If possible, build the sluice before constructing the dike or leave enough space for the sluice, plus its anti-seep boards, to be installed in one piece. In this case you may be able **to assemble and prepare the sluice in good working conditions away from the pond**, and then fix it firmly into place on site.

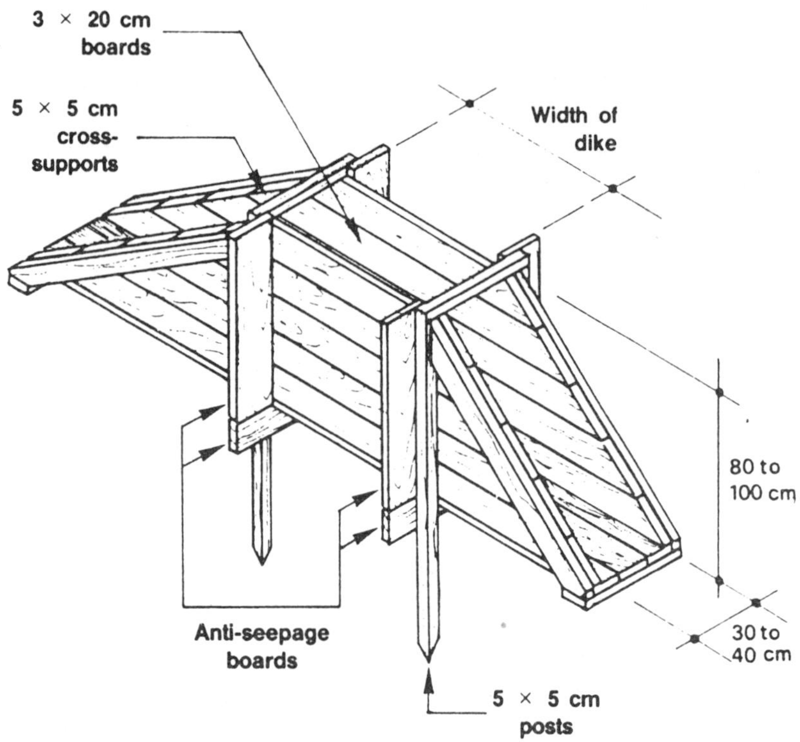

15. If the sluice is too large you will need **to build it on site**.

16. First excavate and clear out the required area. Pack the base area firmly with good pond soil. If pond soils are very soft, either increase the length of the vertical posts or use more of them. Alternatively you can use piling (see Section 105).

Excavate sluice trench, and clean and pack the base

17. Mark out the positions for the main vertical posts. Make up the vertical posts into square frames by bolting, screwing or nailing (one point per corner) the horizontal pieces.

18. Fix the frames into position, and drive them well in. Line the frames up, making sure the horizontal pieces are level, and that they line up along the length of the sluice.

19. Fix the corners of the frame with additional fixings and brackets or angle pieces if needed. Attach the internal planks of the sluice, the external anti-seep boards and the guide rails for the sluice boards.

20. Once the sluice is firmly and squarely assembled and in place, finish any anti-rot treatment. If possible, paint over nail-, screw- or bolt heads with varnish, paint or tar. If necessary, fill any gaps between boards, knot-holes, etc., with clay, tar, putty or mastic sealant. Fill in and pack the dike material around the structure.

Assemble the posts, cross-supports and anti-seepage boards, and drive the frames into the sluice trench

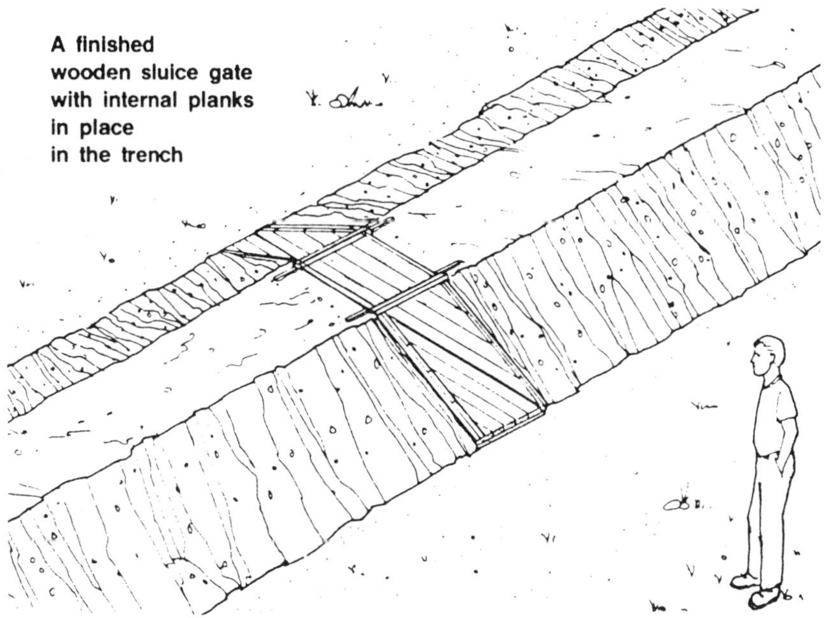

A finished wooden sluice gate with internal planks in place in the trench

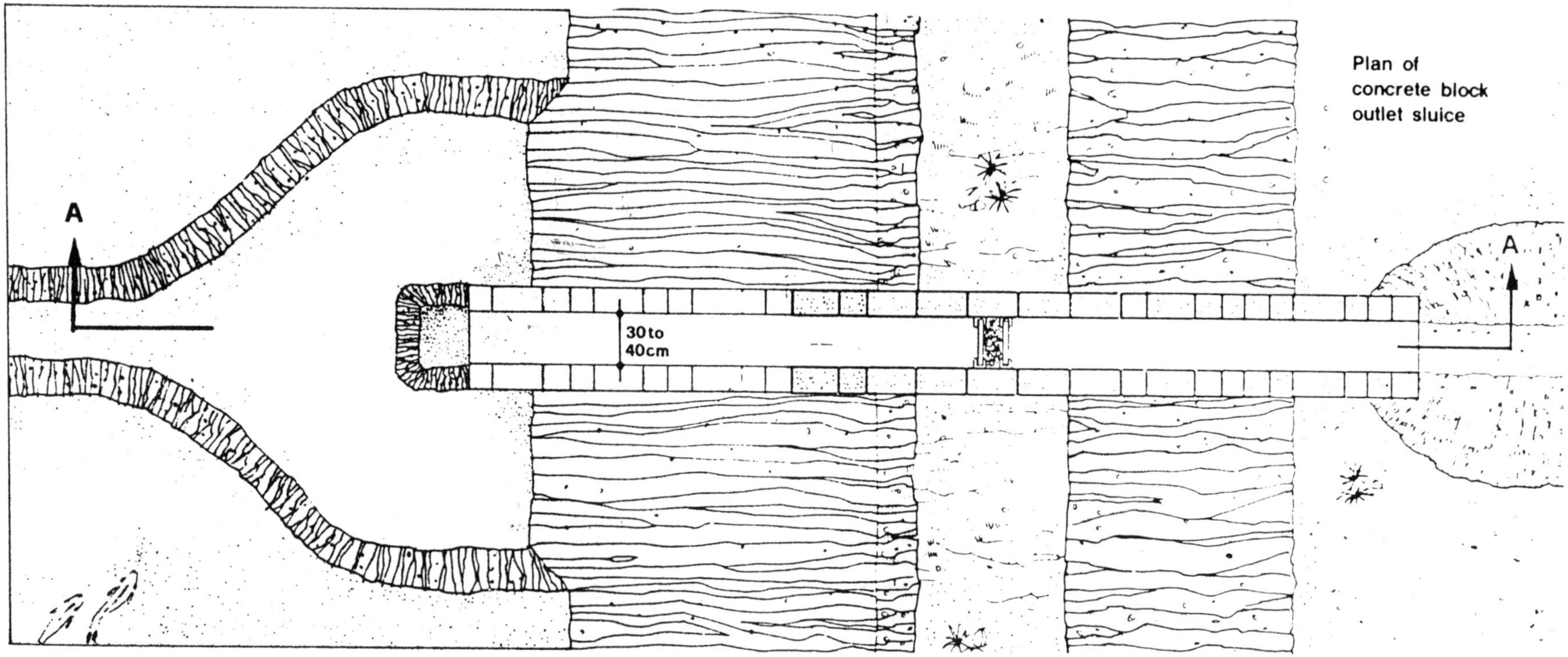
Plan of
concrete block
outlet sluice
A
A
30 to
40cm

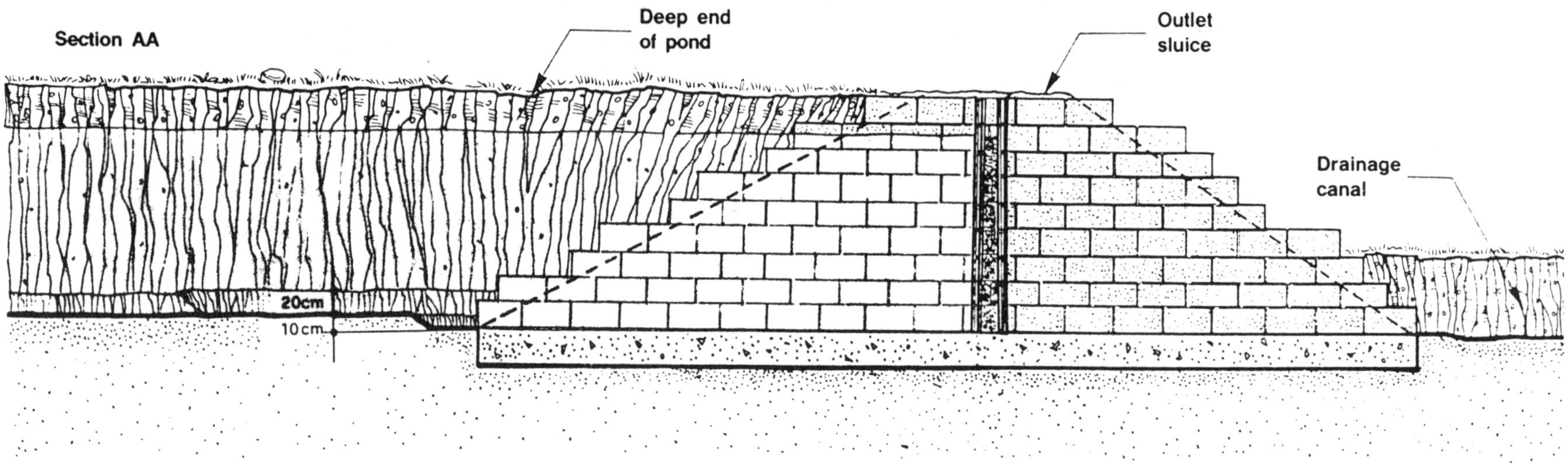
Section AA
Deep end
of pond
Outlet
sluice
Drainage
canal
20cm
10cm

Building a sluice gate with bricks or concrete blocks

21. If you are **using bricks or blocks**, the structure is considerably heavier and needs **good foundations**, usually of concrete or reinforced concrete. You also need good masonry skills. You should proceed as follows (see also Sections 105 and 107):

(a) First prepare, mark and level the site. Dig out the foundation area to the required level. If necessary, fix and level foundation piles and lay and level the bedding material.
(b) Mark out and position simple wooden forms for the concrete base and fix them well. Prepare the surfaces, for example using old motor oil. Fix any reinforcement required. If possible, place some wooden blocks 2 to 5 cm below the upper base level, in the position of the walls. Pour the concrete. Use a lean to medium mix containing 195 to 250 kg cement/m^3 (see Section 34, **Construction, 20/1**).
(c) Once the concrete is well set and cured, prepare and position guide markers for the sluice walls. If you have used wooden blocks, remove these to obtain a "key" for the walls. Otherwise, make a key for the walls using a hammer and chisel to cut the foundation. Build the walls, taking particular care to ensure that inner surfaces are smoothly and cleanly finished off. If you wish, fix in attachment bolts for sluice board guides. For the best finish, it is preferable to plaster the inner walls with a medium mortar.
(d) Once the walls are completed, fix the sluice board guides, using pre-placed attachment bolts or masonry bolts and mortar. Make sure they are parallel.
(e) Carefully place and pack the dike material around the sluice structure.

Building a sluice gate with reinforced concrete

22. You can make a stronger structure, of similar weight, by using **reinforced concrete**. You should proceed as follows (see also Section 105):

(a) Prepare the foundation area as for the brick sluice, fix the base reinforcement, and tie in any wall reinforcement that has to connect into the foundation. Take care that all reinforcement is in position. You may need to make temporary holding forms to support the vertical bars.
(b) Pour the foundation concrete and allow it to set and cure.
(c) Mark out positions, place and fix the forms or shutters for the sluice walls (note that if you are making several sluices it may be worth using steel forms that can be re-used repeatedly). Fix in any additional wall reinforcement needed, and if required, fix in attachment bolts for sluice board guides. Make sure the shutters are well secured at the base.
(d) Pour the wall concrete, tamp it well down and allow it to set and cure. Use medium-rich concrete (250-350 kg cement/m^3). Remove all the shuttering, and clean, pick out and finish off all surfaces.
(e) If sluice board guides are to be added, fix these, using pre-placed attachment bolts or masonry bolts and mortar.

Operating the sluice gate

23. **Before filling the pond**, insert two rows of wooden boards into the set of grooves until you reach a little lower than the maximum water level. Fill the space between these rows with earth or sawdust, compacting well. Place the screen on top of the front row of boards. Fill the pond with water and check that there are no leaks at the sluice gate. If necessary, repeat the compaction process.

24. **To drain the pond**, remove one pair of boards and the compacted material at a time, keeping the screen on top of the front row of boards while the water flows out. When there are only two to three pairs of boards left at the bottom of the sluice gate, remove all the compacting material and the second row of boards. Then finish in one of the following two ways, according to the harvesting method:

(a) Lower the water further and harvest the fish **inside the sluice gate**. If there are two sets of grooves, harvest the fish in front of the screen placed in one of the rear grooves. If you need to remove the screen to clean it, first place another screen in the other rear groove.

(b) Alternately let part of the water, together with fish, flow into **the catch basin** where the latter can be easily harvested; repeat this procedure until all the water and fish have been drained from the pond.

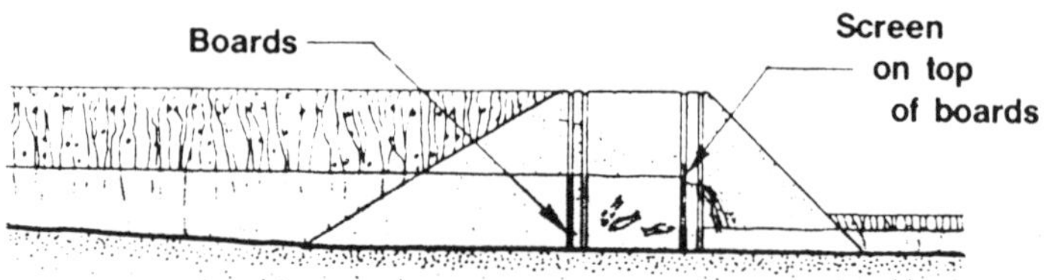

Harvesting inside sluice

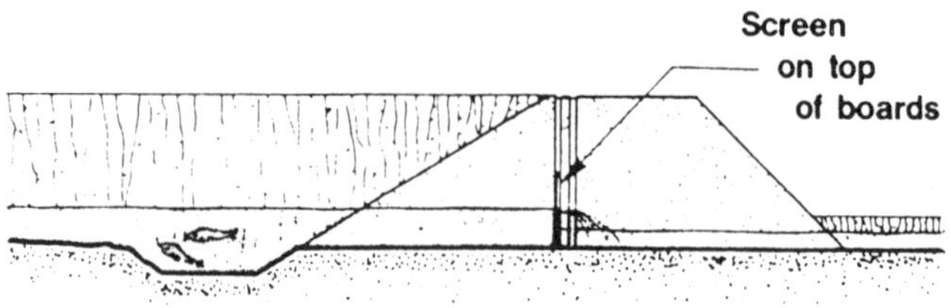

Harvesting in front of sluice

105 The monk outlet

1. The monk is one of the oldest and most common pond draining structures. It consists of **a vertical tower closed with wooden boards** to regulate the water level. The water is discharged through a pipeline buried under the dike. **A screen** keeps the farmed fish from leaving the pond.

2. A monk has **advantages** similar to those of the sluice gate. The pond water level is easily controlled and adjusted. It can function as an overflow. It simplifies the fish harvest. In addition, a monk is more easily protected than a sluice gate, and it is more economical to build if the pond dike is large. However, it has **the disadvantage of not being very simple to construct**, particularly if it is built with bricks or concrete.

3. **The complete monk outlet** consists of:

- **a vertical three-sided tower** (called the monk), usually as high as the outlet dike (see Section 61, **Construction, 20/1**);
- **a pipeline** running through the dike, which is sealed to the back of the tower at its base;
- **a foundation** for the tower and the pipeline; and
- **grooves** to fix the wooden boards and screens which form the fourth side of the monk.

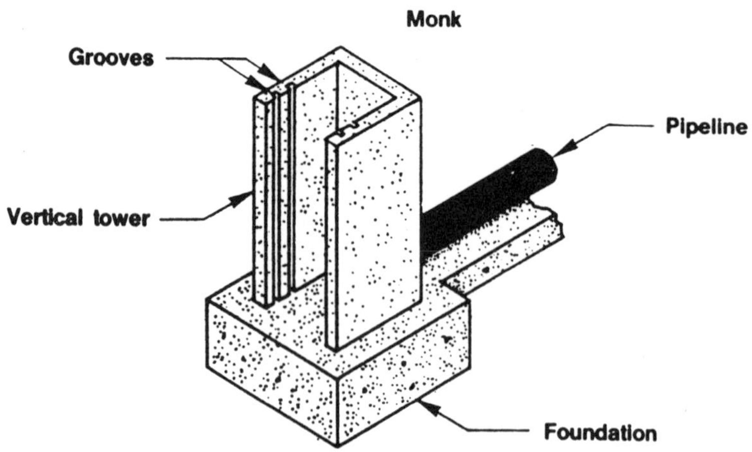

Locating the monk

4. Similar to any other outlet, the monk is generally built **on the side of the pond opposite the water inlet**. It may be placed either in the middle of the dike or, when the water drains, for example, in a catch basin common to two adjacent ponds, in a corner of the dike.

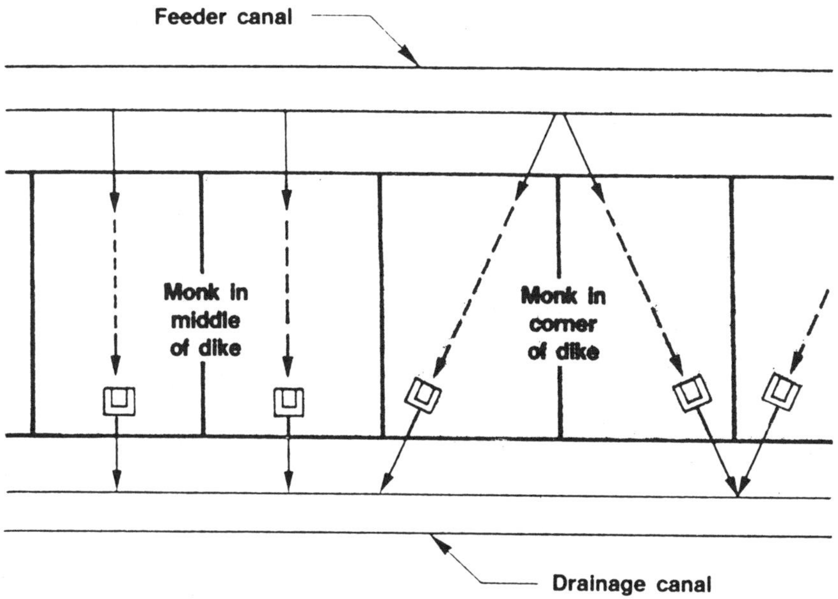

5. The monk can be built either **into the dike or freestanding** some distance into the pond:

- **if the monk is built into the dike**, water infiltration through the dike will be more common and access to the outlet will be easier for poachers. To prevent soil from entering the monk, you will have to build an additional protective wing on both sides, but servicing the monk will be easier;
- **if the monk is built on the pond bottom** in front of the inside toe of the dike, you will need a longer pipeline, but access to the monk will be through a removable catwalk and tampering with it will be much more difficult.

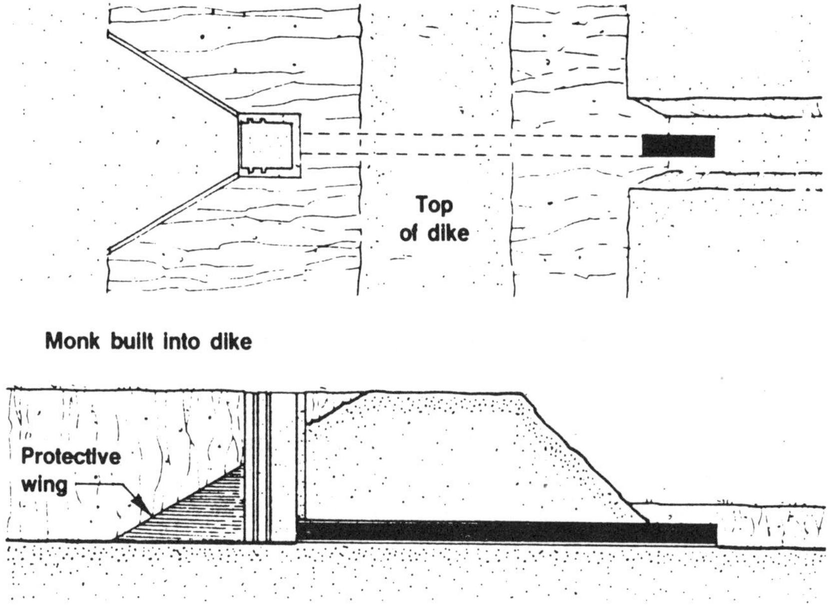

Monk built into dike

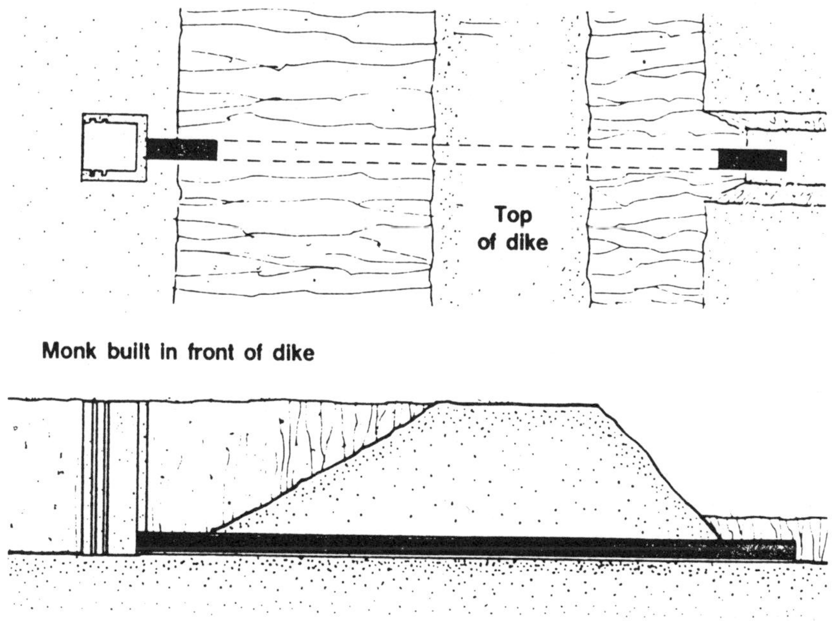

Monk built in front of dike

How much water should a monk discharge

6. The water discharge capacity of a monk depends on **the inside diameter of the pipeline** (see Section 38, **Construction, 20/1**, and Section 100 in this book).

7. To estimate quickly how much water you should be able to discharge from a typical monk, use the pipe flow **Tables 12, 13 and 14** from Section 38, **Construction, 20/1**.

8. **The cross-section of the monk increases as the diameter of the pipeline increases**. Remember that:

- **the internal width** of the tower should be equal to the pipe's diameter plus 5 to 10 cm on each side;
- **the space in front of the first groove** should be about 8 to 10 cm;
- **the gap** between the two rows of boards should be at least 8 to 10 cm;
- the distance from **the last row of boards to the back wall** of the tower should increase as the water discharge capacity increases, up to a maximum value of 35 to 40 cm.

9. To be able to move the boards easily, **try to limit the internal width of a monk to 50 cm at the most**.

Example

Inside dimensions of monks according to pipeline size (in cm)

	Pipeline inside diameter			
	10-15	15-20	20-25	25-30
Internal width	**30**	**33-35**	**40**	**48-50**
In front of groove 1	8	10	10	10
Gap between grooves 1 and 2	8	10	10	10
Distance groove 2 to wall	16	16-20	26	34-37
Width for two grooves	8	8	8	8
Internal length	**40**	**44-48**	**54**	**62-65**

Calculating dimensions for a monk

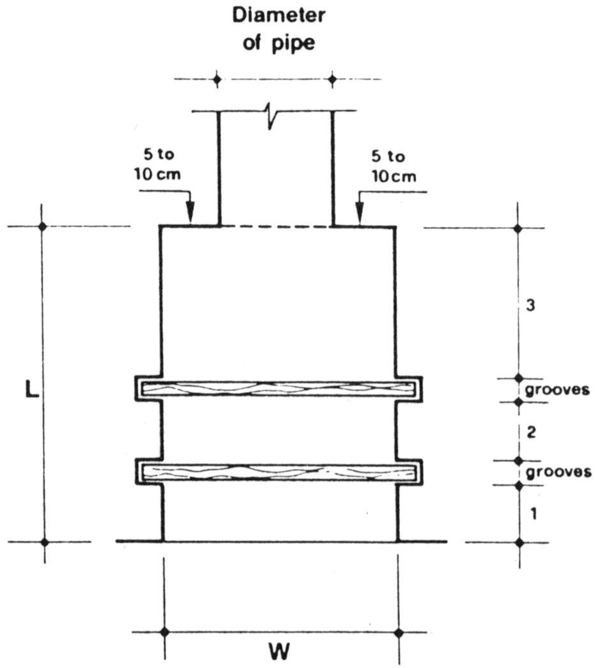

WIDTH (W) = Diameter of pipe + 2 × (5 to 10 cm)

LENGTH (L) = (1) + (2) + (3) + (grooves)

Where (1) = 8 to 10 cm
(2) = 8 to 10 cm
(3) = maximum 35 to 40 cm
(grooves) = 4 cm each

10. **The height of the monk** is related to the maximum water depth in the pond. The monk should be at least 20 cm higher than this depth. Usually, the monk has **the same height as the outlet dike**. Unless specially designed, **the height should not exceed 2.5 m**.

Materials to build a monk

11. Monks can be built in wood, bricks or concrete depending mainly on the availability of materials, their cost, the local technical expertise and the size of the structure.

12. The most difficult type of monk to build is **the brick monk**. It requires a very skilled mason to make it so that it is leak-proof. If not done properly, the mortar surfacing will have to be redone frequently, increasing maintenance costs. Generally, wooden and concrete monks are cheaper and easier to build. You will learn how to build these in the next sections.

Note: the following are some things to remember when **when you build a monk.**

(a) **The pipeline** should be laid down before building the dike and the monk tower.
(b) **Build a solid foundation** to avoid future problems.
(c) **Pay particular attention to**:

- the junction of the monk tower to its foundation;
- the junction of the pipeline to the back of the monk tower;
- the finishing of the monk's grooves.

(d) Give **a reasonable slope** to the pipeline, preferably 1.5 to 2 percent.
(e) If you have **to build several monks** on your fish farm:

- try to standardize their type and size as much as possible;
- for concrete monks, prepare strong forms and re-use them if possible (see Sections 107 and 108).

(f) **Provide a separate overflow** wherever there is danger of uncontrolled entry of flood water into the pond (see Section 111).

106 Wooden monk outlets

1. A simple monk outlet can be built entirely of wood. It is the easiest and cheapest type of monk to construct, although you need to be careful to ensure its watertightness and its durability. **The height of a wooden monk should be limited to 2 m**.

Choosing the wood

2. To build a wooden monk, select **a heavy, durable wood**, which is resistant to water such as iroko or mukulungu (see **Table 6, Construction, 20/1**). To improve its durability, you can treat it with a wood preservative or use discarded motor oil. Remember to wash away surplus preservative before putting your fish in the pond.

3. **Use wooden boards without knots**, 3 to 5 cm thick. For example, for a 2-m-high monk tower, you will require about 0.4 m^3 of wood.

Building the wooden pipeline

4. Instead of using standard plastic or cement pipes, you can build a pipeline entirely of wood. **Simply nail or screw four boards together** in the shape of a box. Fix the structure well over compacted soil, and bury it underneath the outlet dike. Pay particular attention to the compaction of the dike soil around the pipeline (see Section 62, **Construction, 20/1**).

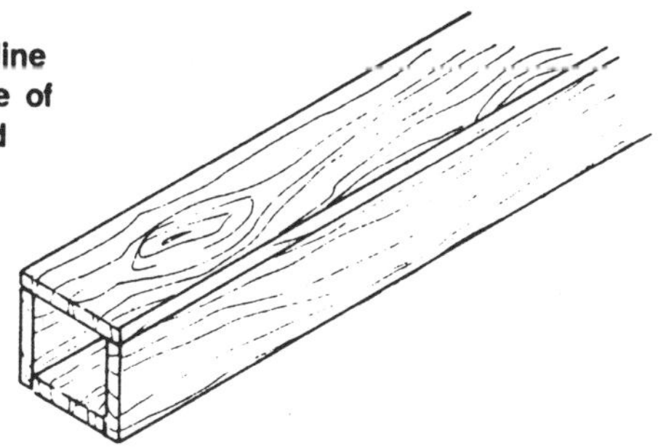

5. In most cases, a foundation is not needed, but in less stable soils such as certain heavy clays, it may be useful to use simple wooden stake piling.

6. Wooden monks require **hardly any foundation**, aş they are very light. It is normally quite sufficient to use light foundation such as paving slabs or simple wood piling, or larger boards placed flat on the pond floor.

7. Both the small- and the medium-size monks are nailed or screwed together so that the side toward the pond is open. Depending on the overall dimensions and on the width of the boards available, assemble the tower as shown in the drawings.

8. It is best to screw **an anchoring post** on each side of the tower First drive these two posts well into the bottom of the pond and then screw them on to the monk.

9. For a stronger structure, you can add **an oblique brace** to each side, supporting the top part of the tower against the pipeline.

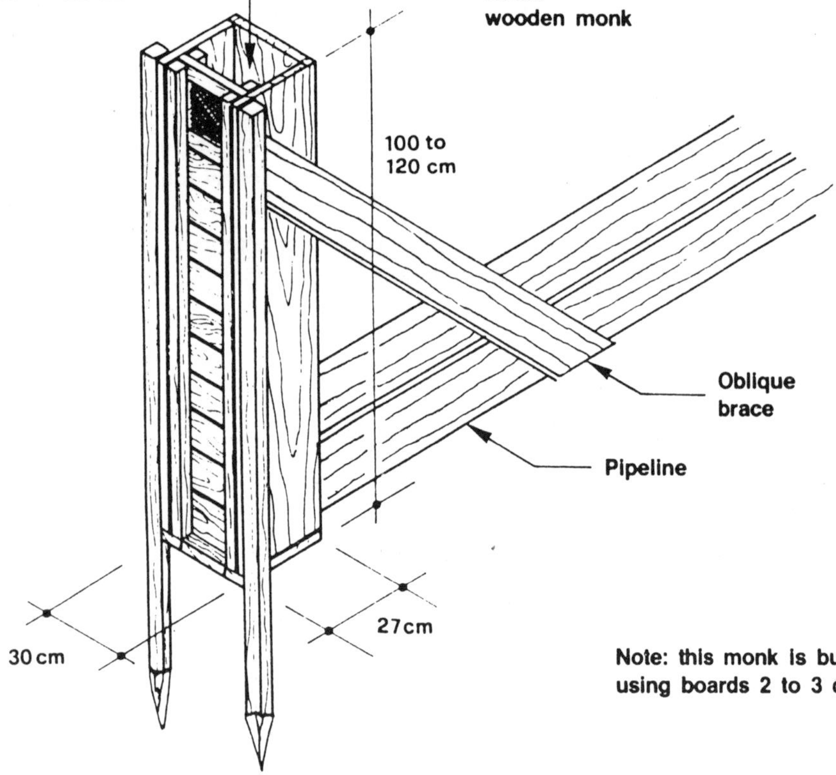

Note: this monk is built using boards 2 to 3 cm thick

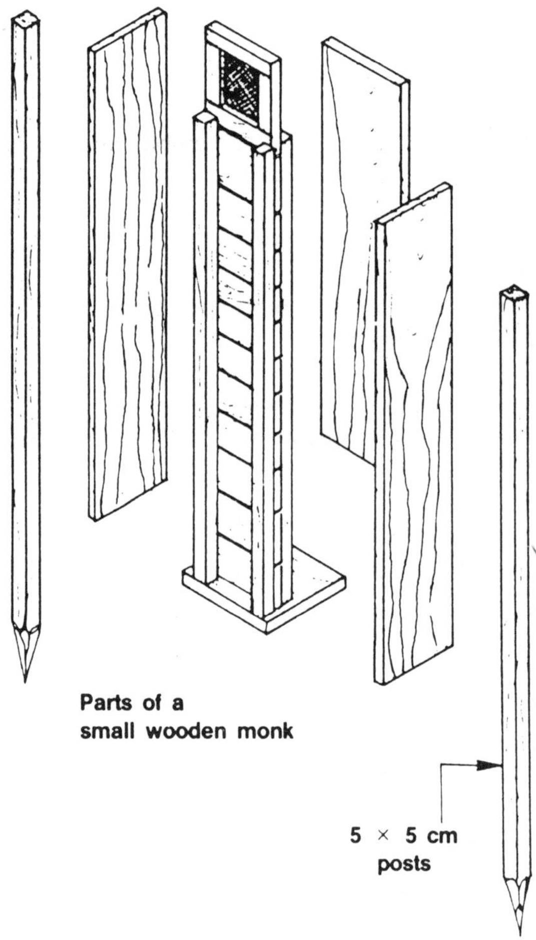

Parts of a small wooden monk

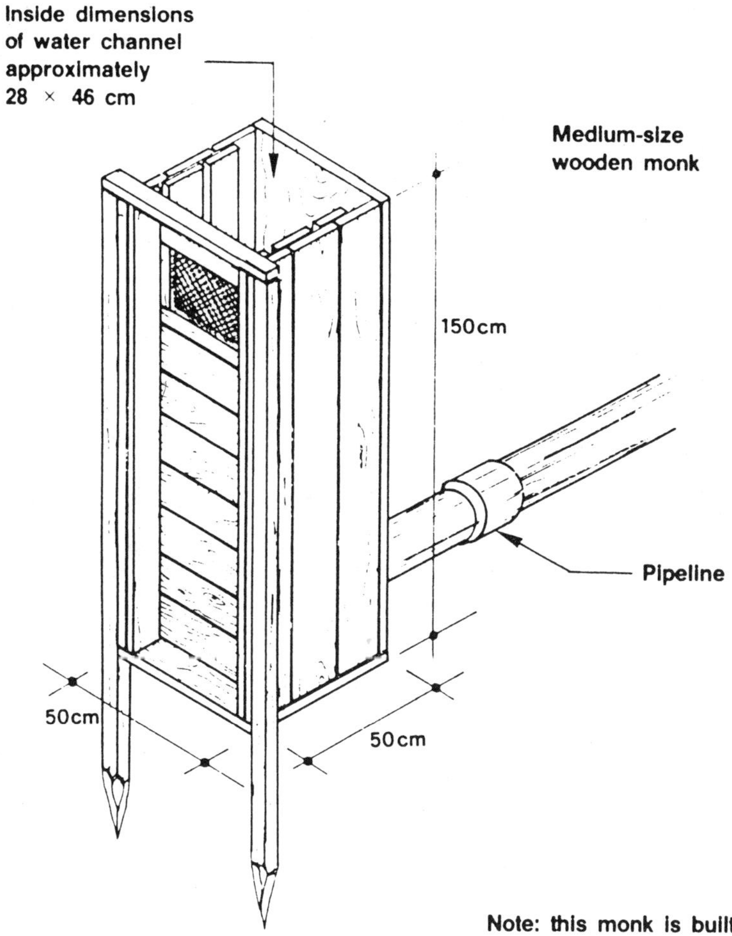

Medium-size
wooden monk

Note: this monk is built
using boards 2 to 3 cm thick

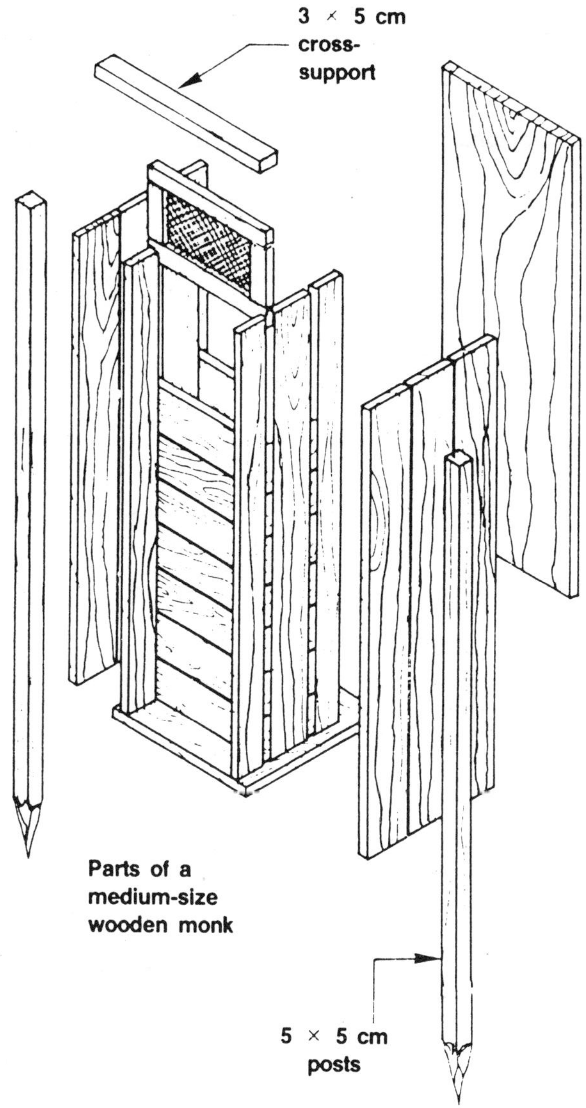

Parts of a
medium-size
wooden monk

107 Small brick, concrete block and concrete monks

1. **Monks of up to 1.5 m in height**, fixed to pipelines up to 25 to 30 cm in diameter, can be built using **single-thickness brick and mortar**. Although taller and wider monks can be built, they require a double-width base and good bracing for stability and strength, and so become too heavy and expensive for most purposes.

2. **Suggested dimensions** for such monks are given in the first part of **Table 48**.

3. Small monks can also be built with **concrete blocks and with reinforced concrete**. In general, the principles of construction are similar (see paragraph 4 onwards), with the following exceptions:

- **brick and block monks** should be well finished internally, using a plaster coat;
- as mentioned earlier, a **skilled mason** should be used; the quality of workmanship should be very high to ensure a durable structure.

TABLE 48

Suggested dimensions for brick, block and concrete monks

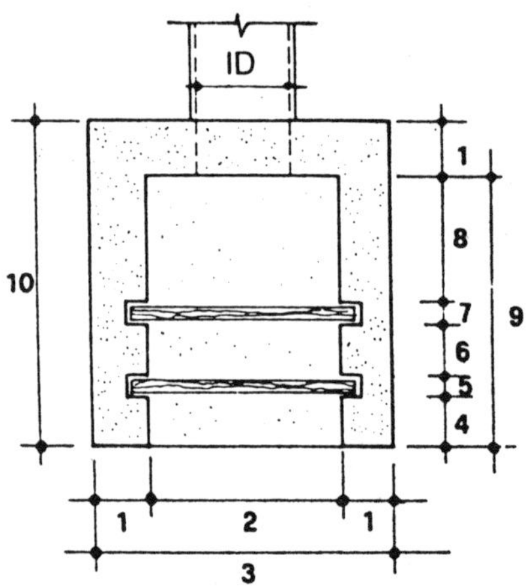

Type of construction	*No concrete reinforcement*			*With concrete reinforcement*	
Height	1 m	1.50 m	1.50 m	1.50-1.80 m	1.80-2.50 m
Pipeline ID (cm)	10-15	15-20	20-25	25-30	30
1. Wall thickness (cm)	7	12	12	12	15
2. Internal width (cm)	30	33	40	48	50
3. Total width (cm)	44	57	64	72	80
4. In front of first groove (cm)	8	10	10	10	10
5. First groove width (cm)	4	4	4	4	4
6. Gap between grooves (cm)	8	10	10	10	10
7. Second groove width (cm)	4	4	4	4	4
8. From second groove to wall (cm)	16	16	26	34	37
9. Internal length (cm)	40	44	54	62	65
10. Total length (cm)	47	56	66	74	80
Estimated volume of concrete (m^3)	0.087	0.261	0.310	0.353	0.630

4. For these monks, **pipelines are usually made of commercially available pipes**, made of either plastic or asbestos cement. For best results, particularly with cement pipes, the pipeline should be supported by **a good foundation, which should be built together with the monk tower foundation**. The joints of the pipes should be well sealed to prevent water seepage.

Concrete pipes

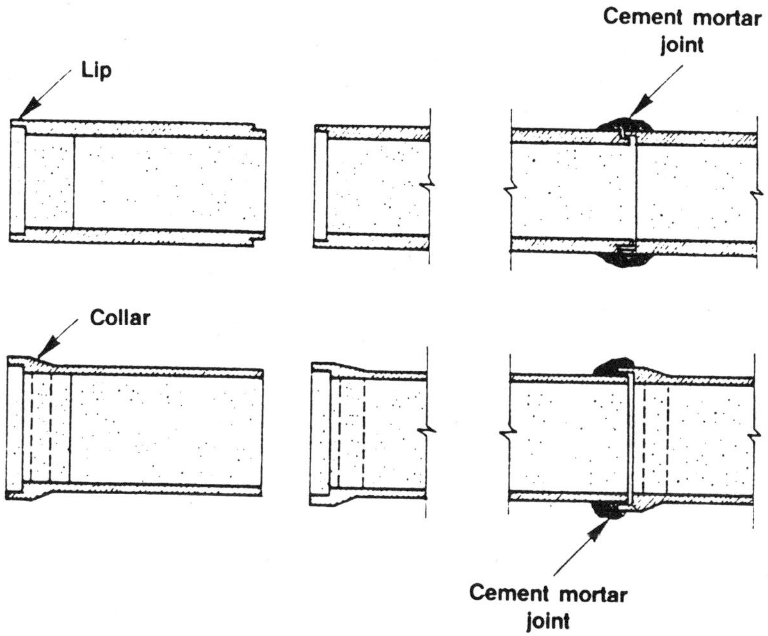

5. To prepare **the foundation for a monk pipeline**, you can proceed as follows:

(a) Excavate and compact well the area along which you plan to build the pipeline. **The level of this area** should be at least 10 cm lower than the lowest point in the pond.

(b) Stake out with a marker line **the centre line of the pipeline**, usually perpendicular to the centre line of the outlet dike (see Section 36, **Topography, 16/1**). The line should extend to 50 cm beyond the pipeline length at each end.

(c) On each side of this centre line measure a distance equal to half the outside diameter of the pipeline plus 10 to 15 cm. Stake it with marker lines, starting at the back wall of the monk tower. These are the **foundation lines**.

(d) Remove the central stakes and line, and between the foundation strings, **dig a trench** whose depth should be:

- in hard, undisturbed soil, 10 to 15 cm depending on the pipe size;
- in soft soil, 20 to 25 cm depending on the pipe size.

(e) Prepare the bottom of the trench carefully, **giving it a slope of 1.5 to 2 percent** toward the outside of the pond (see Section 40, **Topography, 16/1**).

(f) Drive in stakes down the centre of the trench and **adjust their height** to the designed thickness of the foundation and the 1.5 to 2 percent slope.

(g) Prepare a lean concrete (175 kg cement/m^3).

(h) Place the concrete in the prepared trench. **Adjust its surface level** to the level of the top of the stakes. Tamp it well, protect it, and let it cure for two days. (For information on cement concrete, see Section 34, **Construction, 20/1**).

6. When the foundation is ready, lay the pipeline as follows, according to the type of pipe you are using:

(a) **If you are using cement or ceramic pipes**:

- lay the head pipe, **its female end starting about 1 m** from the rear wall of the monk tower, and position it carefully with stones or concrete/mortar;
- lay down the next pipes one by one, connecting the pipes close together, until you reach about 0.50 m beyond the outside end of the foundation, and position these pipes in the same way;
- check the alignment of the pipes and make sure they are well fitted together — it may be useful to fix holding boards and wedges at each end of the pipeline to ensure a good fit;
- fix the pipeline well into its final position with mortar;
- make the joints of the pipes, using a slightly liquid mortar (see Section 33, **Construction, 20/1**).

(b) **If you are using plastic pipes**, there are two main types of connection:

- **push-fit types**, which use simple rubber seals at the joints. These are usually used for low-pressure drainage and can be dismantled later. They are quick and easy to assemble, but because of the possible risk of leakage may be less reliable inside pond dikes;
- **glued (solvent-welded) types**, which are commonly the thicker-walled pressure pipes, and once assembled cannot be dismantled. They require more care and are usually more expensive, but will provide a reliable and long-lasting job.

7. For both types, the overall laying procedure is similar to that for cement pipes, although there is less need for joint anchoring, because the pipe joints are typically 3, 6 or even 9 m apart.

8. In the case of **solvent-welded pipes**, the joints are as strong as the pipe itself, and so the pipe needs less protection from movement. Also, as plastic pipes are flexible and smooth inside, they can deform slightly but still drain well. Therefore it is not necessary to use stiff foundations. It is possible in firm soils to eliminate them altogether, but in soft soils, 5 to 15 cm foundation should be sufficient.

9. **For push-fit pipes**, pay particular attention to the following:

(a) Make sure **joint ends are absolutely clean and free from rough edges** (e.g. that come from sawing the pipe), and that **the rubber seal ring** is not twisted, crushed or broken, and is properly seated.
(b) **Use a suitable lubricant** on the pipe. Silicon grease is best, but even soapy water will do. **Push the pipe in** until it reaches the end of the socket. Do not use too much force, as you may split the socket.

10. **For solvent-welded pipes**, proceed as follows:

(a) Clean the pipe, preferably with the recommended cleaning solution, and apply the cement as instructed. Make sure it is spread all around the pipe.
(b) Push the pipe into the socket. Do not twist it as this may cause "channelling" in the joint, leading to leaks.
(c) Ideally a thin ring of solvent should appear all around the end of the socket; if this is the case, you should have a good weld.

Note: solvent cements are commonly of two types:

- **PVC cement** is for PVC pipes only;
- **ABS solvent** can be used both for PVC and ABS pipes.

11. Check the pipes and the solvent tin carefully to make sure they match.

Using a concrete culvert

12. **A concrete culvert** can also be considered. You can easily build it yourself, on top of the prepared foundation, using wooden forms for the **two lateral walls** and **precast slabs** for the top cover. Build the culvert up to the back wall of the monk tower. Smooth out its bottom surface well with cement mortar.

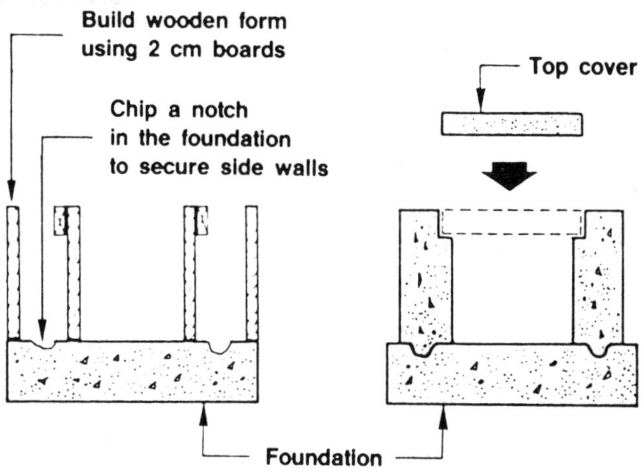

Building the monk tower foundation

13. As discussed before, the monk tower foundation should be built at the same time as the pipeline foundation. However it is generally thicker than the latter. Proceed as follows:

(a) Starting from the pipeline axis, stake out the tower foundation and mark with lines. The tower foundation usually has a square shape. **Its dimensions should be larger than those of the tower**:

- for hard, undisturbed soil, by at least 20 cm on all sides;
- for soft soil, by at least 30 cm on all sides.

(b) Between the marker lines, dig a hole whose depth will vary according to the quality of the soil:

- in hard, undisturbed soil, 30 cm;
- in soft soil, 60 cm;
- if the soil is particularly soft, use a piled foundation (see next page).

(c) Remove all stakes and lines.
(d) Level the bottom of the hole properly.
(e) **Fill this hole with foundation materials**:

- in the bottom half, use rocks and gravel, filling the gaps with a slightly liquid mortar; or use a soft lean concrete (175 kg cement/m^3) (see Sections 33 and 34, **Construction, 20/1**);
- in the top half, use ordinary concrete (250 kg cement/m^3);
- if possible, place in the foundation some pieces of wood of approximately the same width as the monk tower walls, in the position where the walls are to be built. This will help to anchor the walls firmly in place. The pieces of wood should be set about 5 cm below the surface of the foundation.

(f) Tamp these materials well and **adjust the surface level**, as required by the design, so that it is at least 10 cm below the lowest point in the pond.
(g) Protect the surface of the concrete, keep it moist, and let it cure for at least two days.
(h) Remove the wood blocks, leaving the foundation ready for the walls.

Remember: the solidity of a monk tower depends mainly on the stability and strength of its foundation. You should prepare a good foundation for your monk.

Preparing a piled foundation

14. In very soft soils, for example those with significant plastic clay content, the use of **simple piling** will improve the strength of the foundation. In most cases, wood or bamboo can be used, although it is preferable to use durable woods (see Section 31, **Construction, 20/1**). They are typically driven several metres into the ground using a hammer. It is simplest **to test first**, using 2 to 3 m of pile, 6 to 10 cm in diameter, hammering it in until it will go no further. Cut the remainder of the pile, leaving a short stub above the surface. With this as a guide, you can select a suitable standard length of pile. If any pile fails to drive in completely, cut it to leave a stub, as with the test pile. These simple piles are set at approximately 30 to 50 cm from centre to centre. The normal foundation is then built on top of the piles.

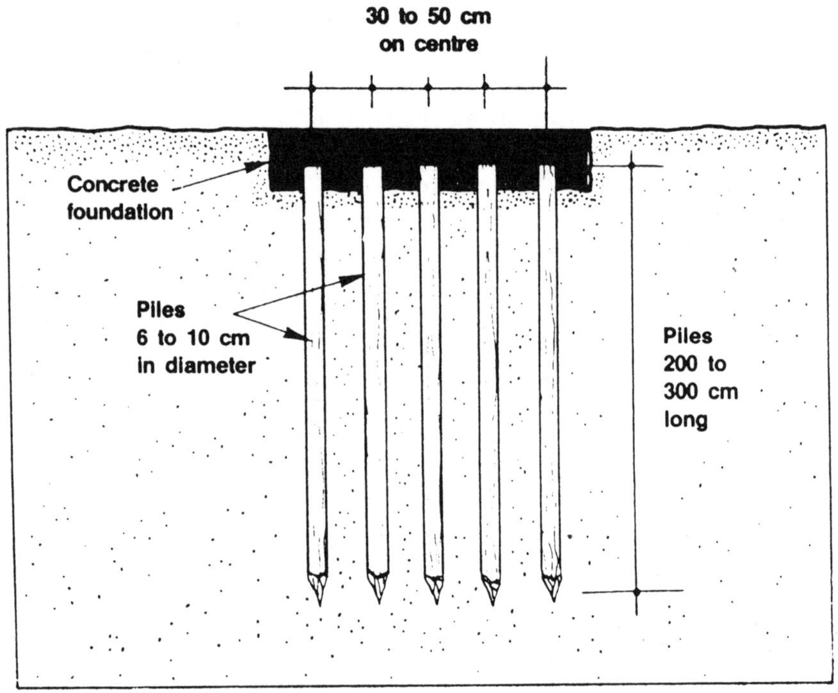

Note: to increase stability, some piles can be driven in at an angle, which has the effect of increasing the base area and tying in the foundation more securely.

Building the concrete monk tower

15. To build the monk tower in concrete, **you need a wooden form** in which to pour the concrete. If you have to build several monk towers of the same dimensions, you can use the wooden form several times, saving time and money. You can also borrow it from or lend it to neighbours and share its cost among a group of people. You will learn more about wooden forms in Section 109. (Refer to Section 34, **Construction, 20/1**, for more details about concrete preparation and placing.)

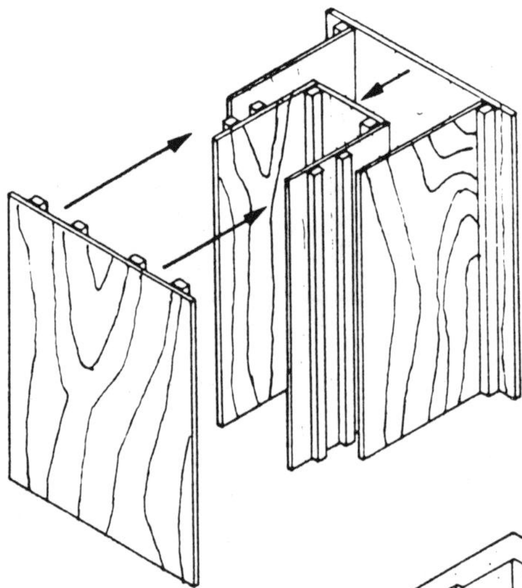

A wooden form for building a small concrete monk

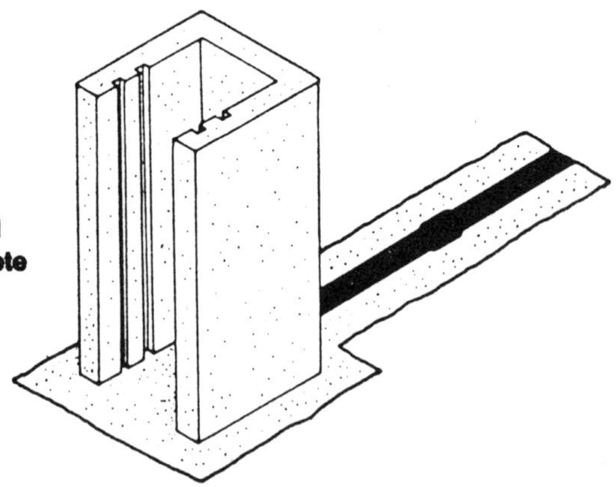

A completed small concrete monk

16. **When you have the wooden form ready on site**, proceed as follows:

(a) With two stakes and a line, mark out **the longitudinal centre line of the monk foundation**. It extends directly from the pipeline centre line.

(b) With nails and a line, mark out on the concrete foundation **the exact position of the three tower walls** to be built, at similar distances from the longitudinal centre line.

(c) If wooden blocks have been used to make the shape of the wall footings, the next stage is quite simple; if not, you will have some more work. Using a hammer and chisel, make deep lines in the concrete surface, about 2 cm from the strings and outside the position of the walls.

(d) **Break the concrete surface between these lines** to a depth of about 5 cm. Clean away the broken concrete. This small gully will be used to reinforce the junction of the tower and its foundation.

(e) Clean the elements of the wooden form well, removing any dry concrete. **Apply used motor oil** to their inside walls so that they can be easily removed from the concrete after it has cured.

(f) Get a piece of straight pipe long enough to run from the pipeline through to the inside wall of the monk. It should have the same diameter as that of the pipeline.

(g) Place it at the back of the tower, on the centre line of the pipeline, making sure that **it overlaps the inside wall slightly**. Block it well in place.

(h) Assemble together the elements of the form; check that the form is well centred, both around the piece of pipe in its back walls and in the gully prepared in the foundation.

(i) **Brace the assembled form strongly** so that it will not move while the concrete is being placed.

(j) Prepare a relatively soft rich concrete (350 kg cement/m^3), (see Section 34, **Construction, 20/1**). **The quantity required to fill the form can be readily estimated** from the dimensions chosen for the tower:

Example

The monk tower is 1.30 m high. The thickness of the walls is 12 cm. Internal width is 33 cm and internal length is 44 cm. **The volume of concrete** required is obtained as:

- volume for the back wall: 0.12 m × 0.57 m × 1.30 m = 0.08892 m^3;
- volume for the two side walls: (0.12 m × 0.44 m × 1.30 m) × 2 = 0.13728 m^3;
- total for the monk tower: 0.08892 m^3 + 0.13728 m^3 = 0.2262 m^3 or **about 0.25 m^3 of concrete**.

Refer to **Table 9** (Section 34, **Construction, 20/1**) for concrete at 350 kg cement/m^3. Mix 90 kg cement with 113 l sand, 200 l gravel and 50 l water.

(k) **Place the concrete into the form** to fill it progressively by layers, tamping each layer well before pouring the next one.

(l) Protect the top of the fresh concrete and **let it cure for at least 24 hours before removing the form**. Be very careful not to break any part of the concrete, particularly next to the vertical grooves.

(m) **Join the pipe piece**, which has been placed in the tower, to the pipeline with mortar.

(n) Using **ordinary mortar**, complete the connection of the tower base with the foundation, both outside and inside. Finish the foundation inside the tower smoothly.

(o) If necessary, **finish the grooves smoothly**. Their quality can be improved by **cementing 4-cm-wide U-irons** into the concrete grooves.

Note: if you choose to use U-irons as grooves, enlarge the grooves made by the concrete forms slightly to about 5 cm.

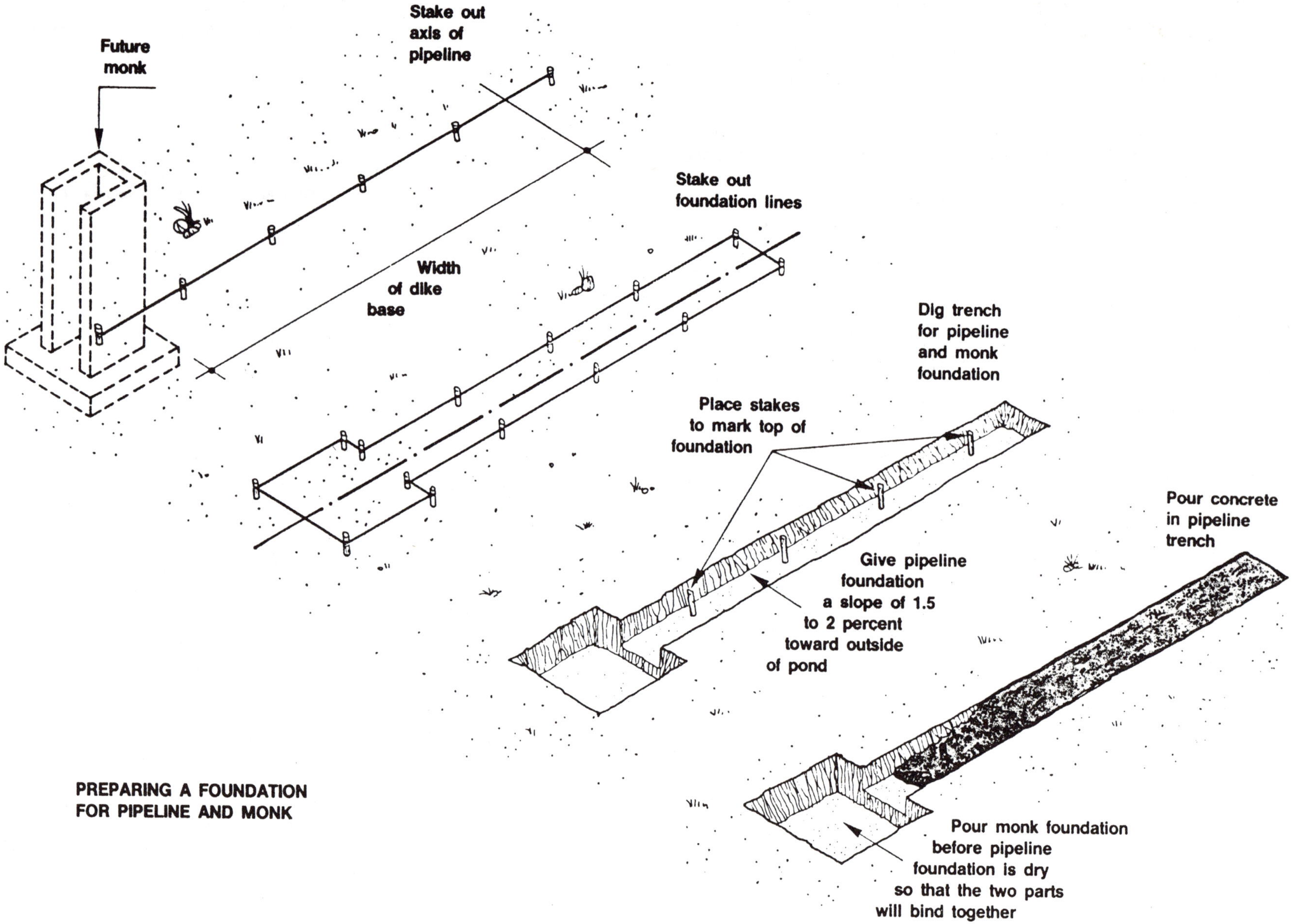

PREPARING A FOUNDATION
FOR PIPELINE AND MONK

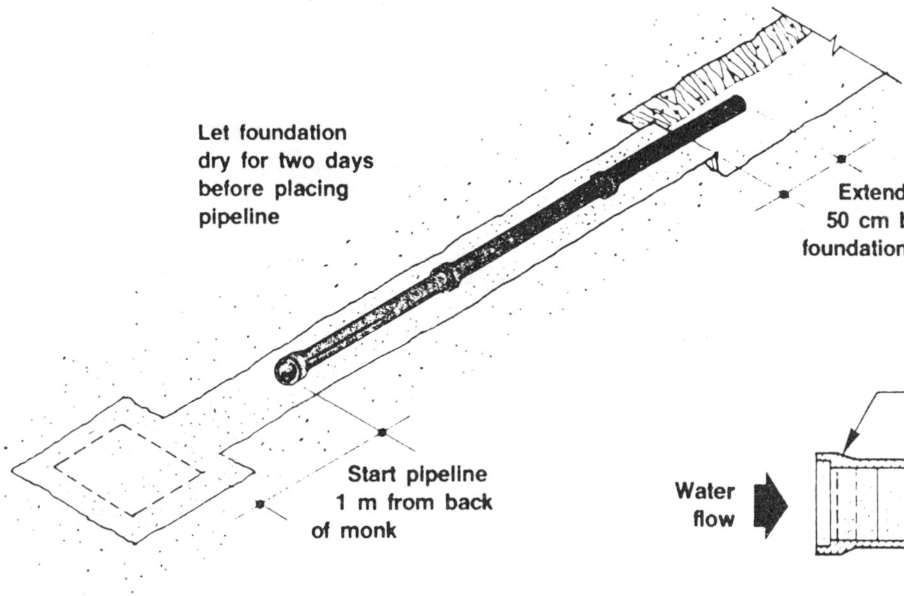

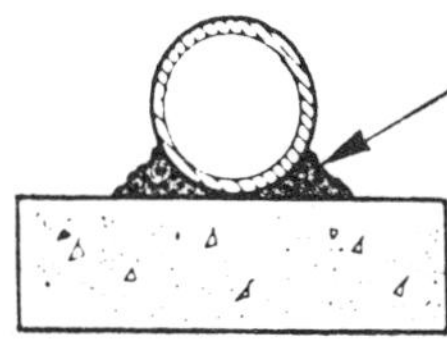

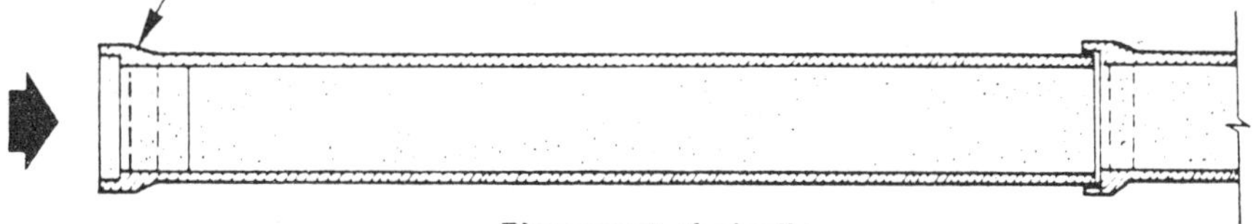

Placement of pipeline

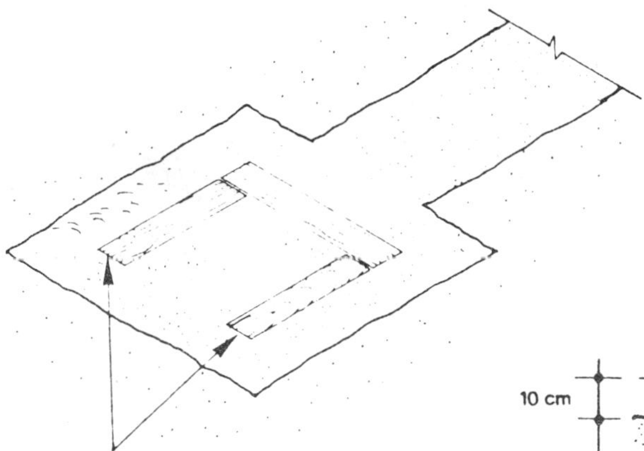

Section of monk foundation

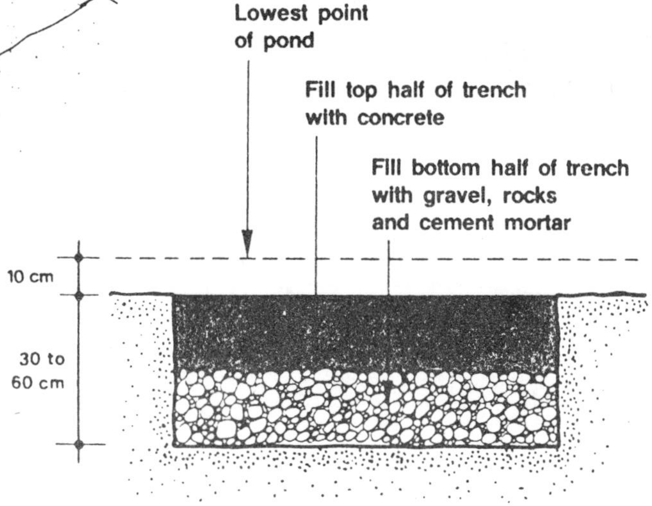

Position of monk on foundation

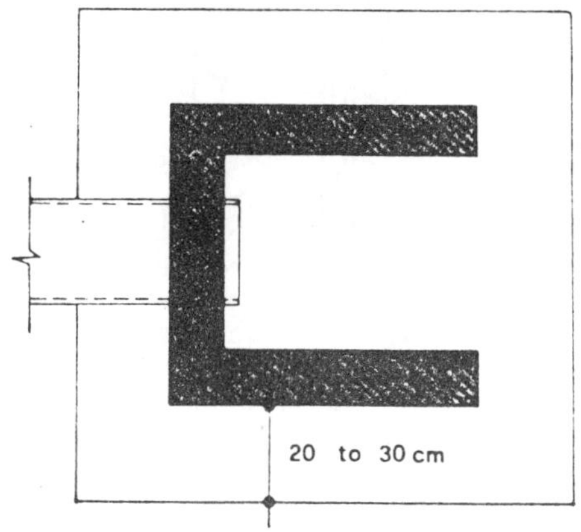

If you did not use wooden strips
to form a base for the monk tower
as shown on page 151,
you will have to
chip out a base
with hammer
and chisel

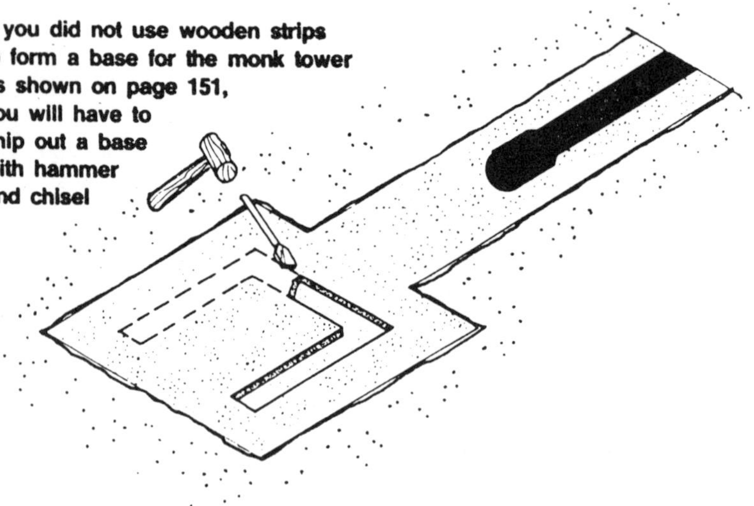

Place a length of pipe
from front end
of pipeline
to inside
of monk

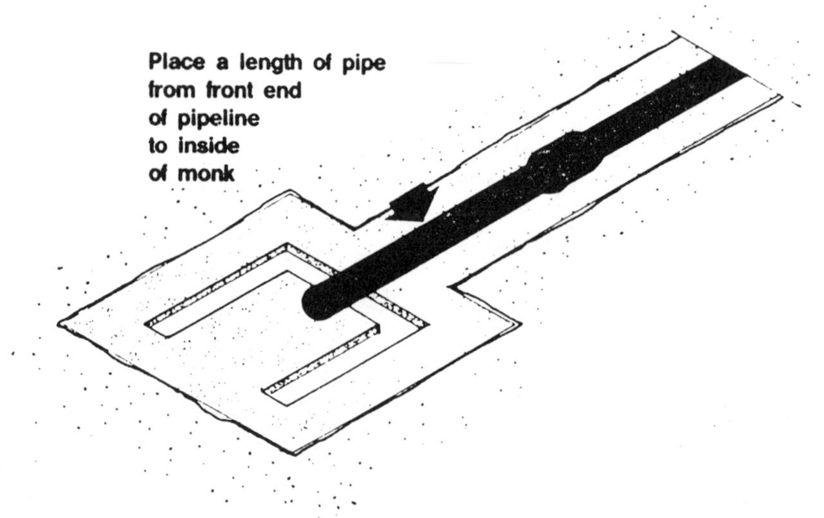

You are now ready
to build or pour
the monk tower

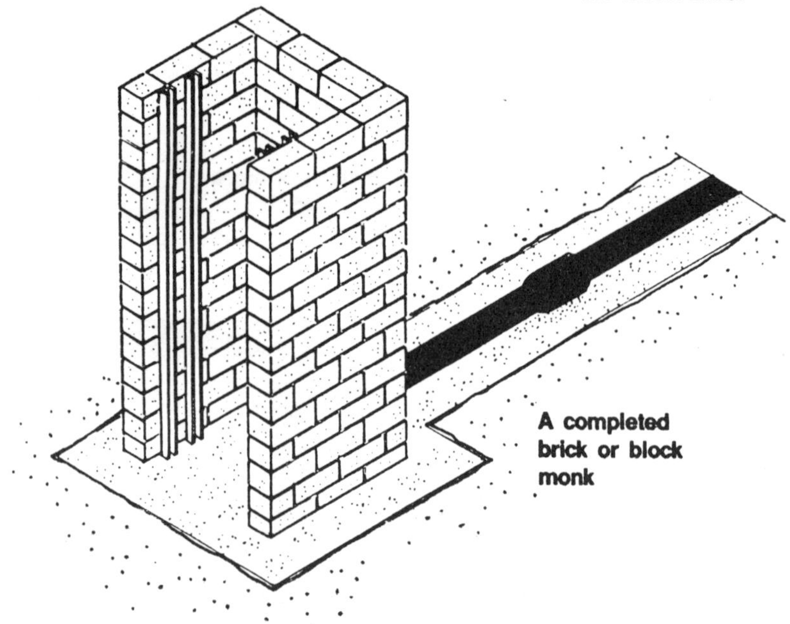

A completed
brick or block
monk

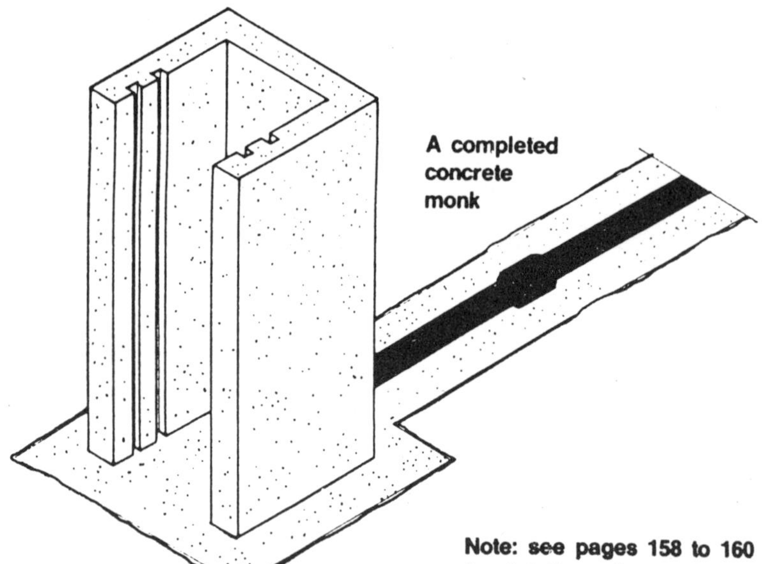

A completed
concrete
monk

Note: see pages 158 to 160
for details on how to build
two kinds of wooden forms
for pouring concrete monks

108 Reinforced concrete monks

1. Larger monks, higher than 1.5 m and including a pipeline with an inside diameter greater than 25 cm, should be built of reinforced concrete.

2. If you use the dimensions shown in **Table 48**, for monks with concrete reinforcement, the result will be a monk tower roughly square in shape (72 × 74 cm or 80 × 80 cm, for example), depending on the size of the pipeline.

Building the pipeline foundation

3. Larger monks are usually equipped with commercially available pipes, either of asbestos cement or concrete, laid down on a good foundation. **Such a foundation should be built together with the monk tower foundation**. To build the pipeline proceed as described in Section 107 with the exception of some dimensions, which should be increased as follows:

(a) **The width of the foundation** should be at least 30 cm wider than the outside diameter of the pipeline.
(b) **The thickness of the foundation** should be:

- in hard, undisturbed soil, 15 to 20 cm depending on the pipe's size;
- in soft soil, 25 to 30 cm depending on the pipe's size;
- in very soft soil, with pile foundations, as described earlier.

4. To build this foundation you can use rock and mortar for the lower half and lean concrete for the upper half.

5. You can build **a thinner foundation** (about 15 cm thick) if you use **reinforced concrete**.

6. **One method is as follows**:

(a) Fill the bottom half of the trench with concrete.
(b) Tamp it well and level properly.
(c) Place the reinforcement (6-8 mm diameter steel bars at about 10 cm intervals, with 6-8 mm cross-bars at about 50 cm intervals tied to these) on top of the wet concrete.
(d) The reinforcing should run the whole length of the pipeline foundation and extend enough to tie in with the monk foundation.
(e) Then, when the concrete below is still wet, cover the reinforcing with more concrete to the top of the trench.
(f) Tamp it well and level properly.

7. **An alternative method is as follows**:

(a) Hang the reinforcing at the midpoint of the trench using lengths of wood and wire hangers.
(b) Fill the whole trench with concrete and at the same time tamp it well.
(c) When the trench is full to the top, level it properly.

Laying the pipeline

8. When the foundation is ready, lay down the pipeline as explained above for smaller monks (see Section 107).

9. If you use 30-cm diameter **asbestos cement pipes**, you can greatly improve their durability by **coating them with 10 cm of soft lean concrete**. Do this after finishing all the joints of the pipeline and before building the monk tower.

Building the monk tower foundation

10. **The monk tower foundation is to be built at the same time as the pipeline foundation**, as described earlier for smaller monks (see Section 107, paragraph 13) except for the following:

(a) **The foundation size** should be larger than the tower base:

- for hard, undisturbed soil, by at least 30 cm on all sides;
- for soft soil, by at least 50 cm on all sides;
- for very soft soil, at least 50 cm, with suitable piles.

(b) **The foundation thickness** should be:

- in hard, undisturbed soil, 50 cm;
- in soft soil, 70 to 90 cm;
- in very soft soil, 70 to 90 cm, plus piles.

(c) **When placing concrete in the top half of the foundation**:

- stop when you reach a level 8 cm below the foundation surface level;
- place **the reinforcement of the monk tower** well on top of the concrete layer (see next paragraphs), with its opening toward the pond, making sure that there is no steel bar in the prolongation of the proposed pipeline;
- pour the last 8 cm layer of concrete without displacing the reinforcement.

Preparing the steel reinforcement

11. You have learned earlier (see Section 45, **Construction, 20/1**) about steel bar reinforcement for concrete. In this section, you will learn more about this through two specific examples of monks.

12. **If you want to build a reinforced monk tower 1.50 m high**, with a draining pipeline of 25-cm inside diameter (**Table 48**, type 4), prepare the reinforcement as follows:

(a) Get about 20 m of 6-mm diameter **steel bar reinforcement; cut the following lengths and shape them as indicated**:

- 3 sections of 3.40 m each to shape into a long U;
- 3 sections of 1.50 m each to shape into a short U;
- 2 sections of 1.95 m each to shape into an L;

(b) **Assemble them as shown in the drawings**, attaching them together at their intersections with 1-mm diameter **soft annealed wire**:

- the three long U shapes are set vertically to reinforce the bottom and the side walls;
- the two L shapes are also set vertically to reinforce the bottom and the back wall;
- the three short U shapes are set horizontally to reinforce the side and back walls, as well as to join together their previous reinforcement.

13. **If you want to build a reinforced monk tower 2 m high** with a draining pipeline of 30-cm inside diameter (**Table 48**, type 5), prepare the reinforcement similarly to the above, as follows:

(a) Get about 25 m of 6-mm diameter **steel bar reinforcement; cut the following lengths and shape them as indicated**:

- 3 sections of 4.45 m each to shape into a narrow U;
- 3 sections of 1.55 m each to shape into a wide U;
- 2 sections of 2.5 m each to shape into an L.

(b) **Assemble them as shown in the drawings**, attaching them together at their intersections with 1-mm diameter **soft annealed wire**.

Building the reinforced concrete tower

14. As explained earlier, you need **a wooden form** in which to pour the concrete (see Section 109). To build the tower, proceed as described in Section 107, paying particular attention to the following points:

(a) **To improve the junction of the foundation and the tower**, break the foundation surface where the tower base will sit and all around the vertical reinforcement bars.
(b) Secure **the small section of pipe well** in the prolongation of the existing pipeline, at the right level in the prepared gully, and within the steel reinforcement.
(c) Assemble **the elements of the wooden form** together without displacing the steel reinforcement.
(d) Make sure that **this reinforcement rises evenly between the sides** of the wooden form — it should not run too close to either side. Secure it well there and **brace the form** so that it will not move when placing the concrete.
(e) **Compact the concrete well** (see Sections 34 and 35, **Construction, 20/1**) without disturbing the position of the reinforcement.
(f) Remove the form very carefully.
(g) **Finish the junction** between the foundation and the tower with ordinary mortar.

BUILDING A REINFORCED CONCRETE MONK

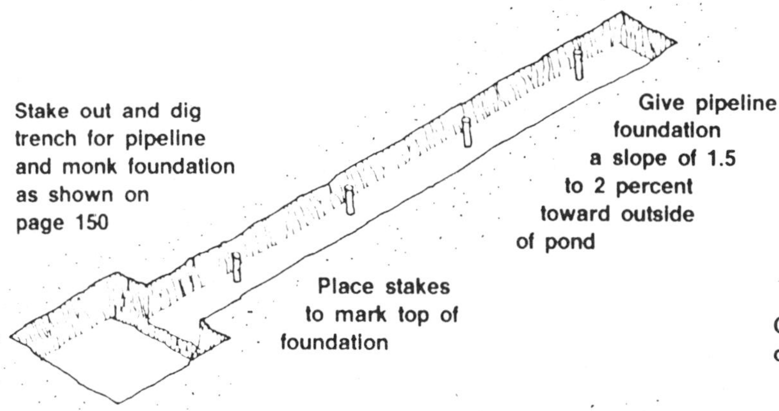

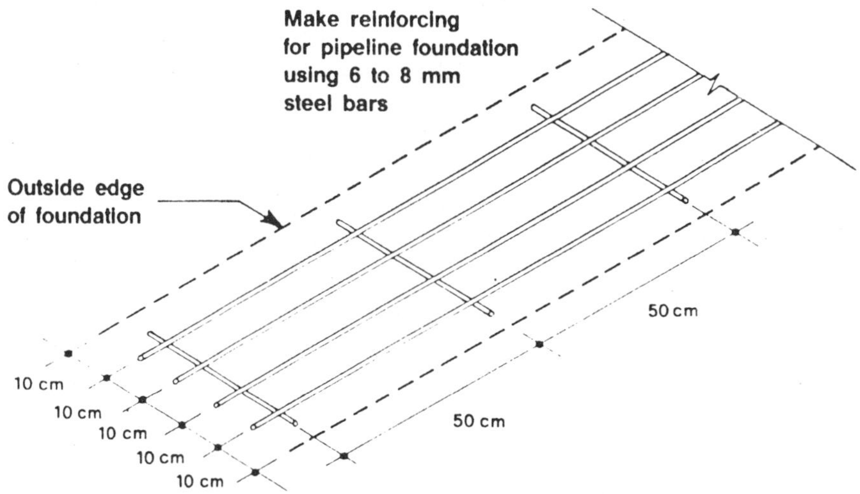

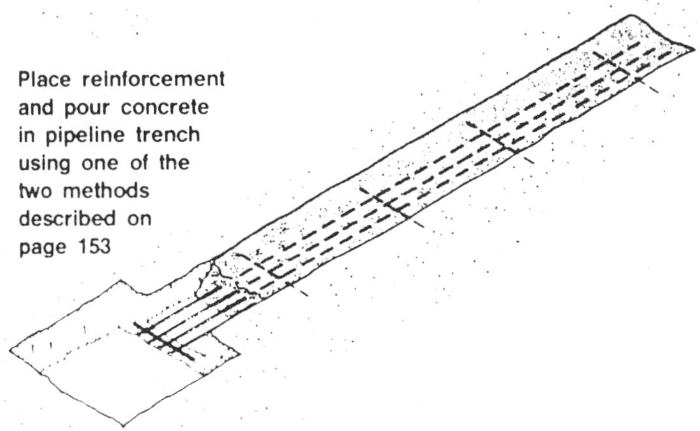

Section of monk and pipeline foundation showing the placing of reinforcement by hanging

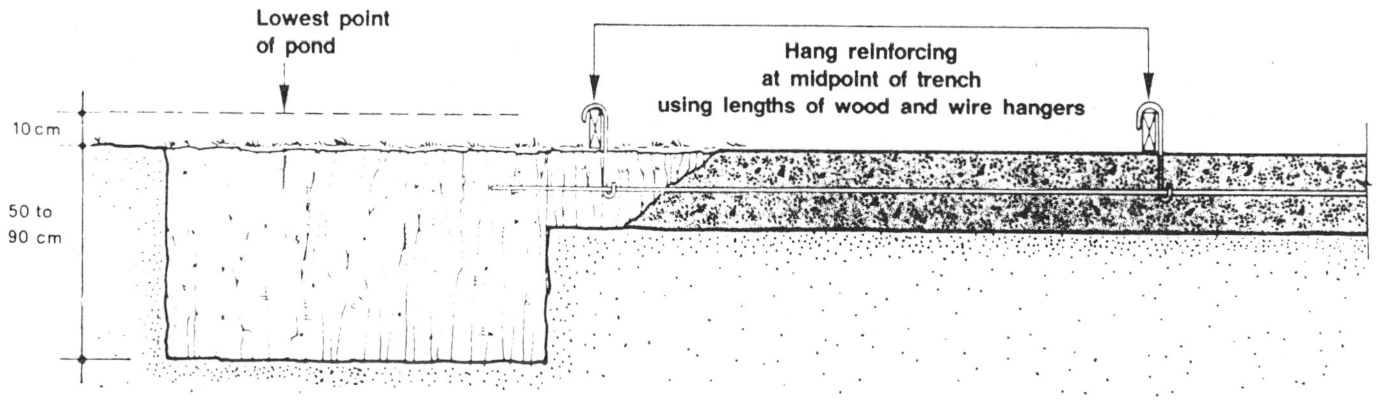

Make reinforcing for monk tower using 6 to 8 mm steel bars

The size of reinforcing structure will depend on size of monk to be built

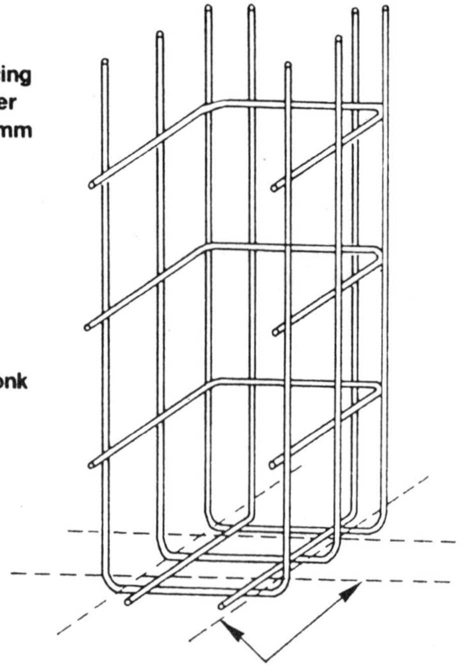

Position of base reinforcing

Note: dimensions for reinforcing steel for a monk of 1.50 m are given on page 154

Plan of monk foundation and position of reinforcing

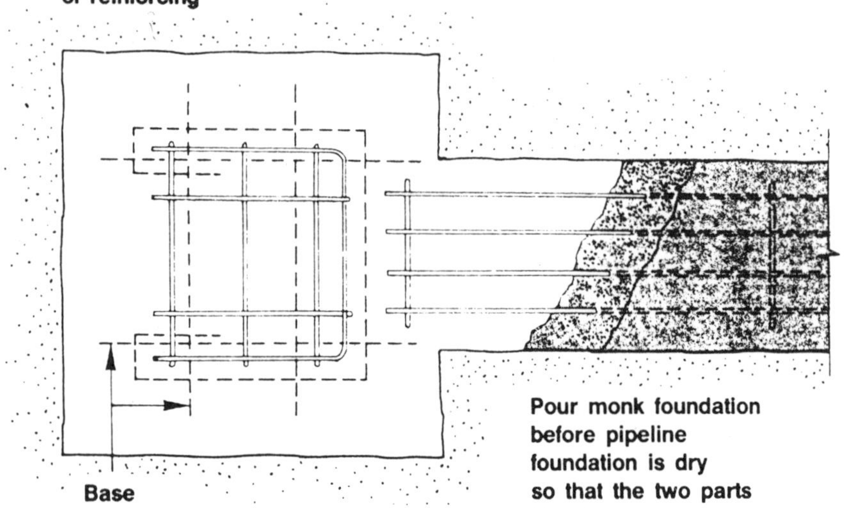

Base reinforcing

Pour monk foundation before pipeline foundation is dry so that the two parts bind together

Place reinforcement for monk tower by hanging

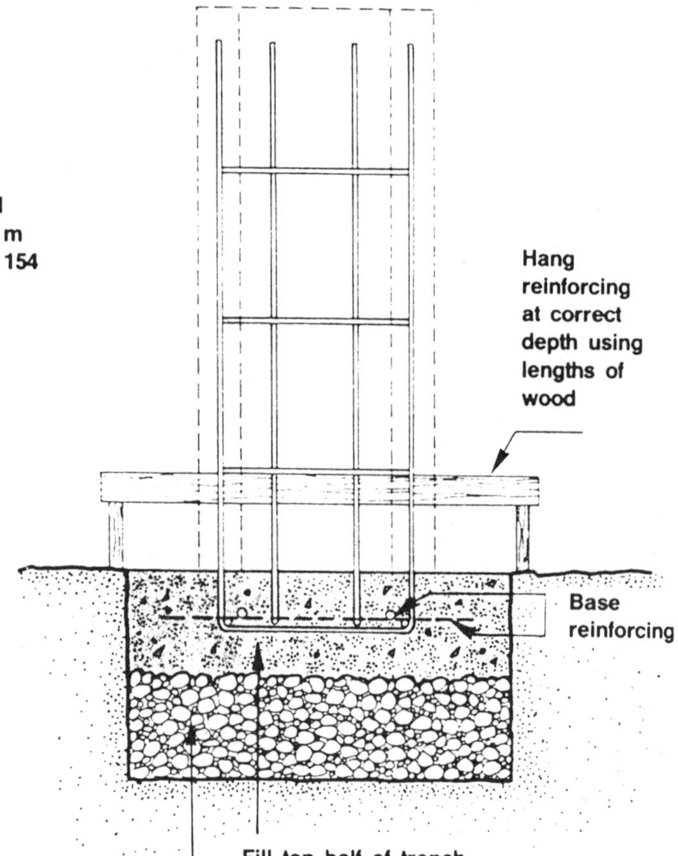

Hang reinforcing at correct depth using lengths of wood

Base reinforcing

Fill top half of trench with concrete

Fill bottom half of trench with gravel, rocks and cement mortar

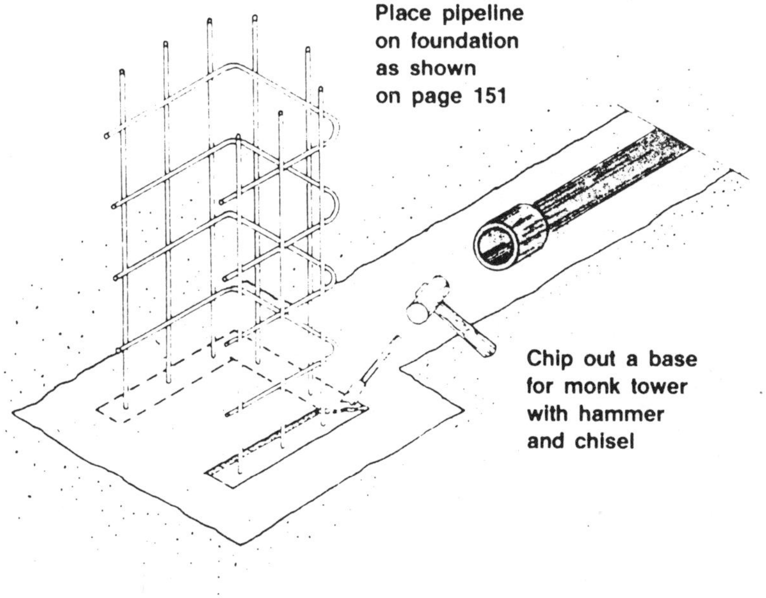

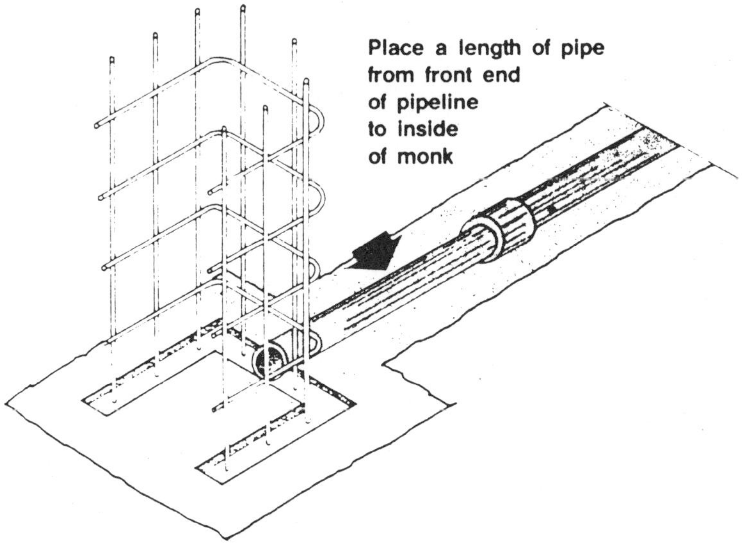

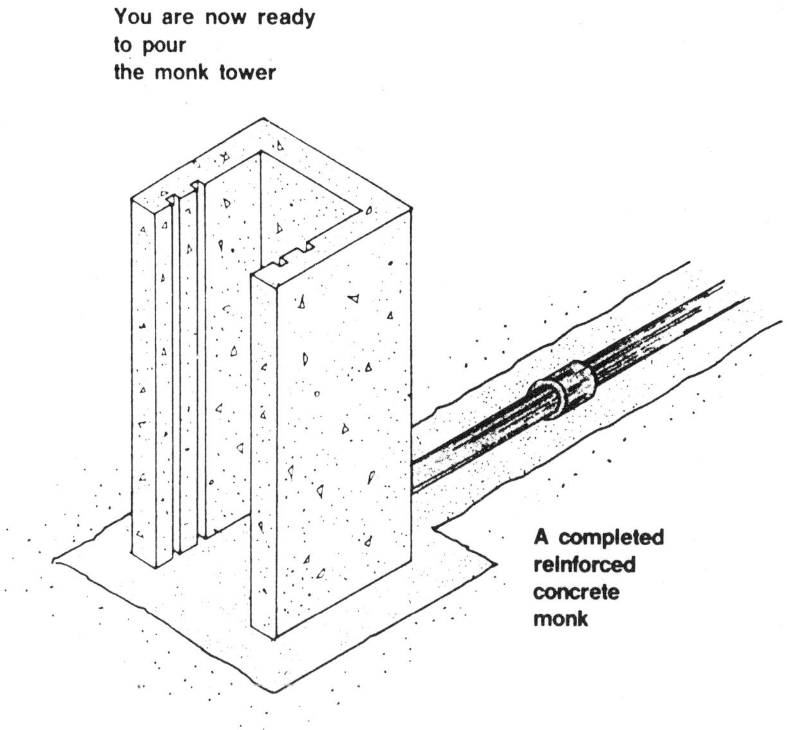

Note: see pages 158 to 160 for details on how to build two kinds of wooden forms for pouring concrete monks

109 Wooden forms for concrete monks

1. In Section 34 of **Construction, 20/1,** we discussed general points on preparing wooden forms for concreting. In this section, you will learn more about the specific design of these forms for building concrete monks.

Preparing a plywood form for small monks

2. For the construction of small monks about 1 m high at the most, you can use a form made of wood strips and **plywood 1.5 cm thick**. The drawings on this page show you how to build this kind of form.

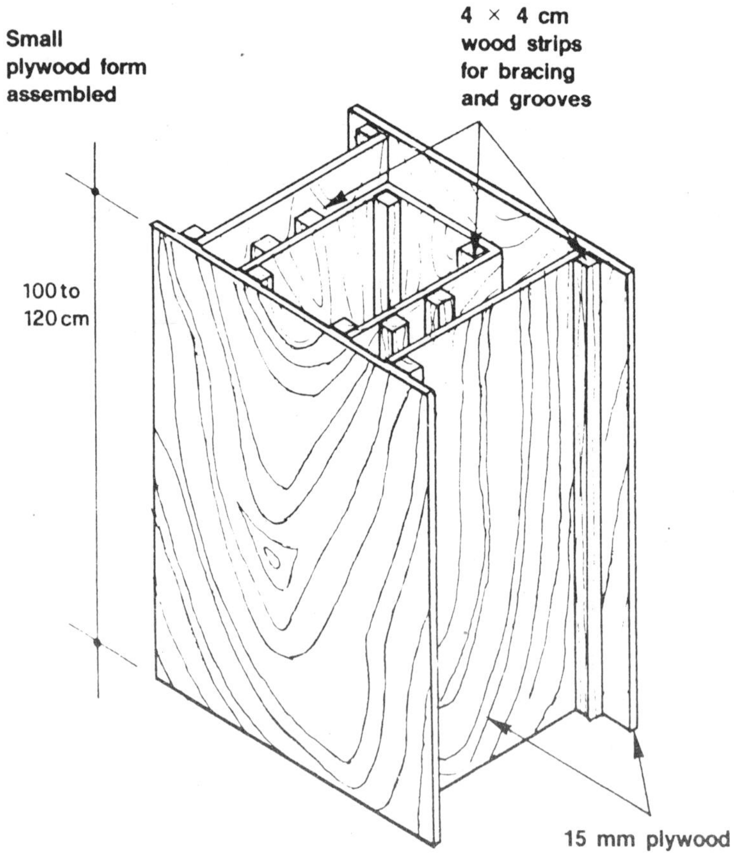

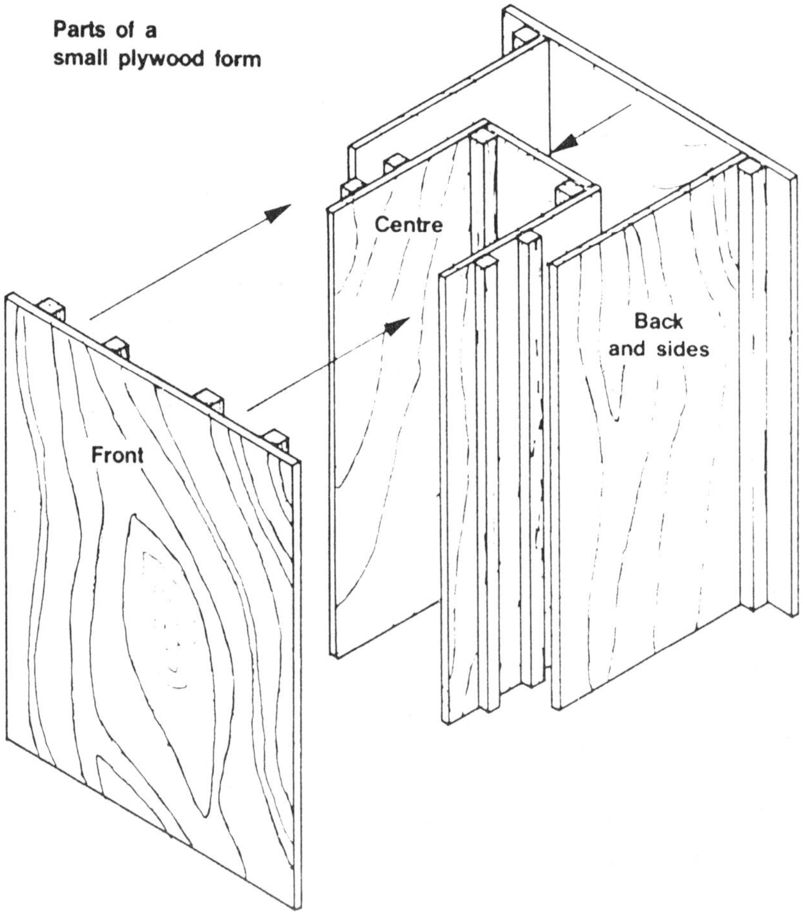

Note: if you want a three-groove monk, add one more groove strip to this design

Preparing a plank form for larger monks[1]

3. You can build a simple form made of **3- to 4- cm thick planks**. The drawings on this page and the next show you how to build this kind of form.

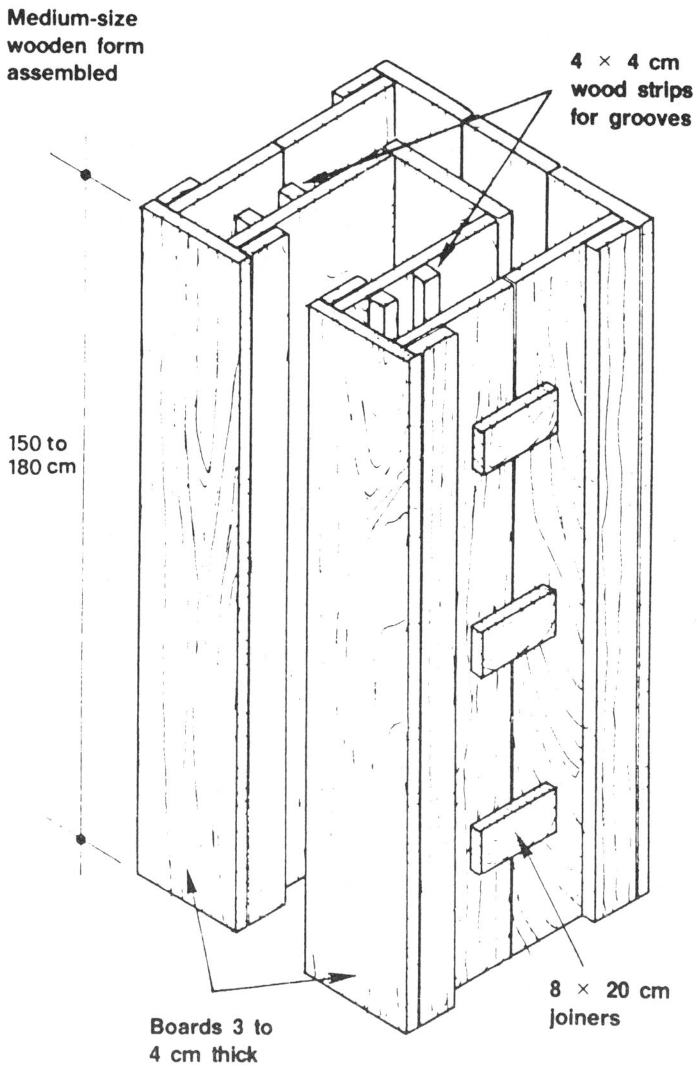

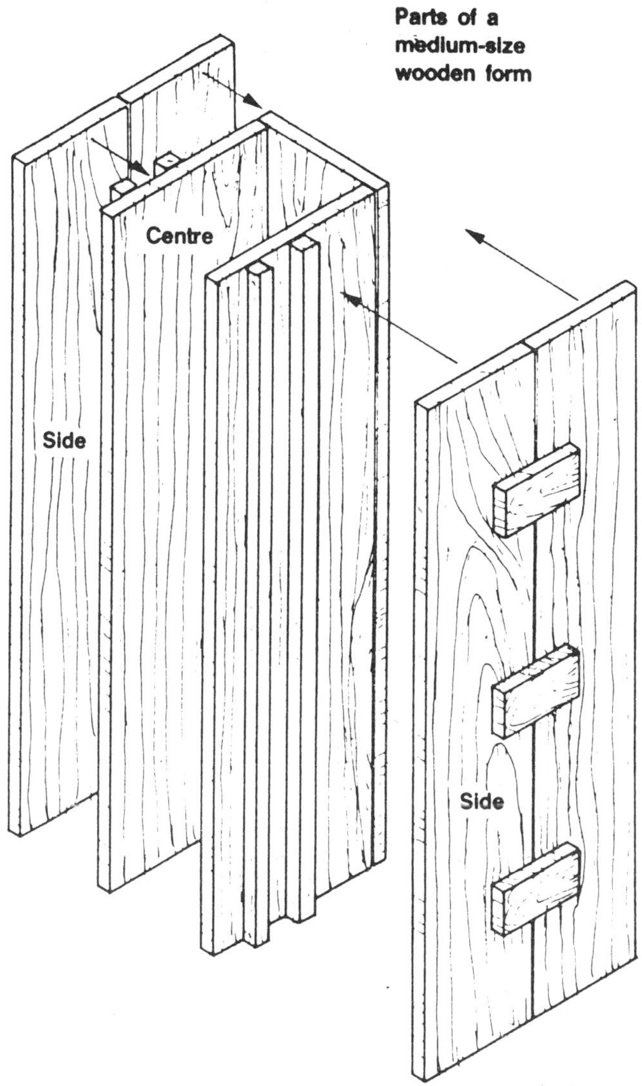

[1] Adapted from a design by J. Miller, FAO Expert in Rural Fish Farming Development

Parts of a medium-size wooden form, continued

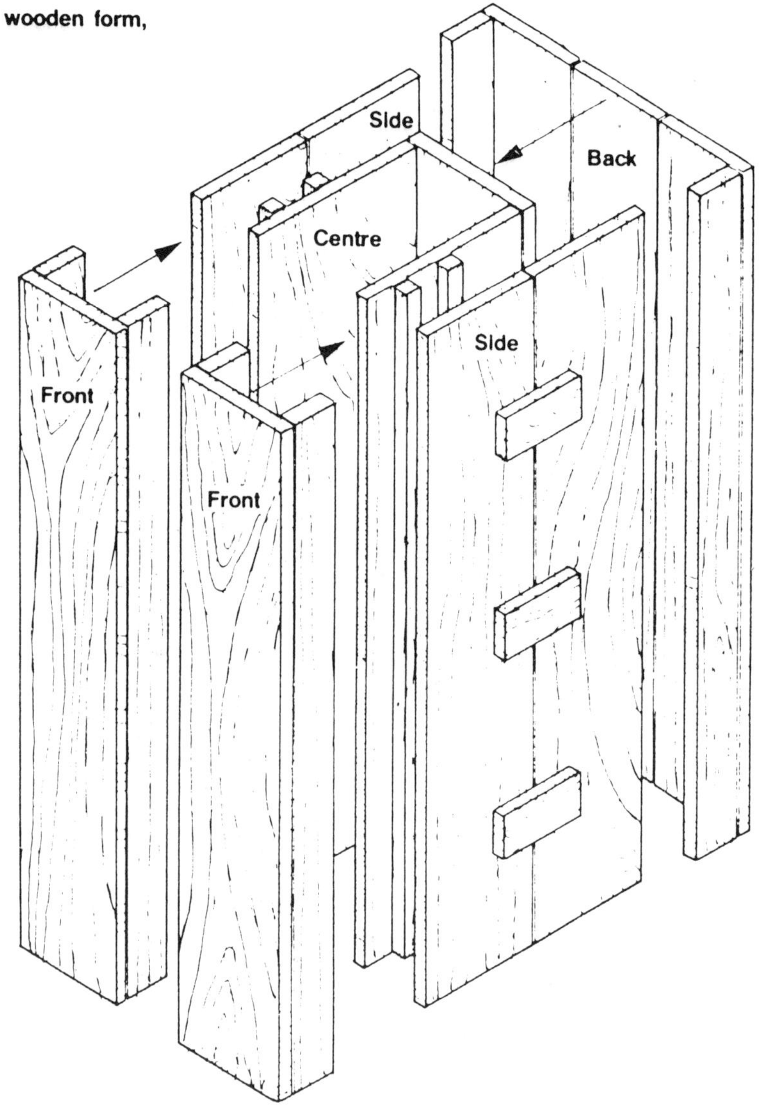

The front and back pieces of this wooden form hold the centre and side pieces in place

Note: if you want a three-groove monk, add one more groove strip to this design

Note: similar wooden forms can be used to construct sluice gates, water distribution and diversion structures in concrete and reinforced concrete

1010 Water control for sluice or monk

Grooves and boards

1. Normally, a sluice or a monk is equipped with **two pairs of grooves**, in which boards are inserted up to the desired water level. To prevent water seepage, the gap between the rows of boards is filled with compressed material such as clayey soil or sawdust.

2. While the pond is in operation, **a screen** should be fitted on the top of the front row of boards to keep the fish from getting out if the water level should rise.

3. **A very small sluice or monk**, however, may have only a single row of grooves and boards. In this case, you can prevent seepage by packing strips of jute burlap between the grooves and the boards. The joints between the boards can also be sealed using polythene or rubber flaps.

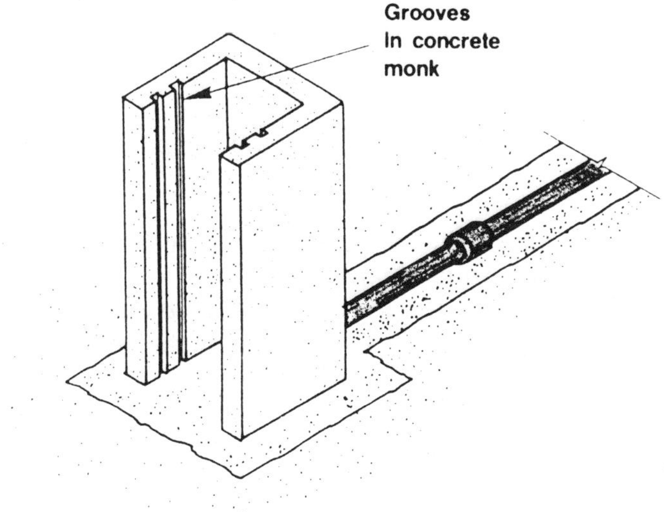

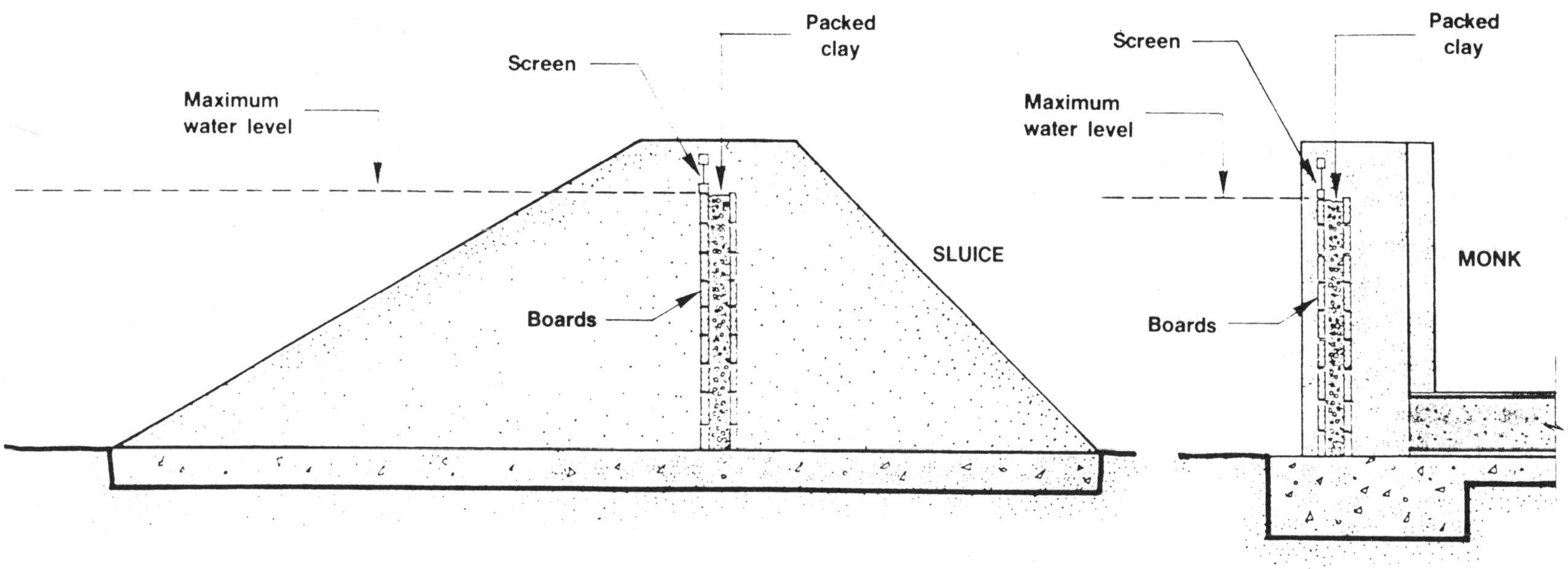

4. **A medium-size sluice or monk** can have two pairs of grooves to be used as follows:

- in the front row of grooves, set a small screen at the level from which you wish to remove water;
- the rear row of grooves is completely filled with boards up to the level of water you want to maintain in the pond.

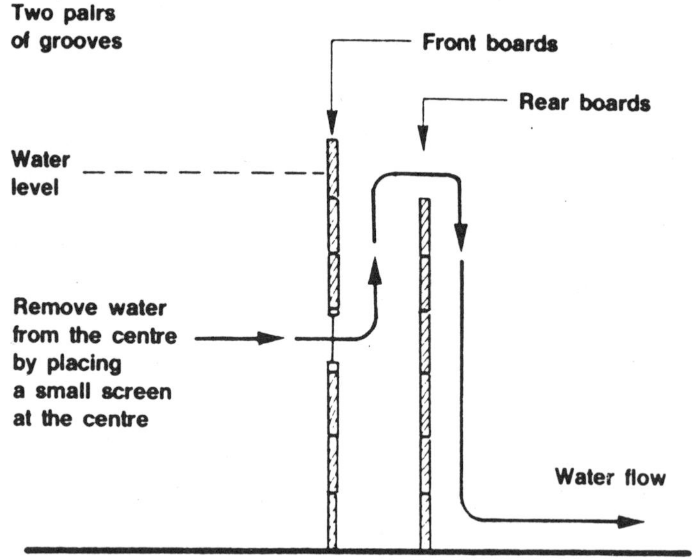

5. **A larger sluice or monk** can have three pairs of grooves to be used as follows:

- in the front row of grooves, set a large screen to filter out the largest debris from the whole water column;
- in the middle grooves, set a small screen at the level from which you wish to remove water;
- the rear row of grooves is completely filled with boards up to the level of water you want to maintain in the pond.

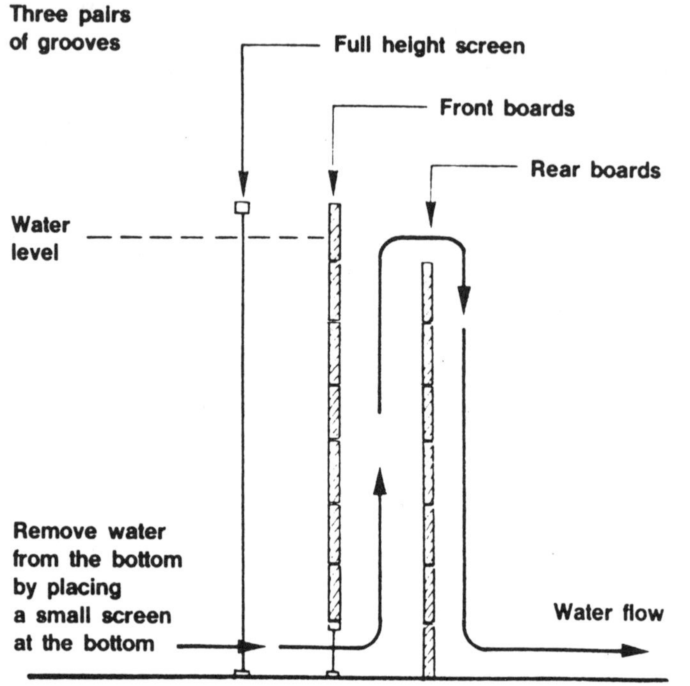

Selecting the boards to use

6. Boards for monks and sluices should be made of **durable wood**, resistant to water and with limited swelling, such as *doussie, iroko* or *mukulungu* **(Table 6, Construction, 20/1)**.

7. Size varies according to the dimensions of the structure but, in general, the following rules should be adopted:

- **thickness**: 2.5 to 3 cm and 0.5 to 1 cm thinner than groove width;
- **height**: 15 to 20 cm (maximum 30 cm);
- **length**: 1.5 to 2 cm shorter than distance between opposite groove ends.

8. It is important that **the wooden boards do not fit too tightly** into the grooves after the wood has swelled under water. If this happens, it will be very difficult to remove the boards when necessary.

9. You can **improve the fitting of the wooden boards** on top of each other and reduce water losses by:

- planing the four sides of the boards well;
- shaping their top and bottom edges as shown;
- avoiding using timber with knots.

10. If your monk is rather high and wide, it will be **easier to remove the boards** from the top of the monk using **a handle with a T welded at its end**. You can easily have one made by a blacksmith. At the back of each board, secure two steel bolts or hooks to lift it.

11. Another way **to reduce water leakage** is to use old inner-tube rubber to make **a simple seal**:

- as a flap below each board, to overlap the previous board and cover the gap between them;
- to put in the groove to help to seal the board and minimize leakage.

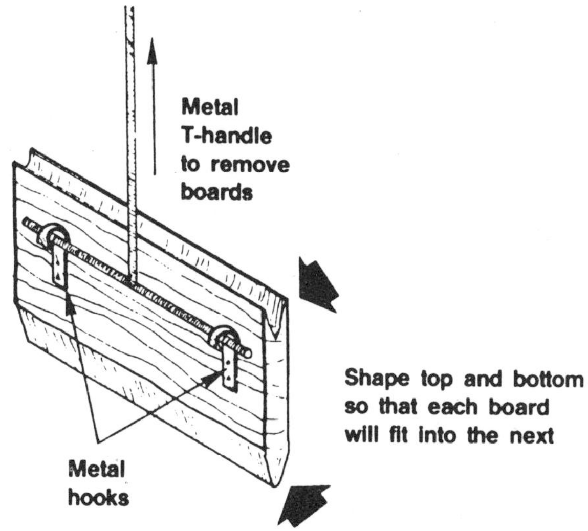

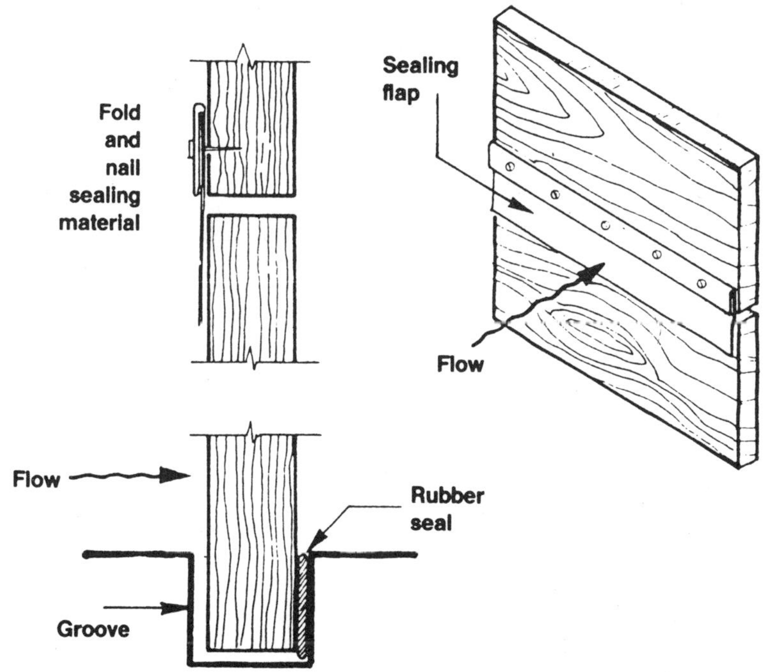

11 FLOOD AND SILT CONTROL STRUCTURES

110 Introduction

1. Your fish farm should be protected against four major causes of fish production losses:

- excessive water supply;
- uncontrolled surface runoff;
- silty water; and
- fast running water.

2. Any **excess water** entering a full fish pond, such as flood water or runoff water, should be immediately and automatically discharged. Depending on the amount of water to be carried away, you can use the pond outlet itself, or additional structures such as **pipe overflows, mechanical spillways and emergency spillways**. You will learn more about each of these in Sections 111 to 114.

3. During heavy rains, **the amount of surface runoff may become excessive**, particularly for barrage ponds or ponds built at the bottom of large sloping areas with little vegetation cover. The runoff water in such cases is also often heavily loaded with fine soil particles that make it very turbid. If the runoff passes across cultivated areas it might accumulate toxic substances such as pesticides. To avoid such water reaching your fish farm, you will have to protect it with one or more protection canals (discussed in Section 115).

4. In certain regions, soil conditions are such that **the water available for fish farming is very turbid**, particularly during the rainy season. To clear such water and improve its quality, you can build **a filter pond or a settling basin** (discussed in Section 116).

111 How to discharge excess water from ponds

1. Excess water should be automatically and safely carried away from a pond. If it is not, the water level in the pond may rise above the maximum designed level and even reach over the top of the dikes. Significant damage may then occur, often leading to the destruction of the dike and the loss of fish.

2. In an earlier manual in this series, **Water, 4**, you learned about the various sources of water and the different factors that control their availability. You learned how **the rainfall and the physical characteristics of the catchment basin** (such as size, slope, soil and vegetation) determine the amounts of water that will reach your fish farm through various ways (such as runoff, groundwater, spring or stream water). Some of this water can exceed your pond requirements. In this case, you will have to discharge it away from your pond.

3. The amount of excess water to be discharged varies according to **the type of pond**:

- **in diversion ponds**, where the water inflow is regulated through the pond inlet, the need to discharge excess water arises only rarely if the inflow structure and the pond are well cared for;
- **in sunk-in ponds** and **in barrage ponds with a diversion canal**, the quantity of excess water is also usually limited except in very large ponds (1 to 2 ha or more) during heavy rains, when runoff may increase considerably and may converge into the pond;
- **in barrage ponds without a diversion canal**, the discharge of excess water may vary in size, may be permanent or seasonal, according to the natural flow regime of the feeding stream; it can become exceptionally high during floods.

4. You may find, in checking through this section, that **the cost of these overflow and protection structures** is high enough to justify either:

- installing a diversion canal (see Section 85); or
- using a different site.

5. There are different ways of **discharging excess water from fish ponds**:

(a) If your fish pond has **a free-flowing outlet**, such as a sluice gate or a monk, and if **the amount of excess water** to be discharged is always small enough, you do not need any other structure. To find out how much water you should be able to discharge through typical outlets, see:

- for sluice gates, **Graph 6** and **Tables 32 and 33** (Chapter 7); and
- for monk pipelines, **Tables 12, 13 and 14** (Section 38, **Construction, 20/1**).

If possible, it is often cheaper to provide a larger than normal pond outlet than to build an additional structure.

Note: remember **to clean the outlet screens regularly** so that the excess water can flow easily through them.

(b) If your pond has **no free-flowing outlet** or if your **outlet is too small**, and if **the amount of excess water** to be discharged is always limited, you can have a **pipe overflow** (see Section 112).

(c) If **the discharge of excess water is relatively large and continuous** for long periods of time, you should, in addition to any pond outlet, build **a mechanical spillway** (see Section 113).

(d) If **the discharge of excess water becomes exceptional** on certain occasions, you should, in addition to any pond outlet and any other discharging structure, build **an emergency spillway** (see Section 114).

6. To assist you in selecting the right type of structure to discharge any excess water from your ponds, consult **Table 49**.

TABLE 49

Structures for discharging excess water from ponds

Type of pond	*Excess water discharge*		
	Small (long-lasting)	Large (long-lasting)	Very large (seasonal)
Barrage pond without diversion canal	Monk, sluice, pipe overflow	Mechanical spillway	Emergency spillway
Barrage pond with diversion canal	Monk, sluice, pipe overflow	—	Emergency spillway
Diversion pond	Monk, sluice, pipe overflow	—	—
Sunken pond	Monk, sluice, pipe overflow	Mechanical spillway	Emergency spillway

112 The pipe overflow

1. To carry away a normal flow of excess water which is reasonably small, you can use a pipe overflow built in the upper part of a dike.

Choosing the correct type of pipe overflow

2. **The number of pipes and their inside diameter** should be selected according to the maximum water flow to be discharged. Usually, no more than two to three pipes are used side by side, and their individual diameter is limited to 15 to 20 cm. Estimate the water discharge capacity of pipes using the chart below to determine the type of pipe overflow you need.

Approximate water discharge capacity of overflow pipes

Inside diameter of pipe (cm)	*Water discharge capacity*			
	(l/s)	(l/min)	(m^3/h)	(m^3/24 h)
5	1.8	108	6.5	155
10	8	480	29	691
15	18	1 080	65	1 555
20	31	1 860	112	2 678
30	70	4 200	252	6 048
40	126	7 560	454	10 886
50	196	11 760	706	16 934
	x	60x	3.6x	86.4x

Note: whenever you use a pipeline to discharge excess water, remember that its capacity depends not only on its **inside diameter**, but also on **the pressure head** (see **Table 12, Construction, 20/1**).

Building a pipe overflow

3. When building a pipe overflow remember the following:

(a) Place the overflow at one corner of the dike.
(b) Fix it well into **the upper part of the dike**.
(c) **Make sure the pipe is long enough** so that its overflowing end discharges the excess water beyond the toe of the dike to avoid eroding it.
(d) Alternatively, if the dike is firm, and you have the material, you can make **a protected area**, using coarse cobble and cement or a large piece of pipe cut to form a semi-circular channel, to take the water down the slope.
(e) Angle the pipe slightly so that:

- its **inside opening** lies at a level 15 to 20 cm below the maximum water level in the pond, to prevent floating debris from obstructing the pipe;
- its **outside opening** lies at the maximum water level and discharges any additional water.

Note: if you wish to remove **cooler and deeper water** from your pond, use an overflow that curves down at its inside end.

Selecting the type of pipe to use

4. Three types of pipe are most commonly used as overflows:

- **bamboo pipes** (see Section 31, **Construction, 20/1**);
- **galvanized iron pipes** (see Section 38, **Construction, 20/1**);
- **plastic pipes** (see Section 38, **Construction, 20/1**).

5. It is best to use **one-piece pipes**, avoiding any joins. If the pipe sags, or extends too far out from the outer side of the dike, it may be useful to put up some simple pipe supports, using for example wood or bamboo.

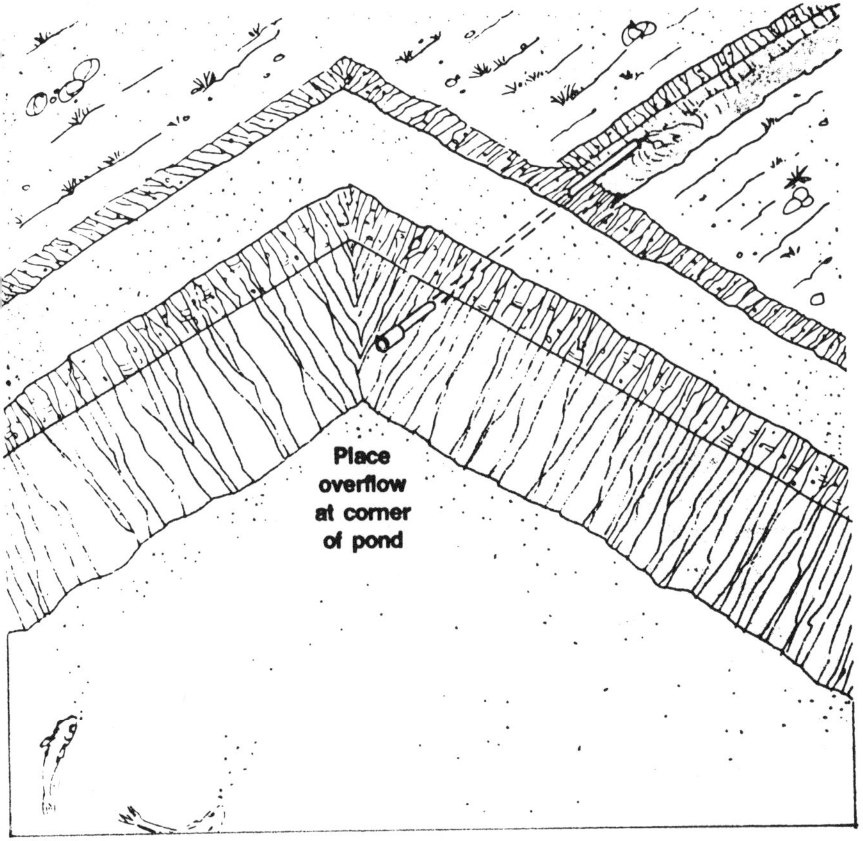
Place
overflow
at corner
of pond

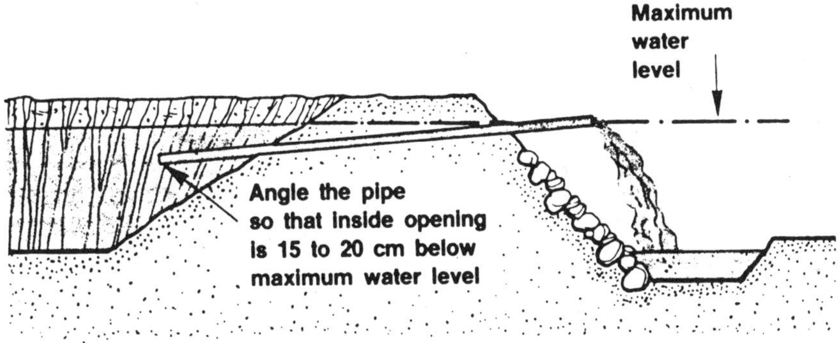
Maximum
water
level
Angle the pipe
so that inside opening
is 15 to 20 cm below
maximum water level

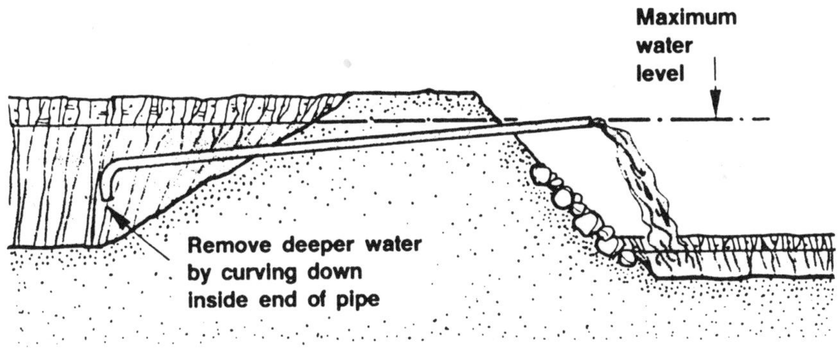
Maximum
water
level
Remove deeper water
by curving down
inside end of pipe

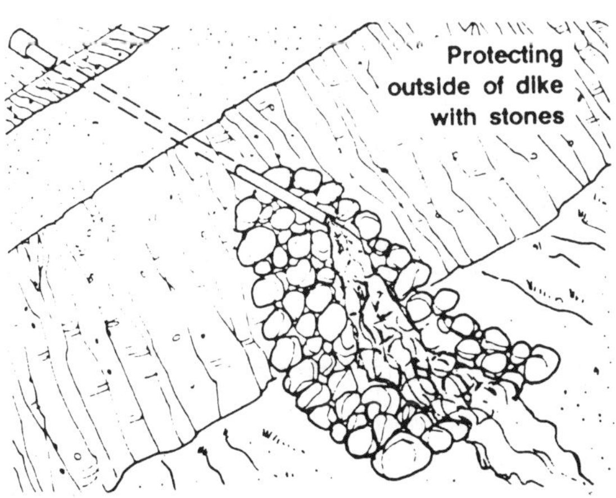
Protecting
outside of dike
with stones

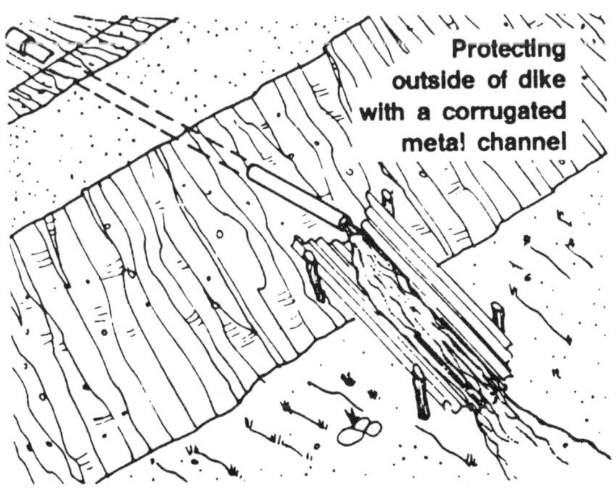
Protecting
outside of dike
with a corrugated
metal channel

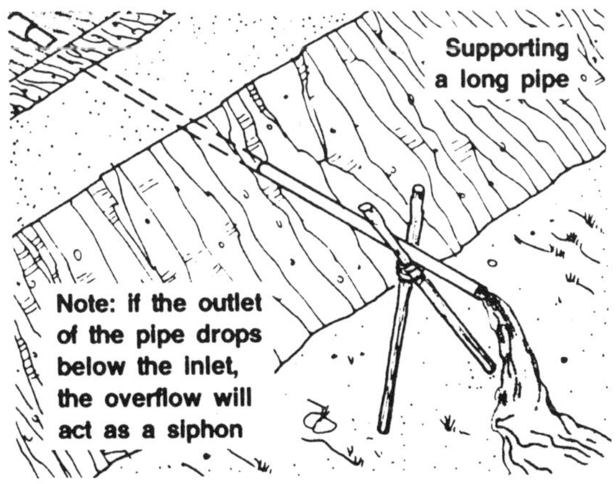
Supporting
a long pipe
Note: If the outlet
of the pipe drops
below the inlet,
the overflow will
act as a siphon

113 The mechanical spillway

1. To regularly carry away **a large flow of excess water**, you should build an open structure, called a **mechanical spillway**, and its draining channel, **the spillway channel**.

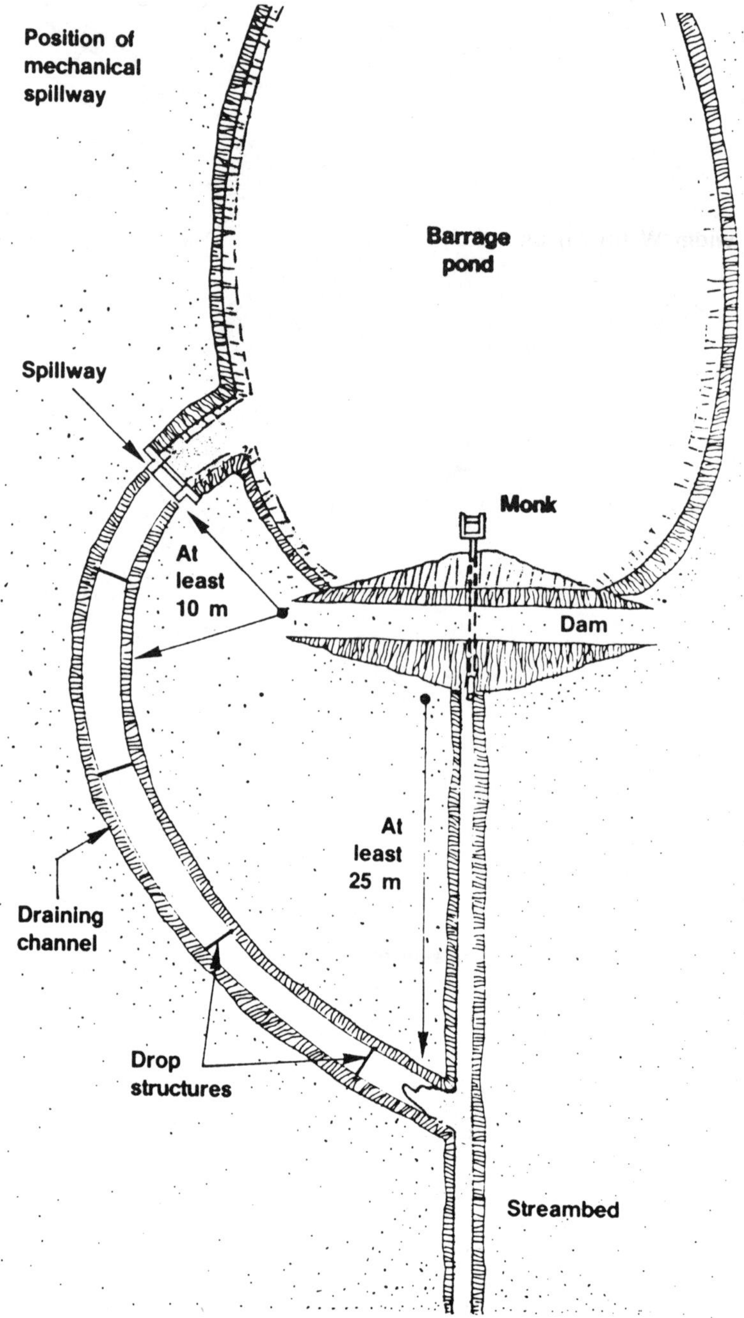

What is a mechanical spillway?

2. A mechanical spillway consists of:

- a horizontal part called **the crest**, across which the water flows;
- two vertical **side walls**, each with a single groove;
- a vertical board and/or coarse **screen** fitting into the grooves.

3. **The level of the crest** controls the level at which the excess water in the pond will start to be carried away.

4. **The width of the crest** and **the height** of the side walls determine the maximum water discharge capacity of the spillway.

5. **The grooves** can be fitted with a **board** to roughly control crest height. The **coarse screen** helps to avoid fish losses, especially when a large quantity of excess water is to be spilled. However, if there is a risk of these blocking with excess debris, it may be better to install an additional, larger screen on the inside. In all cases, the screens should be kept clean.

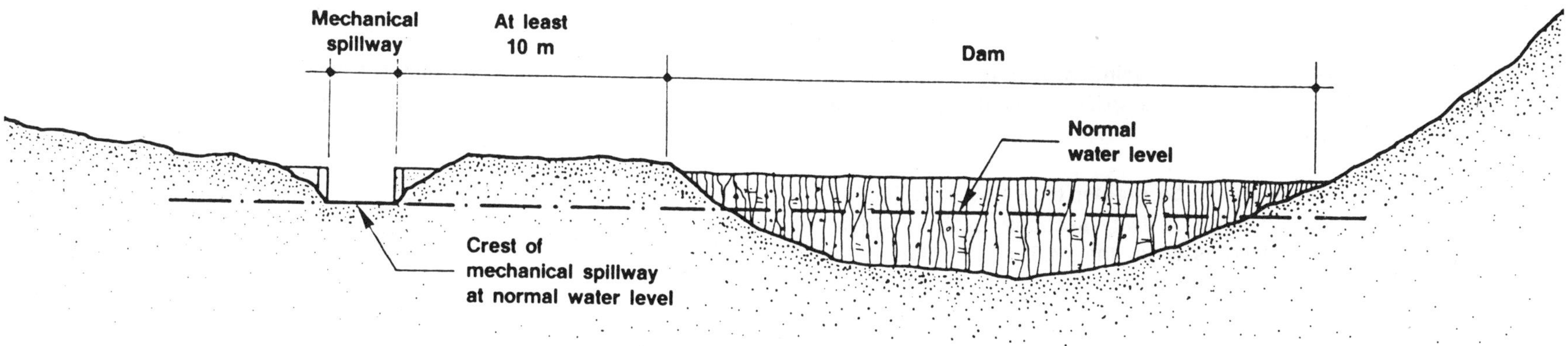

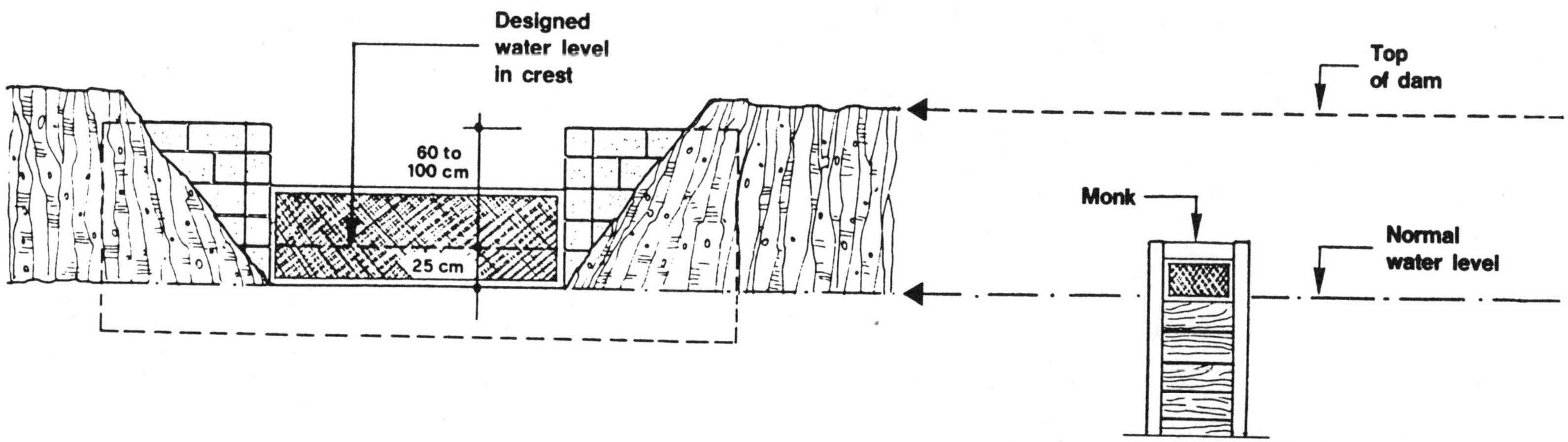

6. **To design a mechanical spillway**, you should first determine **the semi-permanent or permanent discharge Q** (in m^3/s) to be spilled by the structure. This will be equal to the highest normal stream flow minus the water used on the fish farm.

7. Knowing the value of **Q**, calculate **the crest width W** (in m) as

$$\mathbf{W = Q \div F}$$

where **W** (in m) is not larger than 1.5 m;

F is a factor depending on the **maximum height H** (in m) of the water running over the crest of the spillway, usually set at about 0.25 m (see **Graph 13**).

Note: if you determine that **W should be greater than 1.50 m**, it is often simpler to make two or even three spillway sections, each less than 1.50 m. If the mechanical spillway is wider than 4.5 m, you should not design it without the advice of an engineer.

Example

The highest **normal flow** of the stream during the rainy season is calculated as 156 l/s.

- At that time of the year, you will use at least 4 l/s of water for the fish farm.
- The excess water **Q** = 156 l/s − 4 l/s = 152 l/s or 0.152 m^3/s.
- You plan to use a mechanical spillway where the maximum depth over the crest **H** = 0.25 m.
- From **Graph 13**, obtain **F** = 0.18.
- Therefore, the crest width should be: **W** = 0.152 m^3/s ÷ 0.18 = 0.84 m.
- Build a spillway at least 0.85 to 0.90 m wide.

8. Set **the level of the crest** as equal to the maximum water level in the pond.

Note: the crest level plus the maximum water depth over the crest should not exceed the safe water level in the pond. There should still be **at least 20 cm freeboard** left around the dikes.

GRAPH 13

Value of F for spillways of different water depths

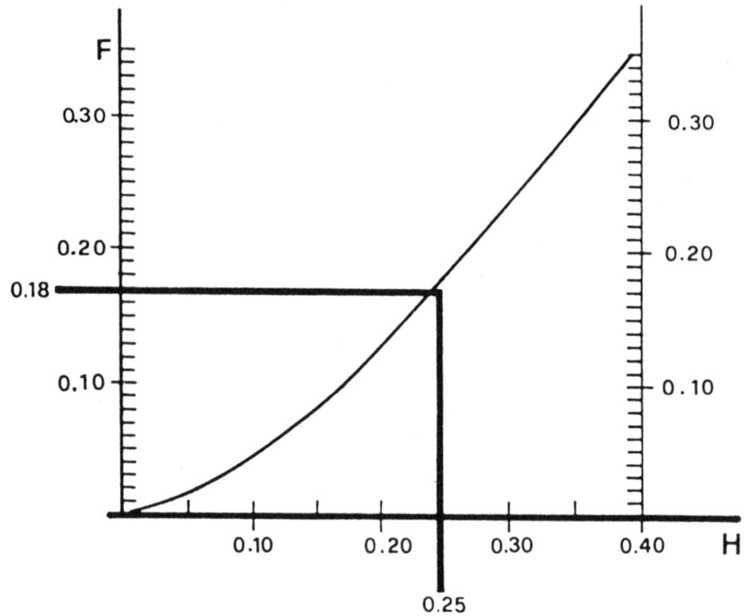

Water depth over the crest (m)

Building a mechanical spillway

9. You should wherever possible build the spillway **on well-settled, natural soil**. Unless it is specially reinforced and constructed, the normal pond dike itself is usually unsuitable, because soil movement will soon cause cracks in the structure and erosion will start damaging it further, together with the dike itself. A specially constructed spillway section within a dike is usually expensive and must be carefully built.

10. It is best **to build the spillway away from the dike**, and **on the least sloping side** of the valley.

11. You can build it with local stones, fired bricks or concrete blocks. You can also use concrete or reinforced concrete.

Concrete block mechanical spillway

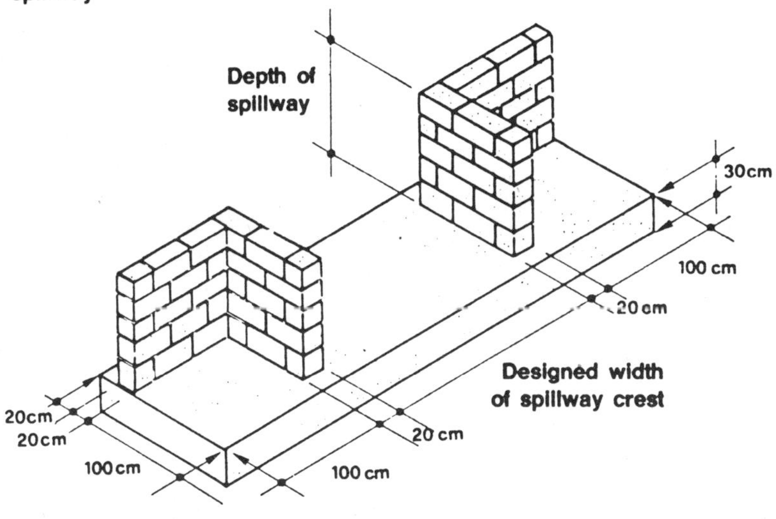

Designing and building the spillway channel

12. The purpose of **the spillway channel** is to carry away the excess water discharged through the spillway and to channel it safely down to the level of the outside toe of the dike.

13. **When designing and building the spillway channel**, remember the following points:

(a) **Lay the channel around the dike**, at least 10 m away from its lateral end and 20 to 25 m away from its outside toe.

(b) **The cross-section of the channel** should be rectangular or trapezoidal and should be equal or greater than the wet cross-section of the spillway at maximum discharge.

(c) **The freeboard** in the channel should be at least 20 to 30 cm, based on the maximum expected flow.

(d) The overall **drop in elevation** of the channel should not normally exceed 1 m per 20 m length. Check the water velocity (see Section 82) to make sure it does not exceed the safe limits for the material you will use.

(e) **If the drop in elevation is greater** than 1 m per 20 m length, it is best to use:

- cobble or concrete-faced channel sections; or
- series of horizontal channels and drop structures (see Section 87); or
- a combination of these.

(f) If you are using a drop structure system, build **the first horizontal section** at or slightly below the level of the spillway crest, for a length of at least 5 m. You can also use a short sloped section immediately next to the crest.

(g) Build **the last horizontal section** at a level such that its water level is about the same as that of the receiving water.

(h) **The drop structures** can be made with rock, brick, block or, more commonly, concrete (see Section 87, paragraphs 21 to 25). You can also build a simple structure with wood, but this requires greater maintenance.

Section through spillway channel with drop structures

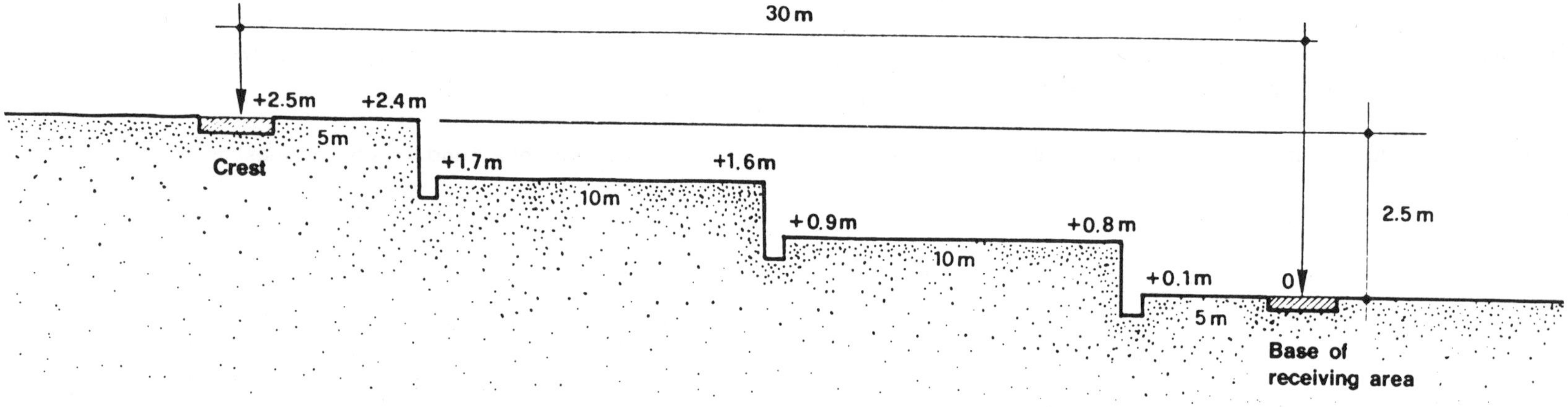

Example

For the previous case, spillway water flow is 0.152 m^3/sec. The difference in height between the maximum pond level (spillway crest) and the base level of the receiving point is 2.5 m.

The overall length of the channel is 30 m. Average fall is therefore: 2.5 m ÷ 30 m or 1:12, which is too steep. A channel with drop structures such as that shown would be preferred. Two or three drops will be sufficient, but a larger number of smaller drops can be used.

You will then need **to check flow and velocity**:

- velocity is greater on steeper parts of the channel such as section A or D, where the gradient is 0.1 m/5 m and **S** = 0.02.
- referring to Section 82, a rectangular channel of 0.85 m width (the same as the spillway width), with roughness **n** = 0.025 **(Table 37)**, would take the flow at a depth of about 0.15 m, with a velocity of 1.3 m/s;
- referring to **Table 35** (depending on the channel materials), if this velocity is too great, it may be better to make the channel more horizontal between drops, with a slightly greater depth or greater width.

Section through crest

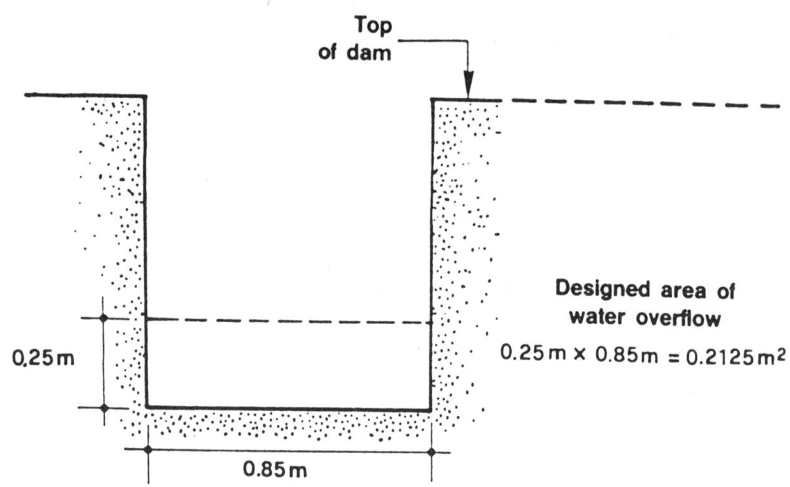

1. The purpose of an emergency spillway is to carry away **exceptional volumes of water** in excess of the flow that can be discharged under normal circumstances by the pond outlet, the pipe overflow and/or the mechanical spillway.

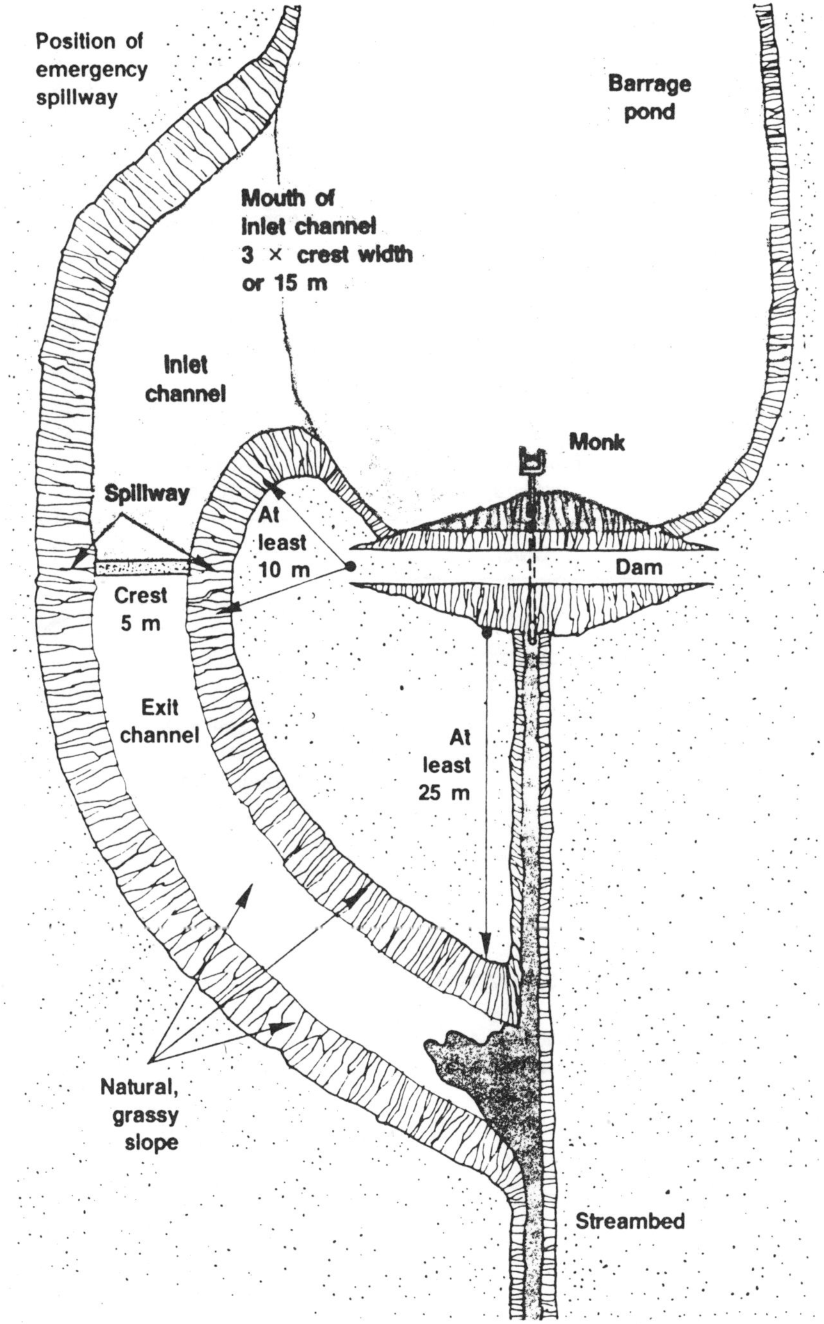

Note: use an emergency spillway only where soils and topography permit, as discussed later in this section. If this is not possible, you may have to use a larger mechanical spillway or even to reconsider the site chosen.

What is an emergency spillway?

2. An emergency spillway is usually an earthen structure which consists of:

- a horizontal portion called **the crest**;
- two sloping **side walls**, giving it a trapezoidal section;
- an **inlet channel**, gently sloping toward the crest and leading the excess water to the structure;
- an **exit channel**, gently sloping away from the crest and leading the excess water safely away from the dike.

3. There are two kinds of emergency spillway:

- **natural spillways**, which use the natural slope of the ground;
- **excavated spillways**, which are dug entirely.

4. Generally, natural spillways should be preferred where possible, and **spillways** cut into rock are ideal.

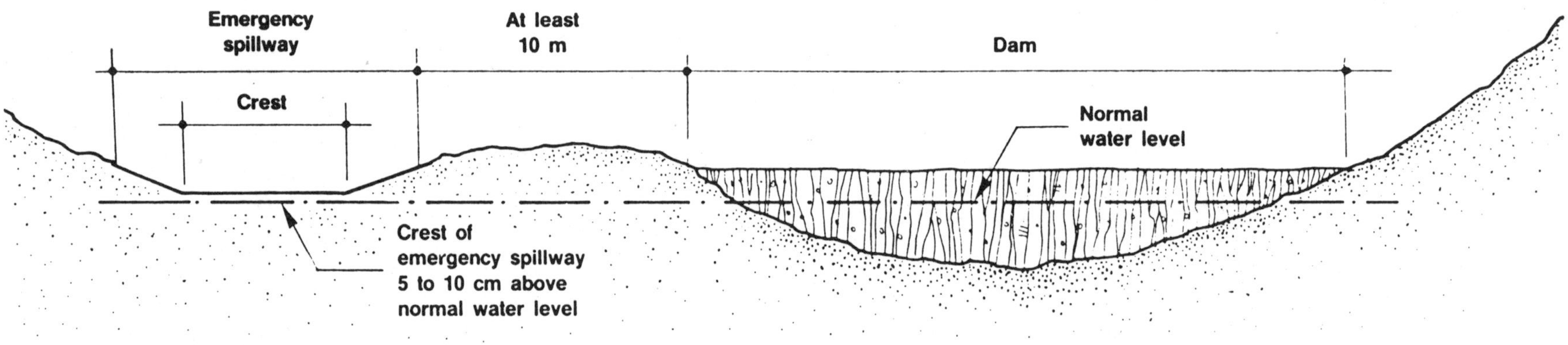

Note: in suitable circumstances you can combine a mechanical spillway with an emergency spillway

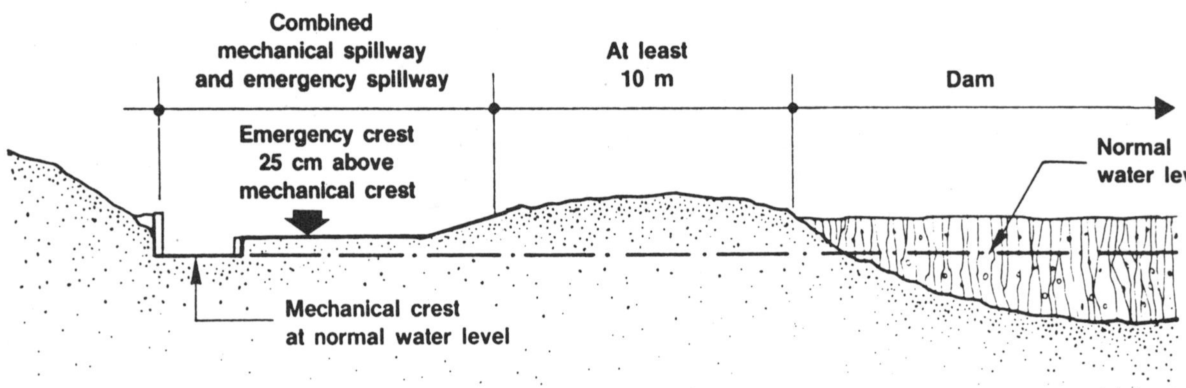

Selecting a freeboard for the dike

5. When an emergency spillway is used, **the minimum freeboard of the dike** is the height of the top of the dike above the crest level of the emergency spillway (see also Section 61, **Construction, 20/1**). This also defines the maximum height of water over the spillway before the dike is overtopped.

6. Select a shallow freeboard. For ponds with a maximum water depth of less than 3 m, **the freeboard should be at least 0.60 m but not more than 1 m**.

Determining the width of the emergency spillway

7. **The width** to be given to the spillway depends on a combination of the size of **the catchment area** that feeds the pond, **the maximum flood** to be discharged and **the freeboard** of the dike. To calculate the width of the emergency spillway proceed as follows:

(a) Determine **the size of the catchment area** (in ha) of your pond. **Water, 4** and **Topography, 16/2** provide useful information on how to do this.

(b) Determine **the flood discharge factor FD** for a freeboard of 1 m maximum from the chart according to the size of the catchment area.

FD = flood discharge factor

	Size of catchment area (ha)								
	50	100	150	200	250	300	400	500	600
FD	18	36	54	72	90	108	144	180	215

(c) From meteorological records, obtain **the mean annual rainfall** (in mm) for the region where the catchment area is found.

(d) Determine **the length of the catchment** (in km).

(e) Determine **the rainfall intensity factor RF** from the chart, according to the mean annual rainfall and the length of the catchment.

RF = rainfall intensity factor

Length of catchment (km)	*Mean annual rainfall*		
	400 mm	800 mm	1 200 mm
0.5	0.98	1.10	1.15
1.0	0.86	0.92	0.94
1.5	0.71	0.76	0.78
2.0	0.63	0.66	0.68
2.5	0.55	0.58	0.59
3.0	0.45	0.51	0.51
4.0	0.40	0.43	0.44
5.0	0.34	0.36	0.38
6.0	0.30	0.34	0.35

(f) According to **the predominant vegetation cover** of the catchment, determine factor **VF** from the chart.

VF = vegetation cover factor

Thick bush	0.05	Medium grass	0.15
Heavy grass	0.10	Cultivated land	0.20
Scrub	0.15	Bare or eroded	0.25

(g) According to **the predominant nature of the soil** in the catchment, determine factor **SF** from the chart. (**Soil, 6**, provides useful information on soils.)

SF = soil factor

Deep, well drained (sandy loam)	0.10	Shallow, with bad drainage (clay)	0.30
Deep, moderately permeable (loam)	0.20	Medium to heavy clays or rocky surfaces	0.40
Slowly permeable (clayey loam)	0.25	Impermeable surfaces or water-logged soil	0.50

(h) According to **the overall slope** of the catchment; determine factor **PF** from the chart.

PF = slope factor

Flat to gently sloping, 0-5 percent	0.05	Hilly or steep, slopes 20-40 percent	0.20
Moderately sloping, 5-10 percent	0.10	Mountainous, slopes greater than 40 percent	0.25
Rolling topography, slopes 10-20 percent	0.15		

(i) Calculate **the topographical factor TF** as **TF = VF + SF + PF**

(j) Calculate **the width of the emergency spillway** (in m) required as

$$\mathbf{W = FD \times RF \times TF}$$

where **FD** is the flood discharge factor;
RF the rainfall intensity factor; and
TF the topographical factor.

Example

A catchment area of 150 ha is 2 km long. The average annual rainfall is 800 mm. The area is heavily grassed and has deep, moderately permeable soils. The overall slope is moderate (5 to 10 percent). The dike is to be built with a freeboard of 0.80 m.

(a) From the charts obtain successively: **FD** = 54; **RF** = 0.66; **TF** = 0.10 + 0.20 + 0.10 = 0.40.

(b) Therefore **W** = 54 × 0.66 × 0.40 = 14.256 m.

- Select an emergency spillway width of 15 m.

Designing the emergency spillway

8. When designing the spillway, remember the following:

(a) **The level of the spillway crest** should be:

- 5 to 10 cm higher than the maximum water level if there is no mechanical spillway;
- about 25 cm higher than the level of the crest of the mechanical spillway, according to the maximum water depth allowed to flow through it (see Section 113).

(b) **The length of the horizontal spillway crest** should be at least 8 m.

(c) **The side walls of the crest** should have a slope of 3:1 or 4:1.

(d) **The inlet channel** should be reasonably short and have smooth, easy curves. Its bottom slope should be at least 2 percent, its side slopes should be at least 3:1, its entrance should be at least 1.5 times as wide as the crest width.

(e) **The exit channel** should be built in undisturbed soil whose natural slope is altered as little as possible. To avoid erosion, the bottom slope should be regular and not too steep. The direction of the slope should ensure that the discharged water will not flow against any part of the dike. The side slopes should be at least 3:1.

Note: whenever possible, select a natural channel for the spillway exit channel.

Locating the emergency spillway

9. The spillway should be built **at one end of the dike**, from which it should be **separated by some undisturbed natural ground**.

10. It should pass **around the end of the dike** and extend downstream on a gentle slope to some **safe channel** well away from the toe of the dike.

11. If necessary, you can **use the same site** for the emergency spillway as for the mechanical spillway and channel. However, you must be sure the channel can carry away the emergency water flow. Thus **the spillway should be sized according to the emergency flows**. As an approximate guide, the cross-sectional area of the channel should be that of the emergency spillway.

12. **The characteristics and topography of the soils** are very important in the selection of the spillway site. The soil should be natural, undisturbed and resistant to water erosion.

13. You should make **detailed soil and topographical surveys** before deciding on the site of your emergency spillway. Avoid loose sands and other highly erodible soils. Look for gentle and regular slopes.

Protecting the earthen emergency spillway

14. To prevent erosion by the turbulent water, you should protect the emergency spillway.

15. **If the soil and climate do not allow the growth of vegetation** on the bottom and side walls of the spillway, protect it as follows:

- strengthen the crest with wooden logs;
- cover the side walls with fixed brushwood or wooden boards; and
- use well-compacted rock, packed in with earth.

16. Whenever possible, protect your spillway with **a dense grass cover**, preferably using perennial grasses as recommended in Section 69.

(a) As soon as possible after construction, prepare the soil surfaces and apply fertilizers.
(b) Sow adapted grasses or transplant them.
(c) On the side slopes, protect the young grass by mulching*.
(d) Water well, if necessary, during dry periods.

Sections through an emergency spillway

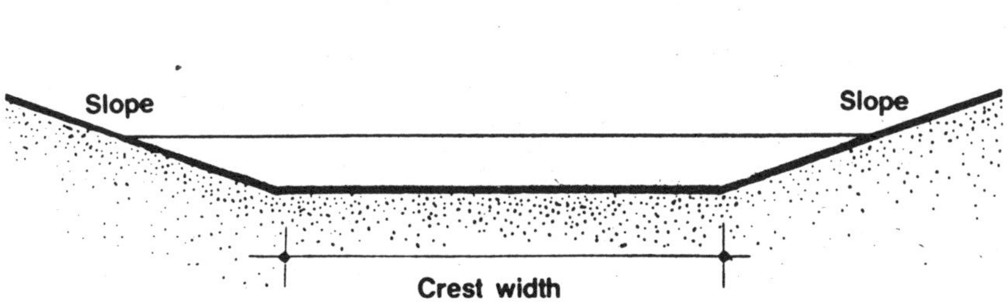

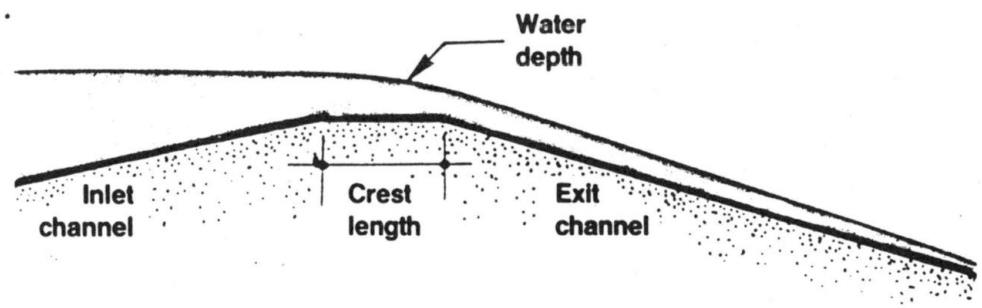

115 Protection canals

1. Protection ditches or canals are built to keep excess runnoff water from a particular area, and are usually built for one of three purposes:

- to percent a roadway or access track;
- to protect a water feeder canal;
- to protect a fish prod.

2. **For the protection of a road or feeder canal**, dig the canal uphill of and parallel to the road or feeder canal you wish to protect. It will collect and carry away the runoff by gravity. Depending on the local topography and the design of your fish farm, the runoff may have to be carried away along a route **crossing the road or feeder canal**. In this case, you can use a short pipeline (see Section 89) or an aqueduct (see Section 88) and have the runoff passing:

- **either above** the feeder canal or road;
- **or below** the feeder canal or road.

3. In the case of a road, you can also use a **simple ford**, or an **angled surface channel**.

4. When used for these purposes, protection canals are usually quite small, typically 0.5 to 1.5 m wide. You should check locally to see what is used for roadside ditches and use these as a guide.

Protecting a water feeder canal

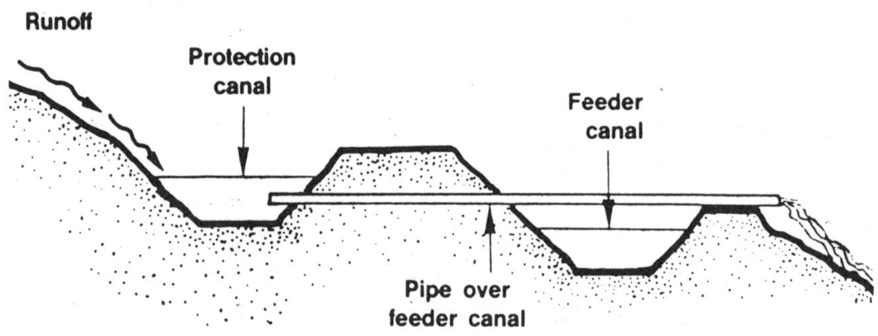

Protecting a roadway

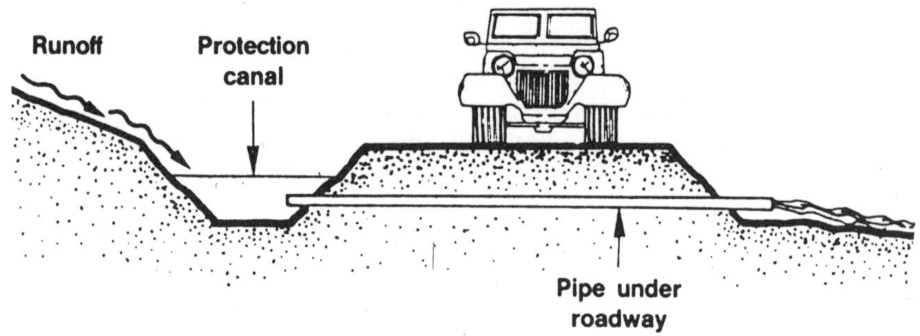

Protecting a pond

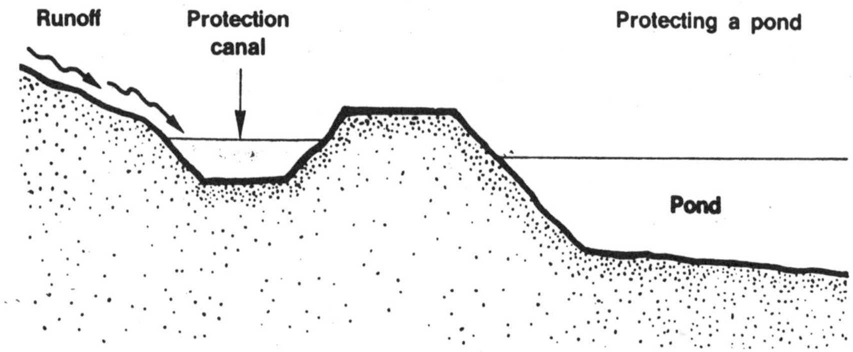

5. **To protect a fish pond**, dig the protection canal uphill of and around the pond.

6. Such a canal can also be used **to store rain-water during dry periods**. The pond can then be supplied with water as necessary, even after its normal supply has dried out. In this case, you should build the protection canal so that its bottom level is higher than the maximum water level in the pond. An overflow should be built at its end to release any excess water.

7. A pond protection canal, if wide enough (3 to 4 m, for example), is also useful for **limiting access to the pond**.

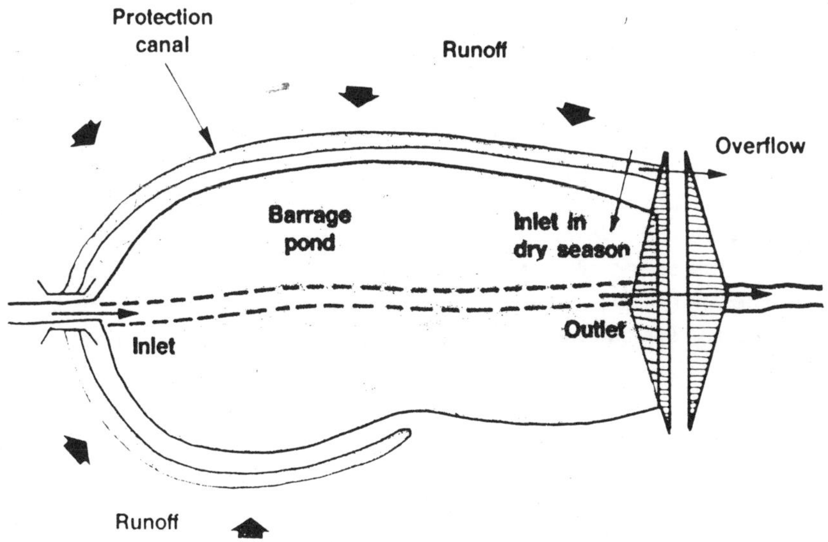

Note: at some sites, it may be useful to build terraced storage ponds.

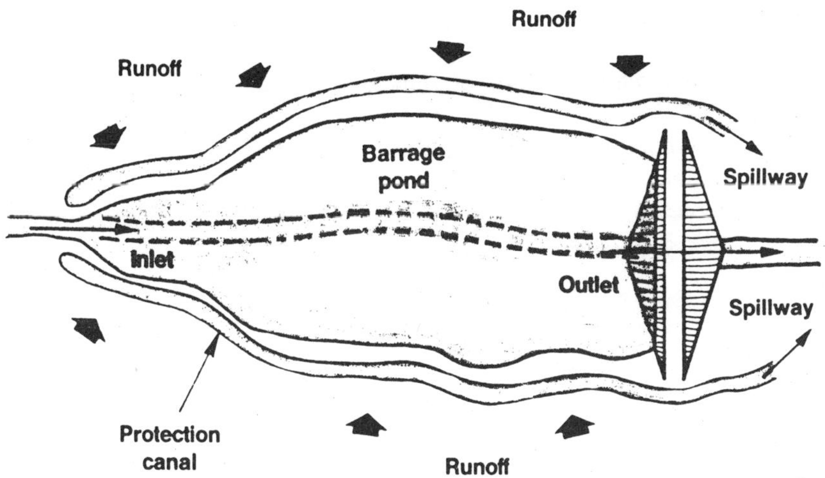

1. A settling basin is specifically designed to improve water quality **by removing the mineral soil particles**, such as fine sand and silt, which can be present in great quantities in certain waters with a high turbidity. This is achieved **by reducing the water velocity** sufficiently to allow the particles to settle. The ability of particles to settle is defined by their **settling velocity Vs** (in m/s), which decreases as the size of the particles decreases. The **horizontal critical velocity Vc** (in m/s), is also important. This is the speed of water flow needed to pick up and carry away a particle after it has settled. This velocity also decreases as particle size decreases. **Typical values for these velocities** are shown in the chart.

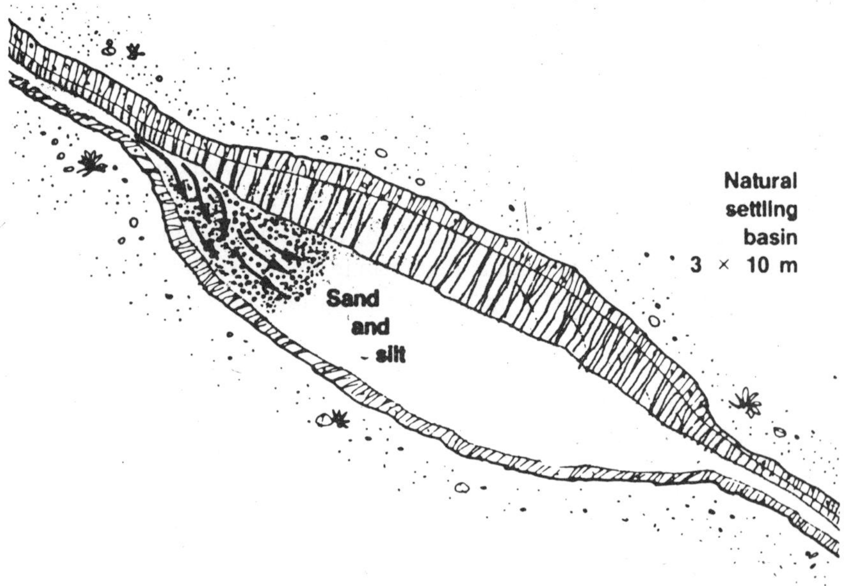

Settling velocity (Vs) and horizontal critical velocity (Vc) for soil particles

	Diameter of the soil particle (mm)										
	0.05	0.1	0.2	0.3	0.4	0.5	1	2	3	5	10
Vs (m/s)	0.002	0.007	0.023	0.040	0.056	0.072	0.15	0.27	0.35	0.47	0.74
Vc (m/s)	0.15	0.20	0.27	0.32	0.38	0.42	0.60	0.83	1.00	1.30	1.90

2. There are different types of settling basins:

- **a simple small pond**, built at the beginning of the water feeder canal;
- **a rectangular basin** built on the feeder canal with bricks, cement blocks or concrete.

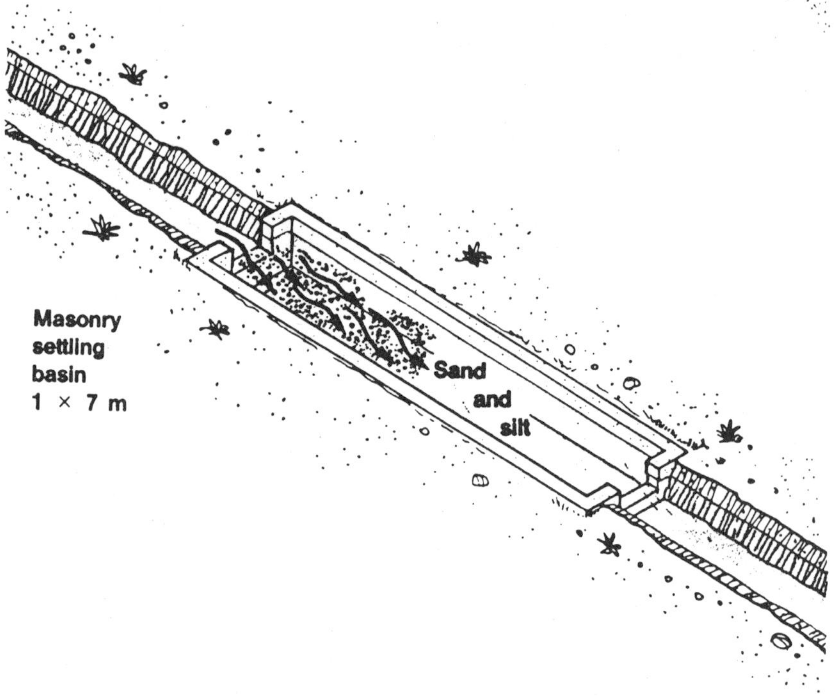

Settling basin

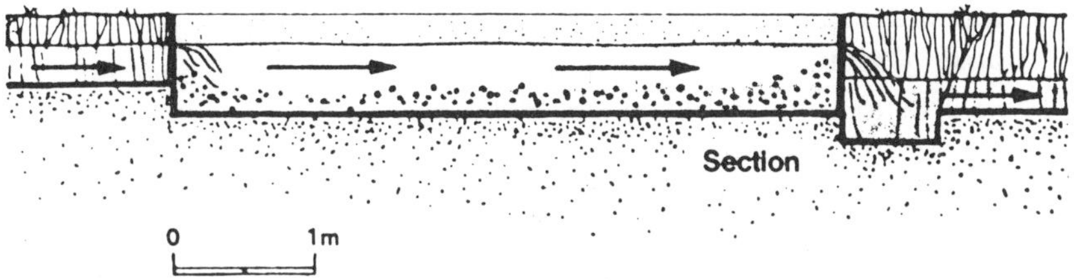

3. If the settling basin is **a simple rectangular basin**, you can determine its size as follows:

(a) **Its minimum horizontal area B** (in m^2) as

$$\mathbf{B = (Q \div Vs) \times E}$$

where **Q** (in m^3/s) is the maximum water flow to be discharged;
Vs (in m/s) is the **settling velocity** for the smallest particle size to be removed (see chart); and
E is the **end allowance factor** for extra space at each end needed to allow the water to run smoothly and evenly. Typically **E** = 1.3 or 1.4.

Example

If **Q** = 30 l/s = 0.030 m^3/s, to settle a particle which has a diameter greater than or equal to 0.1 mm, **Vs** = 0.007 m/s. Therefore the minimum horizontal area of the settling basin **B** = (0.030 m^3/s ÷ 0.007 m/s) × 1.3 = 5.57 m^2. Take **B** = **5.6 m^2**.

Note: in these ideal conditions, 100 percent of particles of 0.1 mm or larger should settle. A smaller proportion of smaller particles will also settle. The smaller the particles, the less the percentage settling.

(b) The **minimum cross-section area A** (in m^2) as

$$\mathbf{A = Q \div V}$$

where **Q** (in m^3/s) is the maximum water flow to be discharged;
V (in m/s) is the **selected water velocity**, which should be smaller than the critical velocity **Vc** (see chart), according to the size of the smallest particle to be removed in the basin.

Example

Following the example above, select for example: **V** = 0.10 m/s, to avoid removing particles with a diameter greater than or equal to 0.1 mm (from chart, for particles 0.1 mm in diameter, **Vc** = 0.20 m/s). Obtain the minimum cross-section of the settling basin as **A** = 0.030 m^3/s ÷ 0.10 m/s = **0.3 m^2**.

(c) **Its minimum width, w** (in m) as

$$\mathbf{w = A \div h}$$

where **A** (in m^2) is the **minimum cross-section area**, and
h (in m) is the **maximum water depth** in the basin.

Example

If, for the example above, **A** = 0.3 m^2 and **h** = 0.25 m, then the minimum width of the settling basin should be **w** = 0.3 m^2 ÷ 0.25 m = **1.2 m**.

(d) **Its standard length, L** (in m) as

$$L = B \div w$$

where **B** (in m^2) is the **minimum horizontal area**; and
w (in m) is the **minimum width**.

Example

If, for the example above, **B** = 5.6 m^2 and **w** = 1.2 m, then the standard length **L** = 5.6 m^2 ÷ 1.2 m = **4.6 m**.

(e) **Its overall dimensions**:

- **inside width** = **w** (in m);
- **inside length** = **L** (in m); and
- **depth** (in m) = water depth (**h** in m) + freeboard (0.20 m) + settling depth (0.10 to 0.20 m).

Example

For the above example, the settling basin characteristics will be:

- inside width, **w** = 1.2 m
- inside length, **L** = 4.6 m
- depth = 0.25 m + 0.20 m + 0.15 m
  = **0.60 m**

Note: the settling basin can be wider, with **a larger cross-section**. This will then allow the standard length to be shorter. As long as the critical velocities are not exceeded, the basin can be shaped to fit local space and to minimize construction costs. As a general guide, **ratios of length: width** are typically between 2:1 and 5:1.

4. **The bottom of the settling basin** is built lower than the bottom of the water feeder canal, to concentrate the soil particles being removed from the incoming water.

Improving the design of the settling basin

5. You can improve the above design in the following ways:

(a) At the entrance, make the water pass over **a wide edge near the basin's surface**, similar to a weir, to minimize disturbances.
(b) At the exit, similarly make **the water spread** over a wide edge near the basin's surface.
(c) Avoid **cross-wind exposure** as this can often agitate the water and resuspend particles.
(d) Within the basin, add **some baffles** to slow down the water further and make it follow a longer zig-zag path. With these baffles, you can **reduce the basin's length** by one third.
(e) Make sure water flows evenly and quietly through the settling basin. Avoid creating areas of turbulence or rapid flow.
(f) Provide **a sloping bottom** (slope = 2 percent) from the downstream end to the entrance of the basin.

6. You should regularly clean your settling basin by **removing the accumulated soil** from its bottom after closing the water supply. You can also remove this soil more regularly using a simple pipe or siphon. Usually, the soil is very fertile, and you should use it in your garden and fields to make your crops grow better.

Improved settling basin

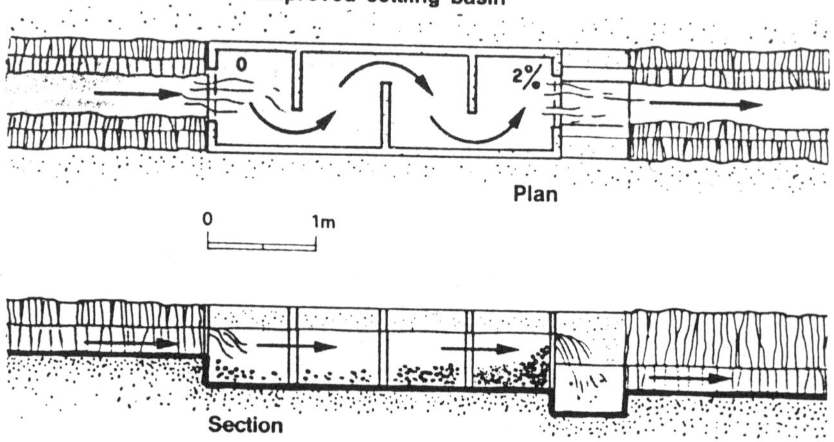

117 Stilling basins

1. Stilling basins are commonly used to slow the water down as it emerges from **the delivery pipe of a pump**; such basins also help to settle out any sand sucked up by the pump. They can be further used as settling basins for finer soil particles, if designed properly.

2. The average velocity of the water discharged from the stilling basin should be reduced below **the maximum velocity permissible** in the feeder canal to avoid eroding it (see Section 82 and **Table 35**).

3. A stilling basin usually has a square base, with a height greater than its base dimension. It can be built with bricks, cement blocks or concrete.

Designing a stilling basin

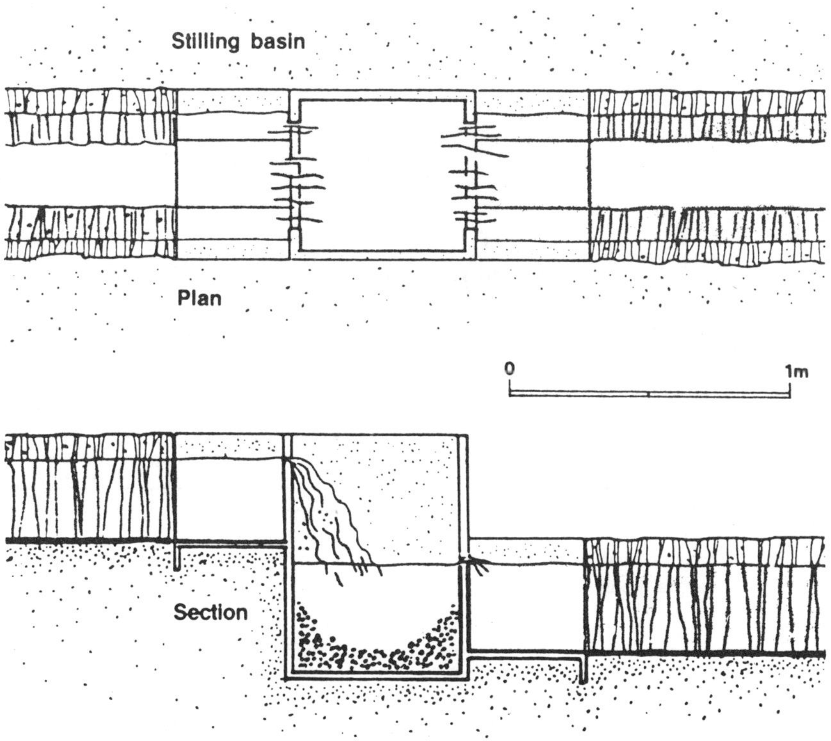

4. When designing your stilling basin, you should proceed as follows:

(a) Determine **the maximum water drop d** (in m) between the water surface upstream of the basin and the water surface downstream, in the feeder canal.

(b) Estimate **the minimum volume V** (in m^3) to be given to the basin as

$$V = (Q \times d) \div 125$$

where **d** (in m) is the **maximum water drop**;
Q (in l/s) is the **maximum water flow** to be discharged.

Note: if you wish the stilling basin **to also function as a settling basin** (see Section 116), divide **(Q × d)** by 40 instead of 125.

(c) Determine **the length L** (in m) to be given to the basin as

$$L \geq 1.5\ d$$

where **d** (in m) is the **maximum water drop** (see above).

(d) Determine **the water depth h** (in m) in the basin as

$$h = h\ (\text{feeder canal}) + 0.10$$

where **h (feeder canal)** is the water depth (in m) in the feeder canal, downstream from the basin.

(e) Determine **the inside width w** (in m) of the basin, which should be greater than the width of the upstream canal, as

$$\mathbf{w = V \div (L \times h)}$$

where **V** (in m^3) is the volume of the basin;
L (in m) is its length; and
h (in m) is the water depth.

(f) Determine **the width of the water inlet** (in m) at the basin's entrance as

$$\mathbf{w\ (water\ inlet) = w - 0.20\ m}$$

where **w** (in m) is the inside width of the basin.

Example

If the maximum water drop **d** = 0.40 m and the maximum water flow to be discharged **Q** = 50 l/s, then:

- **V** = (50 × 0.40) ÷ 125 = 0.16 m^3;
- **L** should be equal to or greater than 1.5 × 0.40 m = 0.60 m (make **L** = 0.70 m for example);
- **h** = 0.30 m + 0.10 m = 0.40 m (with 0.30 m water in feeder canal);
- **w** = 0.16 m^3 ÷ (0.70 m × 0.40 m) = 0.16 m^3 ÷ 0.28 m^2 = 0.57 m; take **w** = 0.60 m;
- **w** (water inlet) = 0.60 m − 0.20 m = 0.40 m.

Improving the design of the stilling basin

5. You can improve the above design by adding to the basin's bottom **a series of angle-irons** placed in alternating rows. These irons should be cemented vertically into the basin's bottom and should extend about 0.30 m above it.

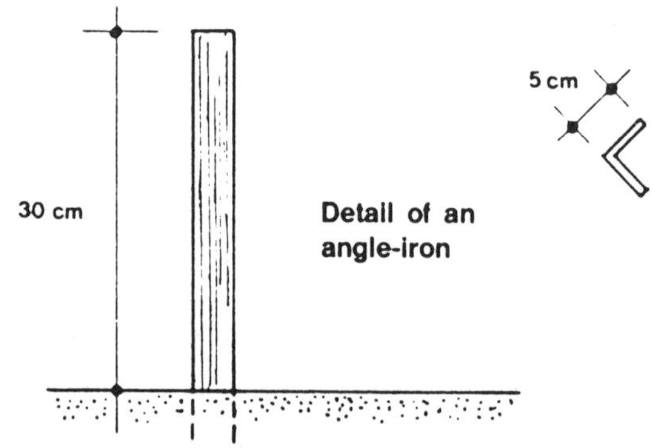

Detail of an angle-iron

Improved stilling basin

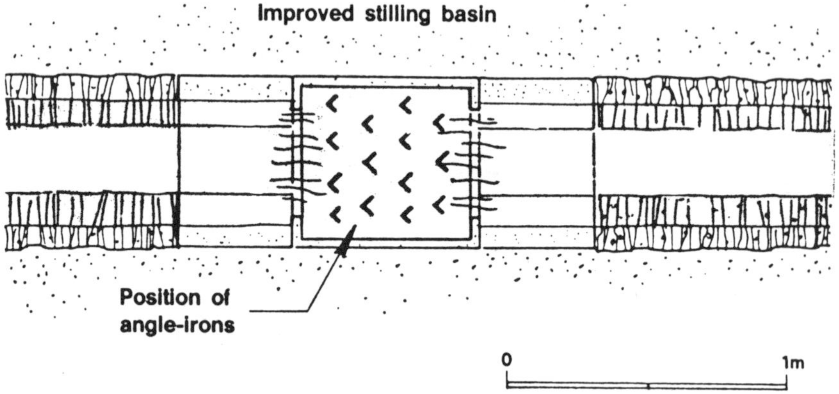

12 DETAILED PLANNING FOR FISH FARM CONSTRUCTION

1. In the previous chapters you learned how to:

- evaluate a potential site for the construction of a freshwater fish farm;
- select the type of pond best adapted to this site and to your needs;
- lay out your fish farm;
- build earthen ponds; and
- construct the various structures required for proper water control and transport.

2. You have **selected a good site** both from the technical and economical aspects (see Sections 22 and 23, **Construction, 20/1**). You have **surveyed this site in detail** for its topography and its soils. On these bases and according to your requirements, you have laid out your fish farm and prepared **a detailed topographical plan** of it (see Section 26, **Construction, 20/1**).

3. Now has come the time to decide on the following important issues:

(a) **When** will the construction begin (see Section 121).
(b) **Who** will construct the farm (see Section 122).
(c) **How** will the construction be done (see Section 123).

4. These decisions may lead to further activities, all of which are looked into in this chapter:

(a) Some more detailed **plans and drawings** may have to be prepared (see Section 124).
(b) A series of **specifications** for the contractor may have to be prepared (see Section 125).
(c) A detailed **schedule of activities** may have to be drawn up (see Section 126).

5. Finally, and in all cases, you will wish to know in advance **how much the construction of the fish farm will cost** (see Section 128).

The selected site

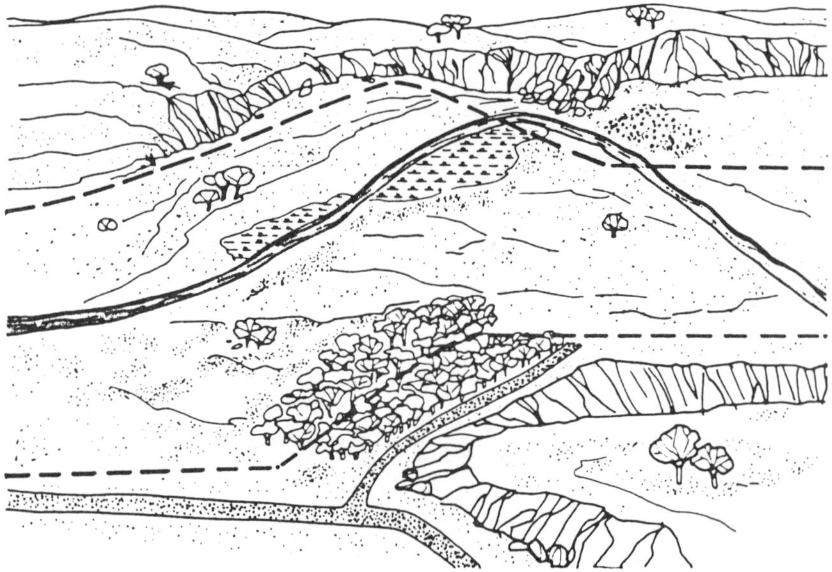

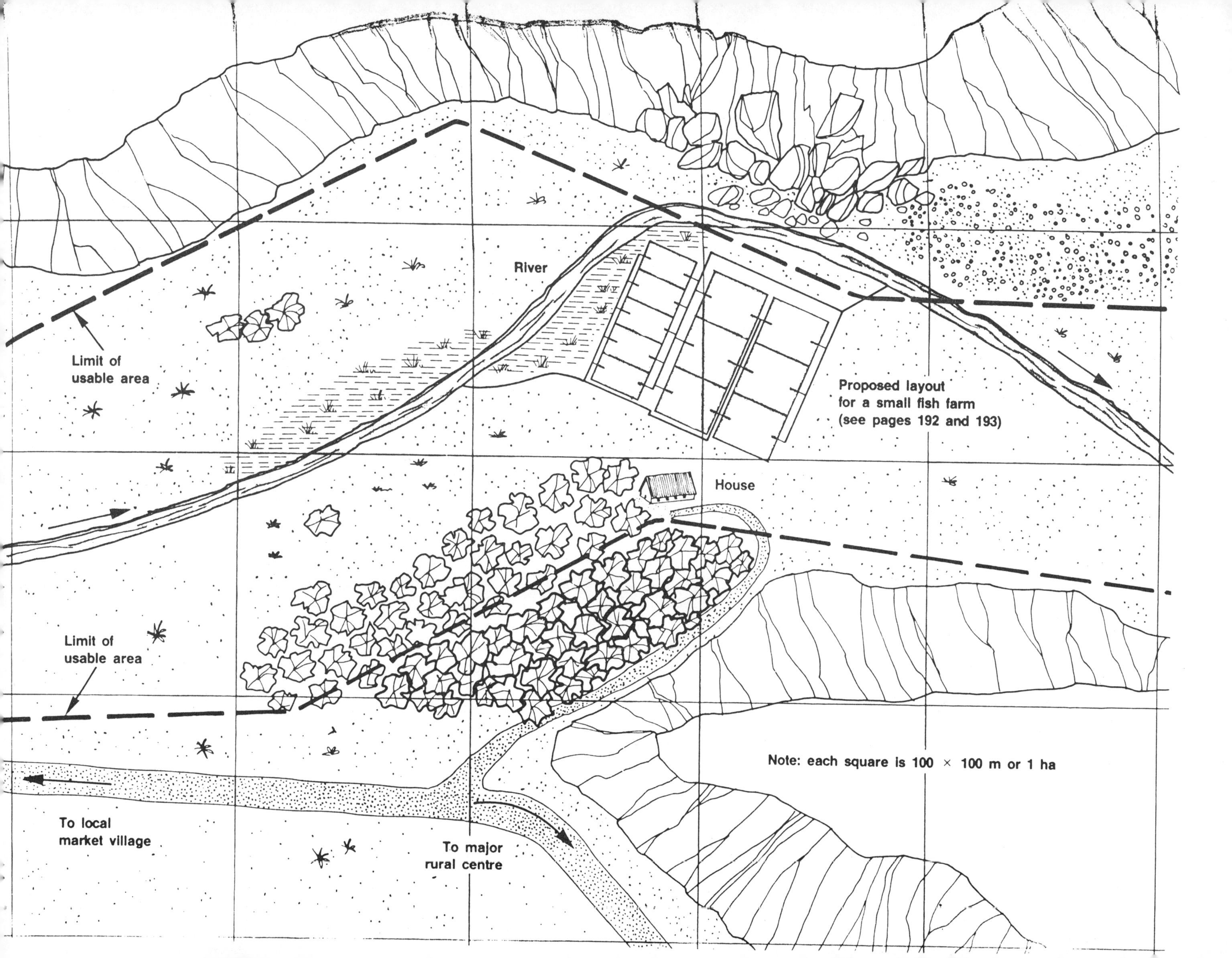

River
Limit of usable area
Proposed layout for a small fish farm (see pages 192 and 193)
House
Limit of usable area
Note: each square is 100 × 100 m or 1 ha
To local market village
To major rural centre

121 When to build the fish farm

1. Before deciding at which time of the year you should build your fish farm, you should ask these questions:

(a) When is **the site** easily accessible?
(b) When is **the soil** relatively dry, soft and easily workable?
(c) Will there be **water** available to fill the pond shortly after building it?
(d) Does it need to be prepared for **stocking with fish** at a particular time?
(e) Will there be **labour, machinery and materials** available at that time?
(f) Should I consider building the farm in stages, over several seasons?

2. **Accessibility to the site** and **workable conditions** are particularly important if you plan to use machinery for the earthwork. Remember that for materials and supplies such as sand, cement, gravel, wood and pipes, access to the site by vehicles may also be required. It is best to **fill the pond** with water shortly after construction to avoid the growth of weeds.

3. Under extreme conditions, select the season for construction using the following guidelines;

- **swampy conditions**: toward the end of the dry season;
- **heavy clay soils**: toward the end of the rains;
- **hard dry ground**: end of rainy season.

4. When building **a barrage pond**, it is best to select the period when the flow of the stream to be dammed is minimal (see Section 65).

122 Who will construct the fish farm?

1. There are three possibilities:

- either you construct the farm entirely yourself; or
- you pay a small, local contractor to do it; or
- you do it partly yourself and partly under contract.

2. You should consider **constructing all or part of the fish farm yourself**, for example when:

- **the fish farm to be built is rather small**, less than 1 ha, and you can rely on some technical assistance from a specialized extension service;
- there is **no qualified contractor** available locally, and you have some of the required experience;
- the interested contractors quote **prices much higher** than your own estimate; check the latter carefully and question the contractors before taking a final decision.

3. **If you decide to construct a larger farm yourself**, before starting its construction you should:

- decide whether you are going to construct it with or without **mechanical means** (see Section 123);
- prepare **the technical specifications** for the earthwork and the structures (see Section 125);
- prepare **a schedule of activities** (see Section 126);
- plan for **the necessary inputs** such as labour, tools, supplies/materials and equipment by working out how much you will need, when you will need these inputs and for how long.

4. **If you decide to employ a contractor** to do part or all of the construction work, it is simplest to discuss **a direct contract** with a known contractor and to agree with him on **an all-inclusive price** for the job. This contractor should not only have the required technical qualifications, but should also have reliable financial credentials.

5. In this last case, you do not have to decide for yourself the best way in which to construct the fish farm according to your plans. This should be done by the contractor. But first, you will have **to prepare plans and drawings** (see Section 124) **and specifications for the contract** (see Section 125). The contractor will base **the price of the contract** on these plans, drawings and specifications. Compare this price with **your own estimate** (see Section 128) and accept it **only** if they do not differ too much.

123 Constructing the fish farm

1. If you have decided to construct your fish farm yourself, you have two choices of ways to proceed:

- either using **manual labour only**; or
- using **machinery** in part and using **manual labour** in part.

2. The choice largely depends upon **the size of the farm** to be built and on **the availability of machinery**. Very small farms up to 1 000 m^2 are usually built using manual labour only. It is also important to have the farm constructed within a reasonable time to bring it into production and reduce the time before earnings are made on your investment.

3. **For manual construction** you will need simple equipment such as picks, hoes, shovels and wheelbarrows. It can be done by yourself and your family, assisted by some friends if necessary. You can also contract someone to dig the pond manually for a fixed price based on the earthwork. The size of individual ponds generally does not exceed 400 m^2. From the volume of earthwork required, you can estimate how long it will take you to build each pond (see Chapter 4, **Construction, 20/1**) and, if necessary, how much it would cost you to subcontract its construction (see Section 128).

Manual construction

4. **Mechanical construction** is done with earth-moving machines such as bulldozers and wheel loaders (see Section 48, **Construction, 20/1** and Section 127 of this volume). Compaction equipment can also be used. This construction method is much faster and not necessarily more expensive than using manual labour only, but it requires the selected **site to be accessible** to the machinery and to have **adequate soil conditions**. It also requires skilled operators. The amount of earthwork should be large enough to justify the costs of transporting equipment to the site. **One way of reducing these costs** would be to join your neighbours and build several fish farms in the same area, using the same machinery.

5. Normally, you would hire the services of **a contractor who owns the necessary equipment**. This is usually done either with a price determined by the nature and amount of work to be done (see Section 122) or for a weekly, daily or hourly rate. While the latter could be cheaper, you should make sure that the operator is skilled, and be careful if there is a risk of the machinery breaking down or getting stuck during construction. Before discussing **the contract**, it is best to estimate yourself the volume of earthwork to be done and, from this, the machinery time needed (see Section 127). You can do similar calculations for other types of construction work, such as site preparation and dike compaction. Remember that it will always be necessary to use manual labour as well, particularly for finishing the construction.

6. A **mixed method** is often the most advantageous way to construct medium-size fish farms. This only involves the use of earth moving equipment, for example an ox-drawn scoop or scraper or a small bulldozer, as discussed in Chapter 4, **Construction, 20/1**, to speed up the main earth movements. All other types of work are done manually.

Mechanical construction

1. At the beginning of this manual (see Section 26; **Construction, 20/1**), you learned how to prepare **a topographical plan** showing the site elevations and the proposed layout of the fish farm, including all its structures.

2. **If the fish farm to be constructed is rather small** (less than 5 000 m^2), you will not usually need to prepare more detailed plans. It is usually sufficient **to mark the main dike boundaries** and **estimate the volumes of earth** required. You will only need to use detailed calculations and markings if the ground is very irregular (see Sections 64 to 68, **Construction, 20/1**). As distances are not great, the planning of earth movements is not so important.

3. The most useful additions to the existing **topographical plan** are the following:

- elevations of the tops of the dikes;
- elevations of the pond bottoms;
- elevations of the feeder and drainage canals;
- characteristics of the dikes (side slope, length, thickness);
- characteristics of the canals (side slope, bottom width);
- characteristics of the roads (elevations, width);
- positions of the other structures such as main water intake, pond inlets/outlets, division boxes on canals, spillways, etc.

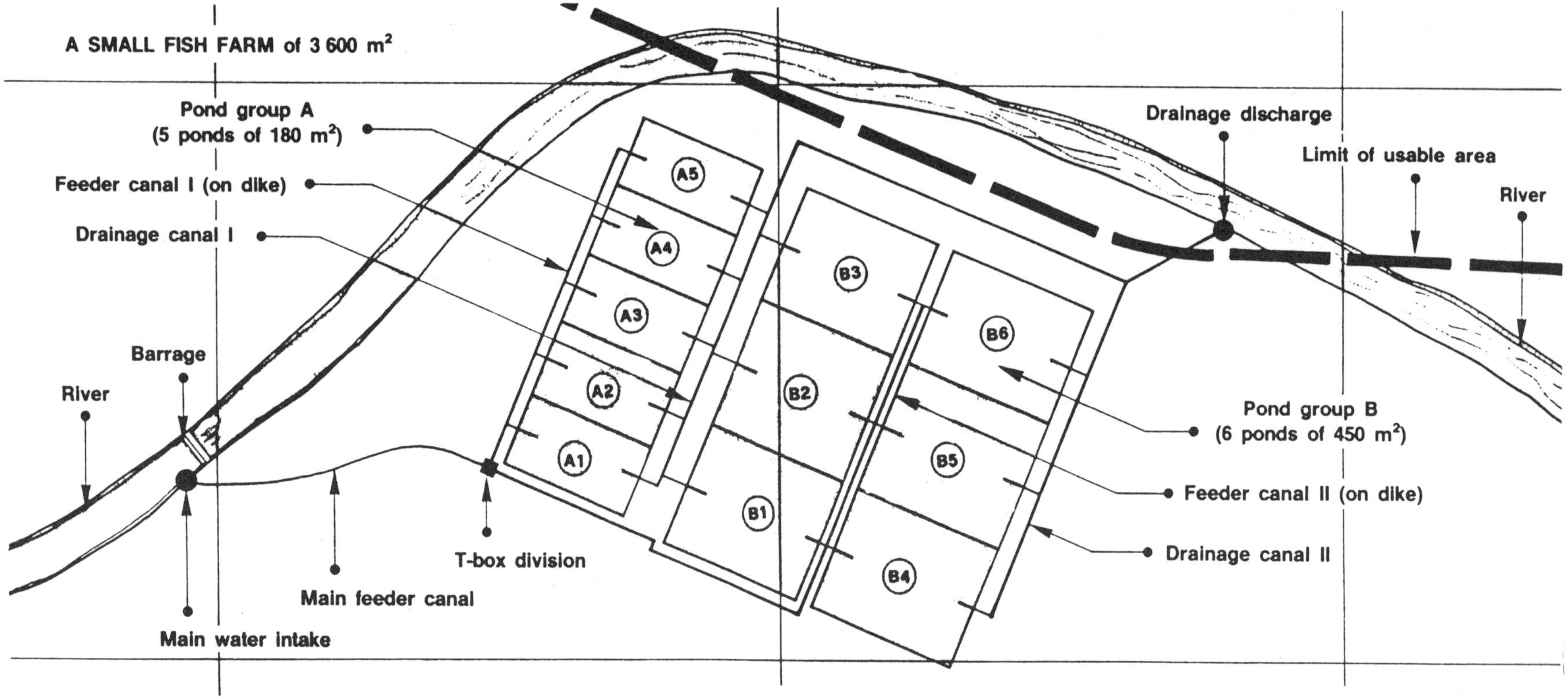

General layout for a small fish farm

KEY

- – – – Feeder canal (FC)
- —·— Drainage canal (DC)
- ▲ Pond inlet
- ● Pond outlet
- ■ T-or X-division box
- ⬇ Elevation measurements

NOTES: Water depths of pond group A: 1.0 to 1.2 m
Water depths of pond group B: 1.1 to 1.3 m
Freeboard: 0.50 m

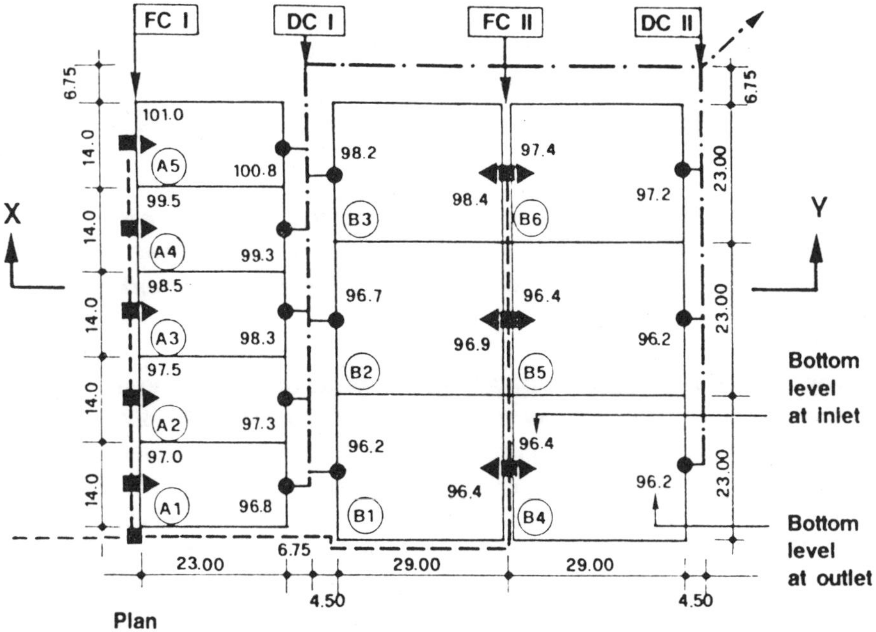

Plan

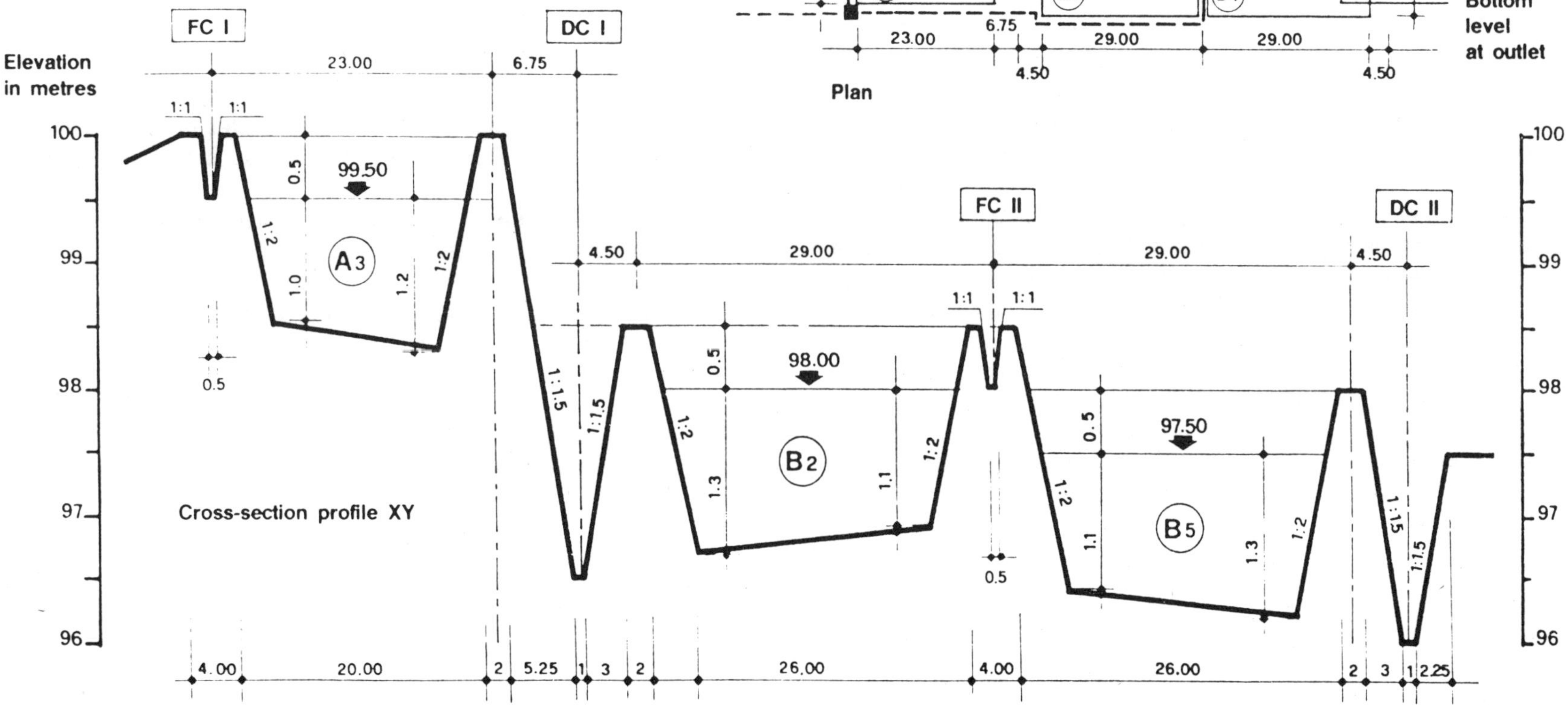

Cross-section profile XY

4. It may also be useful to prepare **detailed drawings of the various structures** as a basis for estimating their individual cost (see Section 128) and for their construction. Elevations should be clearly indicated to avoid any mistakes later.

5. **If the fish farm to be constructed is larger** (greater than 5 000 m^2), it is advisable to prepare more detailed work plans to be closely followed during construction. For each fish pond prepare **a topographical work plan** in the following way:

(a) **Survey the site area** in more detail, either by radiation or using a square grid as discussed in Sections 81 and 114, **Topography, 16/2**. This survey is best done after partial clearing of the vegetation (see Sections 52 to 55, **Construction, 20/1**), and it should provide you with accurate information on the positions and the levels of the pond site area. At this stage you will have pegged out the pond area.

(b) Start **drawing the topographical work plan** of the pond area, including all surveyed points.

(c) Determine the best excavation depth to **balance cut and fill volumes** according to the type of pond (see Sections 64 to 68, **Construction, 20/1**).

(d) Make sure that these calculations tie in with the required **feeder and drainage canal levels**, and that local **soil conditions** such as areas of rock, permeable soil, sources of clay, etc. are taken into account.

(e) Try **to minimize the distances** you move earth within the pond, particularly if you plan to build the pond by hand. As an approximate guide, on flat ground you should aim to transport the earth no more than about one-quarter of the pond's width and on steeper slopes, no more than two-thirds of the pond's width.

(f) Make sure that access roads or tracks, feeder canals and drains are laid out to service the site efficiently and well.

Note: in some cases, particularly for **large sites with many ponds**, it may be necessary to move earth from one area of the site to another. Allow for the extra earth volumes taken away or brought into the pond. Check that the earth volumes balance out over the whole site.

(g) Determine the **key elevations** of the pond.

(h) **Complete the topographical work plan of the pond** by entering the following information on it:

- the elevations of the top of the dikes;
- the positions and elevations of the inlet and outlet;
- the positions and elevations of any other structure to be built in the pond area;
- the characteristics of the dikes.

6. It is also better to prepare **cross-section profiles** of the site and the ponds (see Section 96, **Topography, 16/2**) **in two perpendicular directions**, especially if the site is sloping and if the ponds are to be constructed at different elevations. Indicate on them the key elevations of the ponds and other structures.

7. Add to the existing **topographical plan of the fish farm** all information relating to the feeder/drainage canals, road system, service buildings, etc., which do not appear on the work plans.

8. Prepare **detailed drawings** of the various structures, indicating clearly their key elevations and giving a reference number to each kind of structure.

General layout for a larger fish farm of 16 ponds of 0.25 ha or 4 ha total

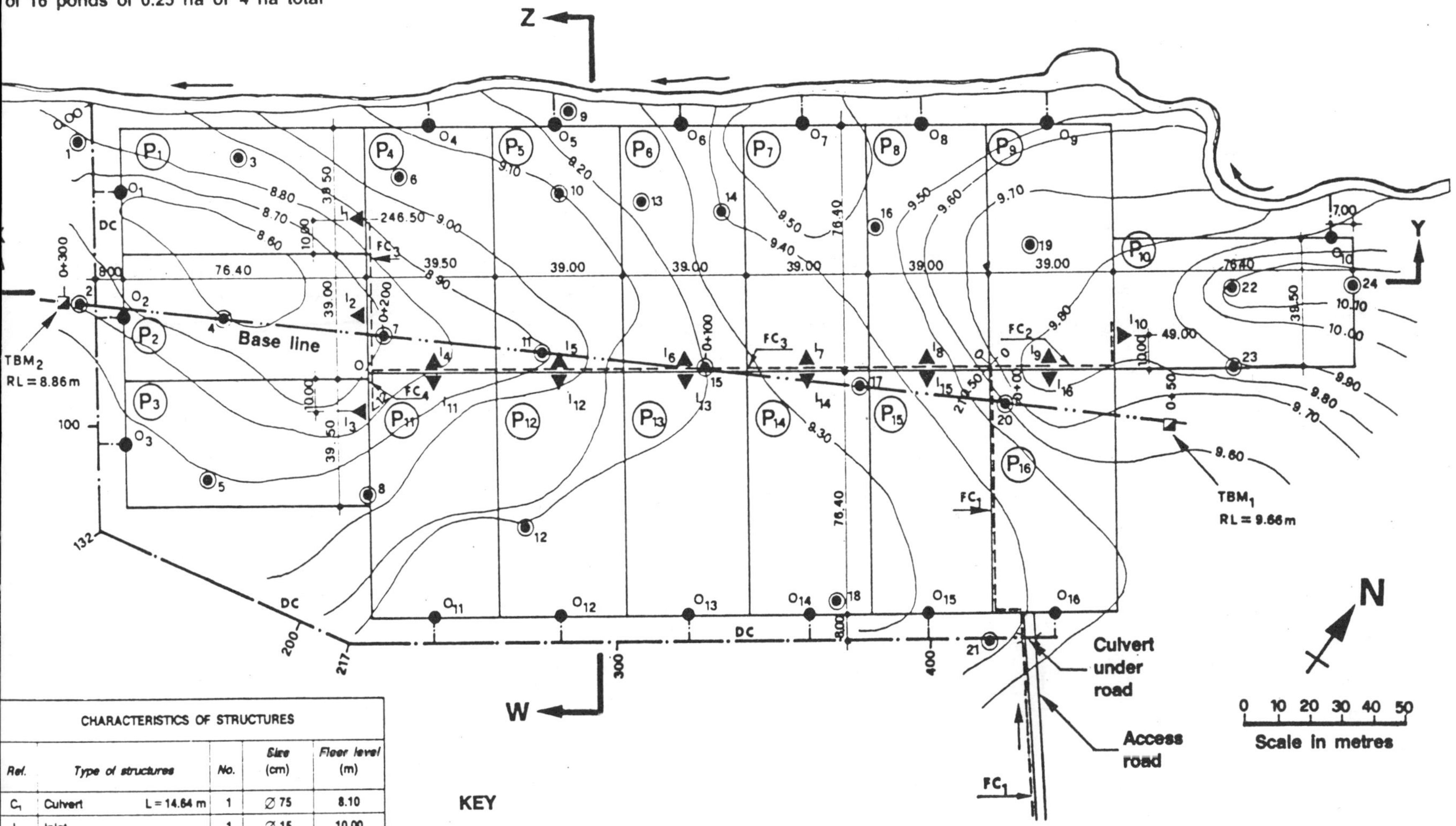

Ref.	Type of structures		No.	Size (cm)	Floor level (m)
C_1	Culvert	L = 14.64 m	1	∅ 75	8.10
I_1	Inlet		1	∅ 15	10.00
I_2-I_9	Inlet		8	∅ 15	10.10
I_{10}	Inlet		1	∅ 15	10.00
I_{11}-I_{16}	Inlet		6	∅ 15	10.10
O_1-O_{16}	Outlet		16	∅ 30	8.70
FC_1	Feeder canal	L = 210.50 m	1	40/60	11.90-10.40
FC_2	Feeder canal	L = 49.00 m	1	40/60	10.20-10.00
FC_3	Feeder canal	L = 246.50 m	1	40/60	10.40-10.00
FC_4	Feeder canal	L = 12.00 m	1	40/60	10.10-10.00
DC	Drainage canal	L = 439.60 m	1	200/275-350	8.20-7.70

Table: CHARACTERISTICS OF STRUCTURES

KEY

- ◉ Soil sampling station
- ◪ Temporary bench-mark (TBM)
- RL Relative level
- —··— Base line
- – – – Feeder canal (FC)
- —·— Drainage canal (DC)
- ▲ Pond inlet (I)
- ● Pond outlet (O)

NOTES:

Water depths in ponds: 1.0 to 1.3 m
Freeboard: 0.50 m
Side slope of dikes: 1:1.5
All measurements are in metres
Contour lines at 0.10 m intervals
For cross-section profiles XY and WZ see page 196

Detailed drawings for a larger fish farm

Typical cross-sections for a larger fish farm

Cross-section profile XY (see page 195)

(see page 195)

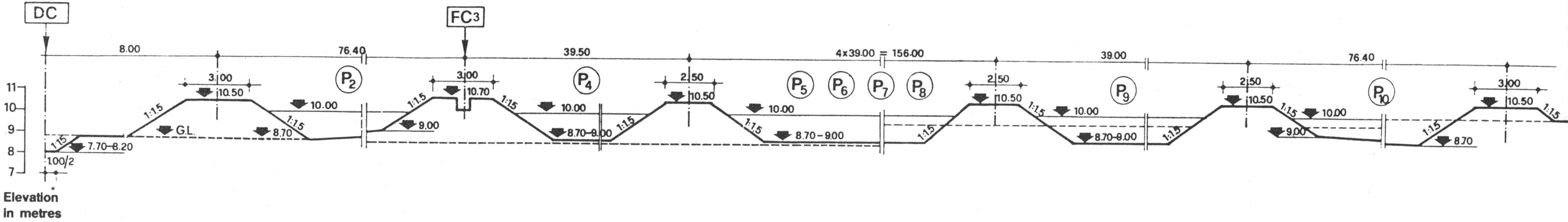

Cross-section profile WZ (see page 195)

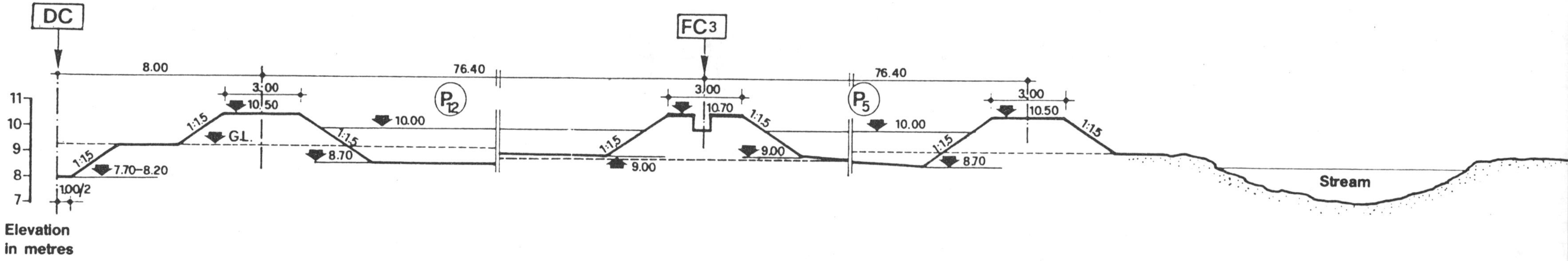

NOTE:

All measurements are in metres

KEY

Elevation measurements

Typical structures for a larger fish farm

Cross-section of a feeder canal

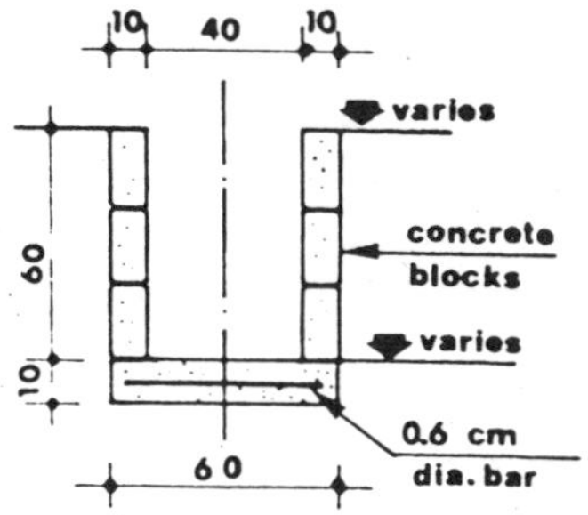

Cross-section of a feeder canal with double inlet

KEY

Elevation measurements

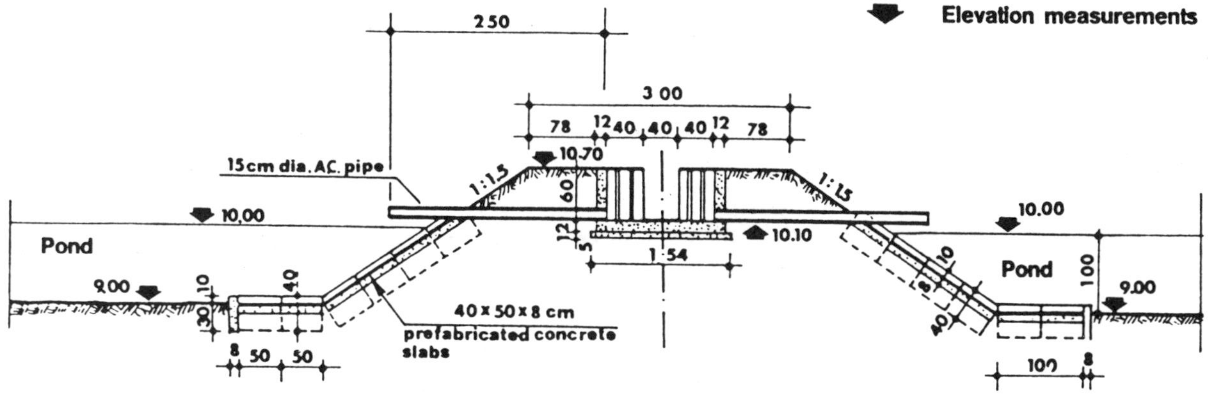

Cross-section of a pond inlet

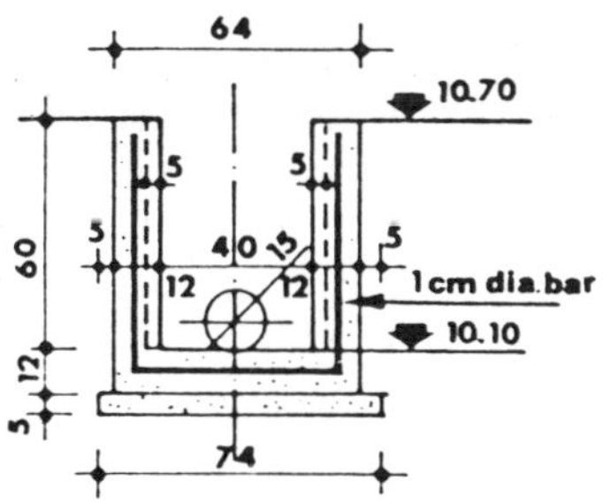

Cross-section of a pond outlet with monk

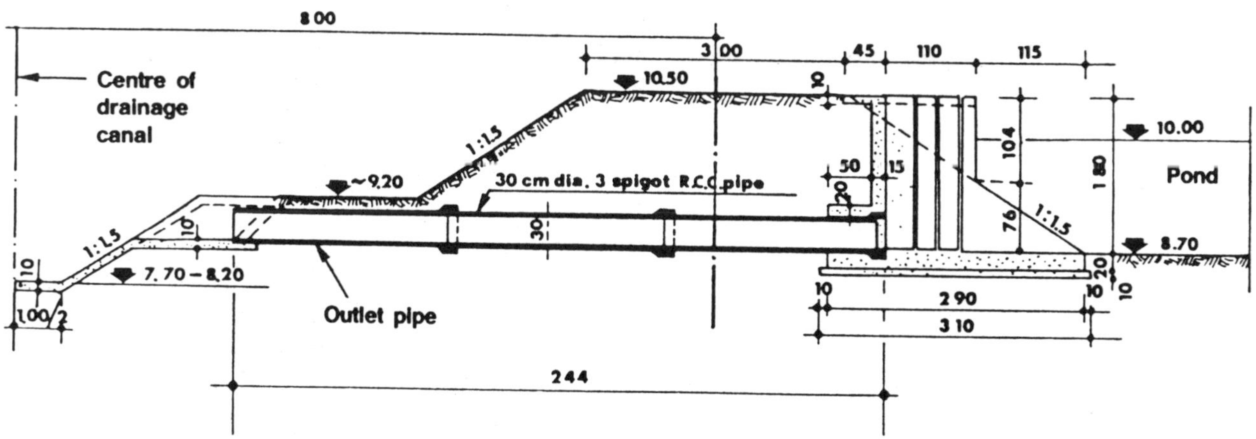

NOTES:

Elevation measurements are in metres
All other measurements are in centimetres

125 Making specifications for construction contracts

1. Preparing the specifications for the construction of a larger fish farm can be a very complicated task, and this should be carried out by a qualified technician or engineer. However, **if your fish farm is small** (less than 5 000 m^2), you may prefer to prepare the specifications yourself for its construction under contract, preferably with some assistance from your extension agent.

2. Make a detailed list of the **non-technical specifications**, under the following headings:

- **general description of the site and work** to be done, including number and size of ponds, canals and structures;
- **description of contractor's responsibilities** for the constructions until completion, delivery and acceptance;
- **description of supervisory, testing and acceptance conditions;**
- **description of payment schedule** according to work progress;
- **time limits for delivery or completion**, and possibilities for extension (for reasons beyond contractor's control) and penalties (for late delivery and loss of production/interest).

3. List the **technical specifications** clearly and in detail, referring to available topographical plans and detailed drawings (see Section 124). These specifications should deal separately with the earthwork and the structures as follows:

(a) **Earthwork specifications**:

- site clearing, including total removal of stumps and roots, handling and placing of cleared vegetation;
- removal of topsoil, including details of area, thickness, storage;
- construction of dikes, including the origin and quality of the soil and its characteristics;
- compaction, covering maximum thickness of layers, soil moisture, content, type of equipment to be used.

(b) **Structure specifications**, listing type and quality of materials to be used in each case, such as:

- reinforced concrete, including mix type, slump test limits, reinforcement quality, curing process, forms;
- wood, listing details of species, treatment, relative humidity, storage conditions;
- bricks or concrete blocks, noting quality, finish, type, weight, storage conditions;
- pipes, listing type, material, storage, handling, laying;
- mortar and plaster mixes, additives, water, etc.;
- paints, specifying number of coats, type.

4. To be able to give **a price quotation** for the construction contract, the contractor will require all the above specifications together with topographical plans and detailed drawings.

126 Making a schedule of activities

1. If you have decided to construct the fish farm yourself, you should prepare a realistic schedule of activities to help you plan for the required inputs.

2. Before starting construction work, in most cases it is necessary to have **a preparatory phase** in which you will:

- **stake out** the position of the ponds, dikes, canals, etc.;
- **clear the vegetation**;
- **remove the surface soil** and store it;
- establish **temporary bench-marks**;
- assemble **the construction materials** required on the site;
- **stake out** in detail the dikes, canals, pond bottom, etc.;
- dig **temporary drains** for carrying away excess seepage or runoff;
- dig **protection canals** (see Section 115).

3. The **temporary bench-marks** (TBMs) that you establish allow you to determine and check by levelling the elevations of the dikes, canals and other structures (see Section 81, **Topography, 16/2**). The main points to consider are that:

- the **number** of TBMs required increases with the size of the farm;
- these TBMs should have the same **reference level** as the existing topographical bench marks;
- the TBMs should preferably be set **around the perimeter dike**;
- the TBMs should be well **fixed and protected** during the whole construction period.

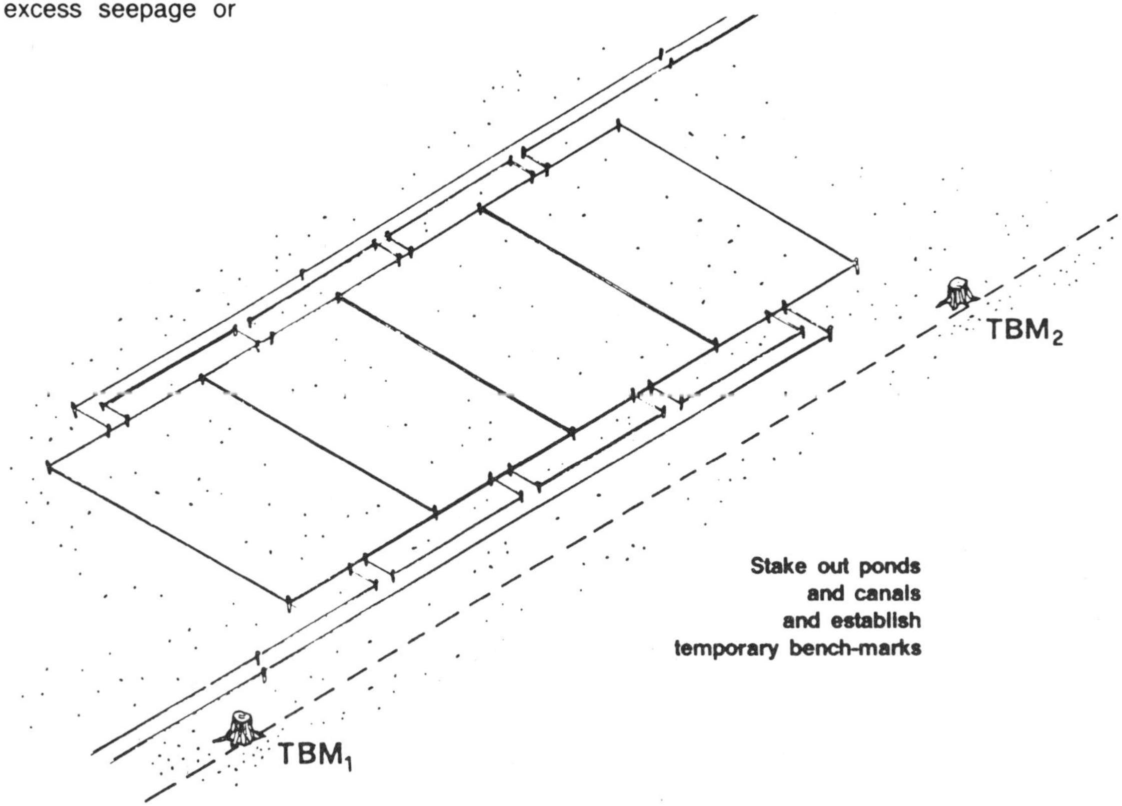

Stake out ponds and canals and establish temporary bench-marks

TABLE 50

Usual order of pond construction phases

Construction phase	*Sunken ponds*	*Barrage ponds*		*Diversion ponds*
		Without diversion canal	With diversion canal	
Diversion canal/main feeder canal	—	—	1	7
Temporary diversion of stream	—	—	2	—
Pond inlet(s)	—	—	3	11
Secondary feeder canal(s)	—	—	—	8
Main water intake	—	—	—	9
Barrage on stream	—	—	—	10
Pond dike(s)				
foundation	3	6	8	1
construction	4	7	9	6
forming	5	8	10	12
grassing	6	9	11	14
Pond bottom levelling	—	—	—	13
Pond outlet(s)				
excavation of pit(s)	—	1	4	2
construction of structure(s)	1*	2	5	4
refilling/compacting pit(s)	—	3	6	5
Spillway(s)				
mechanical spillway	2*	4	—	—
emergency spillway	7*	5	7*	—
Drainage canal(s)	—	—	—	3

* If necessary

4. The construction of one or more ponds and structures can then be initiated, following a sequence which varies according to the type of pond, as suggested in **Table 50**. Not all the steps are necessary in all cases, but depend on the farm design and the kind of structures to be used.

5. You may need **to change the order of construction** in particular circumstances.

(a) **Under swampy conditions or whenever the flooding of the site** is to be avoided, it is best to build the drainage system of the farm first.
(b) **Whenever the ground is flat** (slope less than 0.5 percent), it may be easier to build the drainage canal first, then the inlet canal and then to adjust the levels of the pond dikes and bottom accordingly, at an intermediate elevation.
(c) **For small artisanal ponds**, it may be easier to finish building the dikes before building the pond outlet. Then you need to cut the dike, put in place the outlet structure, and then rebuild the dike on top.

Sunken pond

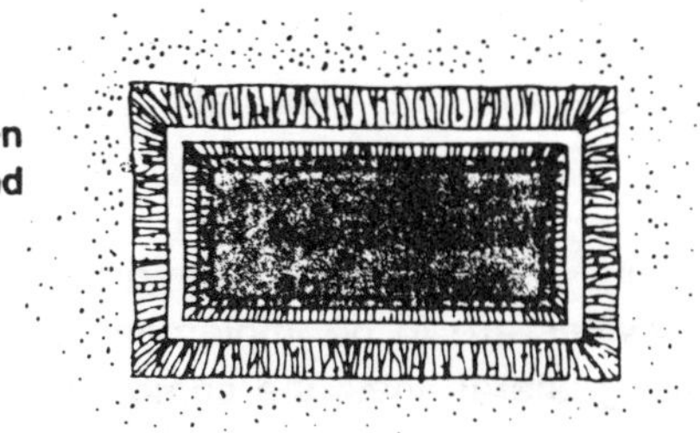

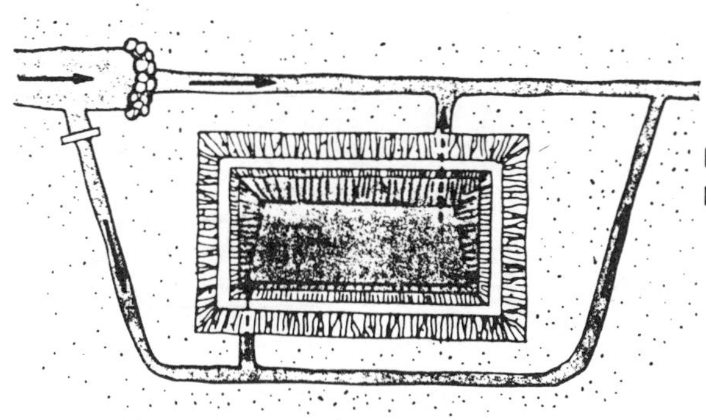

Diversion pond

Barrage pond without diversion canal

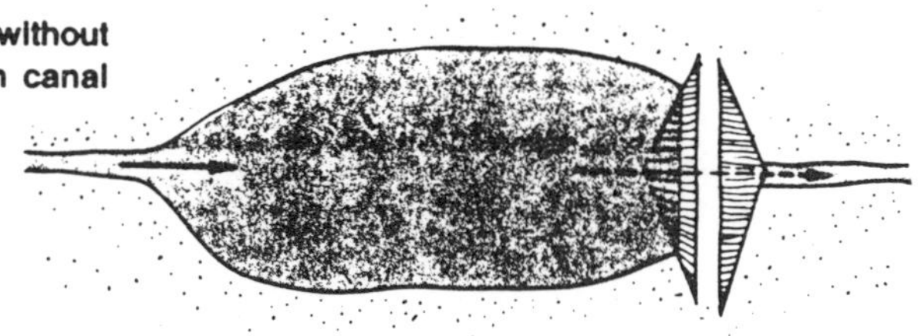

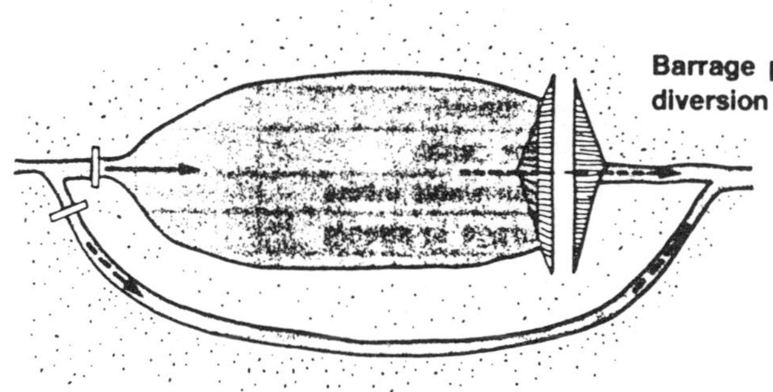

Barrage pond with diversion canal

6. On the basis of this information, prepare **a schedule of activities** showing:

- along a central line, the order in which **the indispensable activities** are to bé implemented; and
- along lateral lines, **the complementary activities**, which can be implemented simultaneously with some indispensable activities.

7. You can **improve this schedule of activities** by adding a background **time scale** (in weeks for example) and by showing for each activity:

- **when** it is planned to take place; and
- **how long** it will take.

8. To do this, use **the working standards** which are given in the next section together with your earthwork estimates and the list of structures to be built.

Schedule of activities

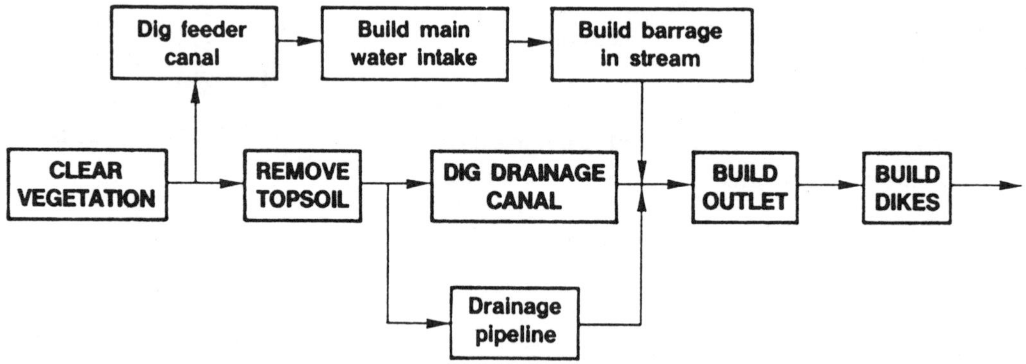

Time scale week by week

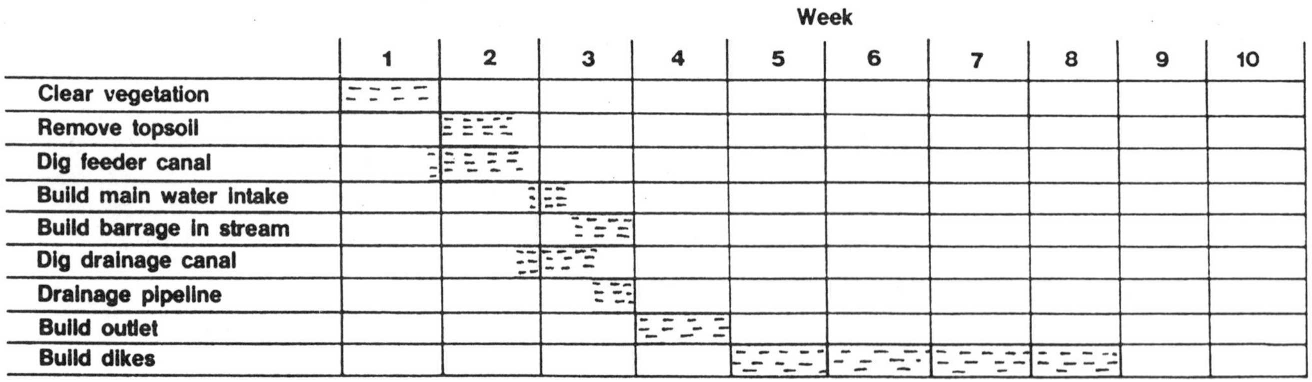

127 Working standards for planning purposes

1. You will use working standards both before construction starts and during its execution, for example:

- to estimate the number of labourers and how long you will need them for each construction phase;
- to select the type of earth-moving equipment and to estimate for how long you will require it;
- to estimate how much the construction will cost you and, if necessary, to discuss with a contractor the price being proposed.

2. In the following paragraphs, general standards useful for fish farm construction are provided.

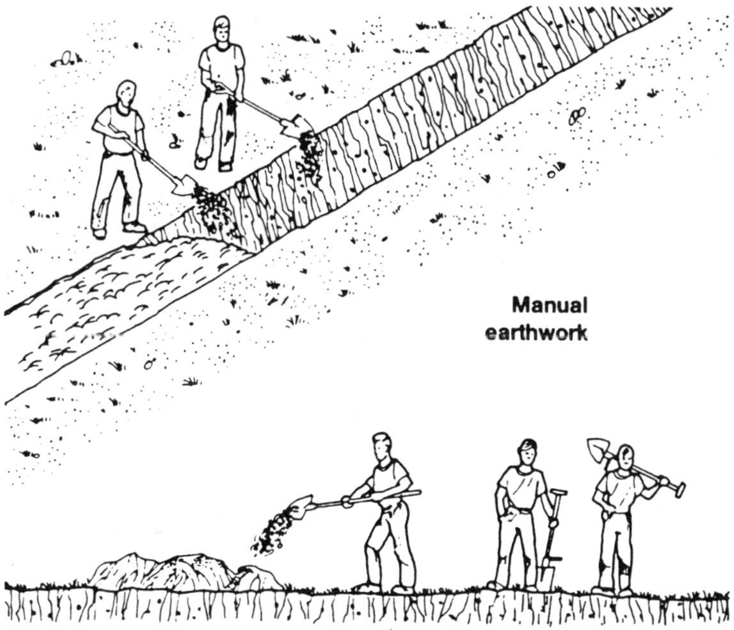

Manual earthwork

Working standards for manual earthwork

3. These working standards will vary mainly with **the nature of the soil**. The harder the soil, the more difficult it is to work it and the lower the work outputs. The presence of excess water also decreases outputs, particularly in heavy and sticky clays.

Example

Working hours for manual construction of rural ponds

	Farm 1*		Farm 2** (working hours)		Farm 3***	
Main water intake with small dam on stream		130		266		130
Feeder canal	(200 m)	*50*	(200 m)	*50*	(270 m)	*70*
Excavation/dikes construction	(150 m^3)	*600*	(400 m^3)	*1 600*	(950 m^3)	*3 600*
Inlet/outlet pipes		5$^+$		4		90$^+$
Total working time		785		1 920		3 890

* One 400 m^2 diversion pond
** Two 200 m^2 diversion ponds
*** Four 400 m^2 diversion ponds and two 100 m^2 diversion ponds
$^+$ Including concrete

4. **When planning earth movements**, you should take into account that to minimize costs, **distances** should be limited as follows:

- horizontal distance for throwing earth: maximum 4 m;
- vertical upward distance for loading earth: maximum 1.60 m;
- oblique distance for loading earth: maximum 4 m.

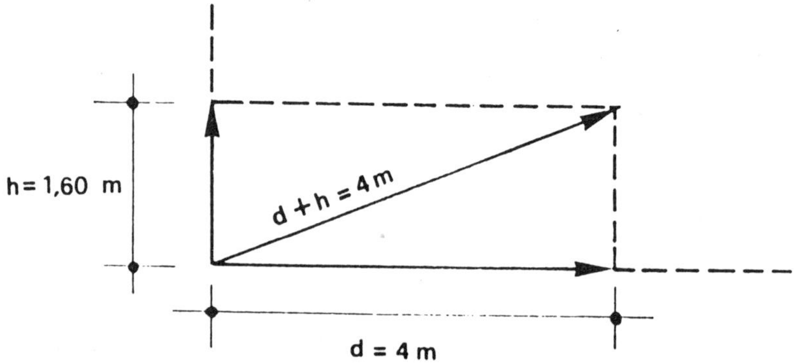

5. **Working standards for earthwork with manual labour** are given in **Table 51**. These are the average outputs to be expected from labourers of average strength working for eight hours at digging and throwing the earth 1 m away. The **minimum values** are for excavating and throwing 1 m away by hoe. The **maximum values** refer to the use of pick and shovel under similar conditions. These outputs should be slightly reduced as the throwing distance increases.

6. **For excavating and forming canals**, the output for a trained labourer varies from 0.8 to 1.2 m^3/day.

TABLE 51

Average output of manual labour for excavations

Nature of soil	*Volume excavated/thrown away* (m^3/8h)	
	By hoe	By pick/shovel
Soft (alluvium, sandy soil)	2.5-3	3.5-4
Moderately hard (loam, light clay)	1.5-2	2.5-3
Hard (heavier clay)	1	2-2.5
Lateritic, moderately hard*	0.5	1-1.5
Water saturated	0.8-1.5	1.5-2

* See Section 18, **Soil, 6.**

7. The nature of the soil determines **the excavation method**. If the soil is soft, it may be possible to work with a shovel only. If the soil is harder, it is best to first use a pick to loosen the soil before shovelling it away. In this case, the following outputs can be expected **for each team of labourers**:

Soil	*Labour and tools*	*Output* (m^3/hour/team)
Normal	2 labourers = 1 pick + 1 shovel	0.8-1.0
Light	3 labourers = 1 pick + 2 shovels	1.5-2.5
Heavy/wet	3 labourers = 2 picks + 1 shovel	0.4-0.6

Working standards for transporting earth with wheelbarrows

8. **A standard metal wheelbarrow** can transport from 30 litres (0.03 m^3) to 60 litres (0.06 m^3) of earth. Preferably, you should limit the **transport distance to 30 m** at the most. For planning purposes, you can then calculate the following:

(a) Estimate the **number N of earth loads** to be transported over **a short distance** per working hour as

N = 60 min ÷ (loading time + transport time)

where **loading time** averages 1.5 minutes per load; and
transport time is based on the total distance to be covered and an average walking speed of 50 m/min on level ground or 40 to 45 m/min on sloping ground carrying the full load uphill. Downhill slopes will similarly increase transport speed.

Note: it is possible to reduce average loading time and even make it zero by using many wheelbarrows.

Example

You transport earth on level ground over a distance of 20 m. Per working hour, you will be able to make:

N = 60 min ÷ [1.5 min + (40 m at 50 m/min)]
= 60 min ÷ (1.5 min + 0.80 min)
= 60 min ÷ 2.30 min = **26 trips**

(b) Estimate **the amount of excavated earth** you can load in one wheelbarrow, for example 0.045 m^3.

(c) Estimate how **many effective working hours** there will be for each working day, taking into account resting periods, which normally total about 25 percent. For example, each eight-hour working day may provide six effective working hours.

(d) Estimate **the earthwork volume** to be moved daily for each wheelbarrow from the above figures.

Example

- Number of trips per effective working hour = 26
- Effective working hours per day = 6
- Excavated earthwork per wheelbarrow = 0.045 m^3
- Earthwork volume moved daily per wheelbarrow = 26 × 6 × 0.045 m^3 = 7.02 m^3 or about **7 m^3**

(e) **To service each wheelbarrow for transport distances up to 30 m**, you will need at least:

- one worker to dig and fill the wheelbarrow; and
- one worker to push the wheelbarrow.

(f) You might need **additional workers** at particular spots:

- at the dumping site, to assist emptying the wheelbarrow completely;
- along the transport road at climbing points, to help bring the wheelbarrow up the slope.

Average output of various types of machinery

9. **Working standards for the most commonly used earthworking machinery** are given in **Table 52** (see also Section 48, **Construction, 20/1**). These machines are particularly useful if relatively large areas of land are involved.

Bulldozer

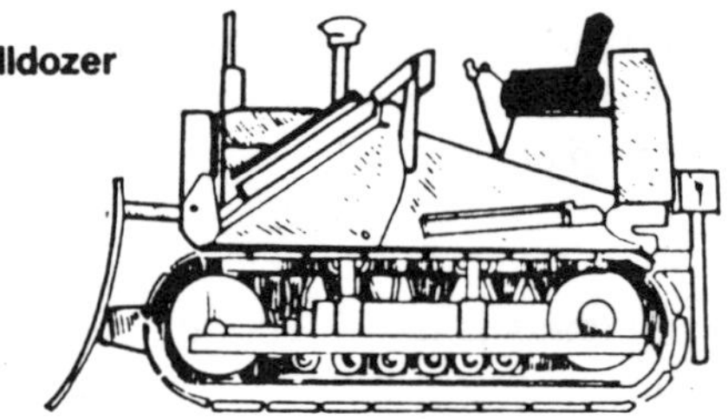

10. **Bulldozers'** output increases as the **engine power** increases, as shown in **Table 53**. These are approximate outputs for normal soil conditions and for **a maximum transport distance of 50 m by pushing**. Lower outputs should be used as soil conditions worsen, for example in heavy/sticky clays.

Example

One contractor calculated the average time necessary to construct one 2 500 m^2 diversion pond, as part of a 5 ha commercial farm as given here.

Item	*Machinery/labour*	*Working time* (hours)
Ripping/removing topsoil	Bulldozer D4	13
Excavation/dikes construction	Bulldozer D8	56
Levelling dike tops	Bulldozer D4	8
Levelling pond bottom	Grader	6
Compaction of dikes	D4 + roller	12
Finishing dikes	6 men	32
Monk construction	4 men + concrete mixer	32
Inlet construction	4 men + concrete mixer	32

Wheel loader

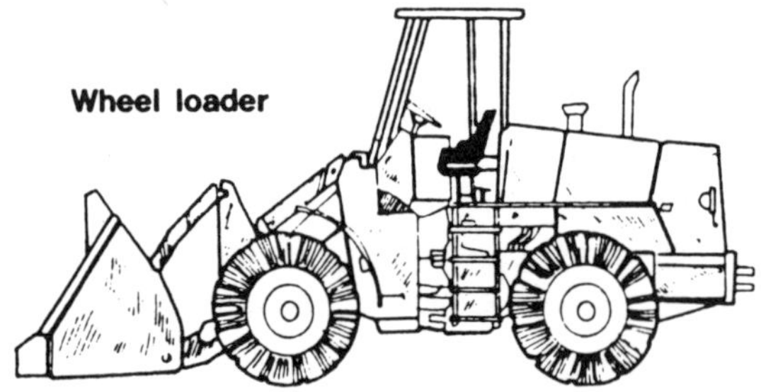

Working standards for construction of structures

11. To build **a brick wall** (half-stone pattern), you will require about 50 bricks per m^2. A good mason can place at the most 600 to 800 bricks per eight-hour day. Note, however, that walls should not be built up more than about 50 cm/day.

12. **To mix and cast concrete**, the required labour varies from eight work-hours/m^3 (without reinforcement) to ten work-hours/m^3 (with re-inforcement).

Compactors

TABLE 52

Average output of various machinery per working hour

Machinery and use	Unit	Haul distance		
		0-20 m	50 m	100 m
BULLDOZER				
Clearing bush, diam. 0-6 cm	m^2	—	300	—
diam. 6-10 cm	m^2	—	200	—
Felling trees, diam. < 20 cm	piece	30	—	—
diam. 20-40 cm	piece	10	—	—
diam. > 40 cm	piece	2-5	—	—
Cutting/moving surface soil				
thickness 10 cm	m^2	—	400	300
thickness 11-20 cm	m^2	—	200	150
Levelling, thickness 10 cm	m^2	800	—	—
thickness 11-20 cm	m^2	400	—	—
Cutting/moving firm earth	m^3	40	20	10
Moving loose earth	m^3	90	50	20
Building dikes	m^3	100	—	—
WHEEL LOADER				
Loading loose earth	m^3	60-100		
COMPACTORS (25-cm layers)				
Sheepsfoot roller	m^2	1 000		
Steel-wheel roller	m^2	2 000-5 000		
Rubber-tyred roller	m^2	5 000-15 000		
Platform vibrator	m^2	300-600		
Frog	m^2	30-150		

TABLE 53

Approximate output of bulldozers for earthwork

Approx. power rating (horsepower)	*Approx. blade capacity** (m^3)	*Excavation/transport*** (m^3/h)	*Spreading loose earth* (m^3/h)
40	1.2	13-17	18-24
70	2.5	22-29	30-39
90	3.6	32-40	42-54
130	4.0	46-71	60-76

* When completely filled. In practice, filling is usually 30 to 60% of this, depending on site conditions
** Excavations by layers less than 0.5 m thick. Transport over 50 m at most, in good site conditions

Note: it is possible to estimate output by defining the time for each trip based on: excavation of blade load 0.5-1 min; pushing, 2 km/h; returning 4-5 km/h; turning, positioning, gear change 0.5-1 min; allow a maximum of 50 minutes per hour utilization. These output figures decrease considerably in difficult site conditions such as sloping ground.

CONTENTS, continued

CONTENTS, continued

CONTENTS, continued

CONTENTS

HOW TO USE THIS MANUAL

The material in the two volumes of this manual is presented in sequence, beginning with basic definitions. The reader is then led step by step from the easiest instructions and most basic materials to the more difficult and finally the complex.

The most basic information is presented on white pages, while the more difficult material, which may not be of interest to all readers, is presented on pages with a grey or a light blue background.

Some of the more technical words are marked with an asterisk (*) and are defined in the Glossary on page 213.

For more advanced readers who wish to know more about fish-farm construction, a list of specialized books for further reading is suggested on page 214.

THE AQUACULTURE TRAINING MANUALS

The training manuals on simple methods for aquaculture published in the FAO Training Series are prepared by the Inland Water Resources and Aquaculture Service of the Fishery Resources and Environment Division, Fisheries Department. They are written in simple language and present methods and equipment useful not only for those responsible for field projects and aquaculture extension in developing countries but also for use in aquaculture training centres.

They concentrate on most aspects of semi-intensive fish culture in fresh waters, from selection of the site and building of the fish farm to the raising and final harvesting of the fish.

FAO would like to have readers' reactions to these manuals. Comments, criticism and opinions, as well as contributions, will help to improve future editions. Please send them to the Senior Fishery Resources Officer (Aquaculture/Publications), FAO/FIRI, Viale delle Terme di Caracalla, 00100 Rome, Italy.

The following manuals on simple methods of aquaculture have been published in the FAO Training Series:

Volume 4 — Water for freshwater fish culture
Volume 6 — Soil and freshwater fish culture
Volume 16/1 — Topography for freshwater fish culture: Topographical tools
Volume 16/2 — Topography for freshwater fish culture: Topographical surveys
Volume 20/2 — Pond construction for freshwater fish culture: Pond-farm structures and layouts

Three final volumes are being prepared:

Volume 20/1 — Pond construction for freshwater fish culture: Building earthen ponds
Volume 21/1 — Management for freshwater fish culture: Ponds and water practices
Volume 21/2 — Management for freshwater fish culture: Farms and fish stocks

"This book was originally published by the Food and Agriculture Organization of the United Nations (FAO) as Pond construction for freshwater fish culture. Pond-farm structures and layouts, FAO Training Series/20/2.

First Indian Edition 1996
ISBN 81-7035-157-X

Rs. 840

Published by: **Daya Publishing House**
1123/74, Deva Ram Park,
Tri Nagar, Delhi-110035
Phone: 7220593
Fax: (011) 7215653

Printed at: Kirpal Print-N-Pack, Delhi

FAO TRAINING SERIES 20/2

simple methods for aquaculture

POND CONSTRUCTION

for freshwater fish culture

pond-farm structures and layouts

Text: A.G. Coche
J.F. Muir
Illustration designs
and layout: T. Laughlin

Reprinted by arrangement with the
Food and Agriculture Organizaton of the United Nations

by

DAYA PUBLISHING HOUSE
DELHI–110035